REMOTE SENSING WITH POLARIMETRIC RADAR

BICENTENNIAL 1807 WILEY 2007 BICENTENNIAL

THE WILEY BICENTENNIAL–KNOWLEDGE FOR GENERATIONS

Each generation has its unique needs and aspirations. When Charles Wiley first opened his small printing shop in lower Manhattan in 1807, it was a generation of boundless potential searching for an identity. And we were there, helping to define a new American literary tradition. Over half a century later, in the midst of the Second Industrial Revolution, it was a generation focused on building the future. Once again, we were there, supplying the critical scientific, technical, and engineering knowledge that helped frame the world. Throughout the 20th Century, and into the new millennium, nations began to reach out beyond their own borders and a new international community was born. Wiley was there, expanding its operations around the world to enable a global exchange of ideas, opinions, and know-how.

For 200 years, Wiley has been an integral part of each generation's journey, enabling the flow of information and understanding necessary to meet their needs and fulfill their aspirations. Today, bold new technologies are changing the way we live and learn. Wiley will be there, providing you the must-have knowledge you need to imagine new worlds, new possibilities, and new opportunities.

Generations come and go, but you can always count on Wiley to provide you the knowledge you need, when and where you need it!

William J. Pesce

WILLIAM J. PESCE
PRESIDENT AND CHIEF EXECUTIVE OFFICER

Peter Booth Wiley

PETER BOOTH WILEY
CHAIRMAN OF THE BOARD

REMOTE SENSING WITH POLARIMETRIC RADAR

HAROLD MOTT
The University of Alabama

IEEE PRESS

WILEY-INTERSCIENCE
A John Wiley & Sons, Inc., Publication

Published by John Wiley & Sons, Inc., Hoboken, New Jersey.
Published simultaneously in Canada.

Library of Congress Cataloging-in-Publication Data is available.

ISBN-13: 978-0-470-07476-3
ISBN-10: 0-470-07476-0

Printed in the United States of America.

10 9 8 7 6 5 4 3 2 1

To my sister Aileen

CONTENTS

PREFACE

The author's purpose in writing this book was to present the principles necessary for understanding polarized radiation, transmission, scattering, and reception in communication systems and polarimetric-radar remote sensing. The book can be used as a text for an undergraduate or graduate course in these topics and as a reference text for engineers and for scientists who use remotely sensed information about the earth. Chapters 1, 2, 4, and 5 are at an introductory level and, together with Chapter 6 and selected material from Chapter 3, can be used for an undergraduate course in electrical engineering. Chapters 3 and 6–9 are at a more advanced technical and mathematical level and provide suitable material for graduate study. Student deficiencies in antennas and radar can be corrected with selections from Chapters 2, 4, and 5. Problems, ranging from straightforward in the introductory chapters to more challenging in the advanced chapters, are provided for pedagogical purposes. Scientists who can profitably study the book are agronomists, geographers, meteorologists, and others who use remotely sensed information. For those who wish to go beyond this discussion of principles to learn of the achievements of polarimetric radar remote sensing and see predictions of future technological developments, I recommend a complementary book, *Principles & Applications of Imaging Radar: Manual of Remote Sensing*, Volume 2, by F. M. Henderson and A. J. Lewis, Wiley, 1998. A comprehensive description of earth-survey sensors is given by H. J. Kramer, *Observation of the Earth and Its Environment: Survey of Missions and Sensors*, Third ed., Springer, 1996.

The material in Chapters 3 and 6 is at a higher mathematical level than the introductory chapters. That in Chapters 7–9 is still more detailed and original and will require more diligent study for a complete understanding. The reader with an understanding of calculus, vector analysis, matrices, and elementary physics can readily comprehend the material, however.

In the author's earlier books, *Polarization in Antennas and Radar* and *Antennas for Radar and Communication: A Polarimetric Approach*, published by Wiley-Interscience in 1986 and 1992, a polarization ratio was defined for an antenna functioning as a transmitter and another for it as a receiver. After reflection, it was thought best to describe the antenna in the same way, regardless of its function, and that has been done in this text. In the earlier books, there was no distinction made between coherently and incoherently measured target matrices, but the distinction is an important part of this book.

ACKNOWLEDGMENTS

The following persons read chapters of this book and provided valuable suggestions: Dr. Ernst Lüneburg of EML Consultants, Wessling, Germany, Jerry L. Eaves of the Georgia Tech Research Institute, Professor Emeritus Ronald C. Houts of the University of West Florida, Assistant Professor John H. Mott of Purdue University, Professor Jian Yang of Tsinghua University, Beijing, Professor Robert W. Scharstein of the University of Alabama, and Dipl. Ing. Andreas Danklmayer of the Deutschen Zentrum für Luft-und Raumfahrt (DLR), Oberpfaffenhofen. Many of their suggestions were incorporated in the text, and I am most grateful for their help.

The late Dr. Ernst Lüneburg encouraged me and provided assistance in the writing of the book over a period of several years. His meticulous and selfless analysis of Chapters 3, 6, 7, and 9 led to extensive changes in the presentation and mathematical developments of the chapters. He read Chapters 1 and 2, also, and exerted a strong influence over still others. Dr. Lüneburg was one of the most able electromagnetics theoreticians and mathematicians of our time and a most generous and helpful friend and colleague. It was a privilege to know and work with him.

Professor Emeritus Wolfgang-Martin Boerner of the University of Illinois, Chicago, maintained his interest in the text from its beginning and provided valuable help and encouragement at critical times. My second book was dedicated to him, and I am thankful for his continued friendship and assistance. I also wish to thank the University of Alabama for providing me with an office and assistance during the years since my retirement as Professor of Electrical Engineering.

This book is dedicated to my sister Aileen in remembrance of her help in my early years and for her love throughout my life.

HAROLD MOTT

Tuscaloosa, Alabama

CHAPTER 1

ELECTROMAGNETIC WAVES

The Maxwell equations,

$$\nabla \times \tilde{\mathbf{E}} = -\frac{\partial \tilde{\mathbf{B}}}{\partial t} \tag{1.1}$$

$$\nabla \times \tilde{\mathbf{H}} = \tilde{\mathbf{J}} + \frac{\partial \tilde{\mathbf{D}}}{\partial t} \tag{1.2}$$

$$\nabla \cdot \tilde{\mathbf{D}} = \tilde{\rho} \tag{1.3}$$

$$\nabla \cdot \tilde{\mathbf{B}} = 0 \tag{1.4}$$

represent the physical laws that are the electromagnetic basis of radar remote sensing. $\tilde{\mathbf{E}}$, $\tilde{\mathbf{H}}$, $\tilde{\mathbf{D}}$, $\tilde{\mathbf{B}}$, and $\tilde{\mathbf{J}}$ are real vectors that symbolize the space- and time-dependent physical quantities of electric field intensity, magnetic field intensity, electric flux density, magnetic flux density, and electric current density. They are in a bold typeface, as are all vectors and matrices in this book. The parameter $\tilde{\rho}$ is a real scalar function of space and time representing electric charge density. The operations indicated are the curl and divergence and the partial time derivative. The rationalized meter-kilogram-second (SI) unit system is used throughout this work.

Note: Only referenced equations are numbered, and numbered equations are not more important than unnumbered ones.

Remote Sensing with Polarimetric Radar, by Harold Mott

Electric current density $\tilde{\mathbf{J}}$ has value because of the flow of electric charge and is related to the rate of change of electric charge density in a region by

$$\nabla \cdot \tilde{\mathbf{J}} = -\frac{\partial \tilde{\rho}}{\partial t}$$

This relationship expresses the conservation of charge, and the equation is the *equation of conservation of charge* or the *equation of continuity*. It can be derived from the Maxwell equations, or, conversely, the divergence equations 1.3 and 1.4 can be derived from it and the curl equations 1.1 and 1.2.

1.1. THE TIME-INVARIANT MAXWELL EQUATIONS

The sources and fields vary in a sinusoidal manner in many phenomena of electromagnetics, and the Maxwell equations can be written in a more tractable form by the substitution, shown here for the electric field intensity, but applicable to all field and source terms,

$$\tilde{\mathbf{E}} = \mathrm{Re}\,(\mathbf{E}\, e^{j\omega t})$$

The tilde is used in this formulation to represent quantities that vary with space and time. Quantities without the tilde are functions of space only. With this substitution,

$$\nabla \times \mathbf{E} = -\mathbf{M} - j\omega\mathbf{B} \tag{1.5}$$

$$\nabla \times \mathbf{H} = \mathbf{J} + j\omega\mathbf{D} \tag{1.6}$$

$$\nabla \cdot \mathbf{D} = \rho \tag{1.7}$$

$$\nabla \cdot \mathbf{B} = \rho_M \tag{1.8}$$

In forming these equations, a magnetic charge density ρ_M and magnetic current density $\mathbf{M}$ were added to the equations formed directly from (1.1) and (1.4). They correspond to the electric sources $\mathbf{J}$ and ρ and make the Maxwell equations symmetric. Physical quantities corresponding to these additions do not exist, but it is convenient, when considering some antenna or scattering problems, to replace the actual sources by equivalent magnetic sources having properties that ensure the fields obey (1.5)–(1.8). (Elliott, 1981, p. 32). The equations 1.5–1.8 are called the *time-invariant Maxwell equations* or the *complex Maxwell equations*.

For linear, isotropic media the field terms are related by the constitutive equations,

$$\mathbf{D} = \epsilon\mathbf{E}$$

$$\mathbf{B} = \mu\mathbf{H}$$

$$\mathbf{J} = \sigma\mathbf{E}$$

where the constants are respectively the permittivity, permeability, and conductivity of the medium in which the electromagnetic field exists.

1.2. THE ELECTROMAGNETIC TRAVELING WAVE

The nature of solutions to the Maxwell equations is brought out more completely in Chapter 2, but a simple development that illustrates the characteristics of certain solutions of importance in remote sensing is shown here.

In a lossless region, with current and charge densities zero, the time-invariant Maxwell curl equations are

$$\nabla \times \mathbf{E} = -j\omega\mu_0\mathbf{H} \tag{1.9}$$

$$\nabla \times \mathbf{H} = j\omega\epsilon_0\mathbf{E} \tag{1.10}$$

E and **H** are functions of **r**, the vector distance from an origin to the point at which the fields are determined, but an important class of electromagnetic fields is that for which they depend, locally, only on the scalar distance from a point. Figure 1.1 shows the coordinates.

In the vicinity of point P we set

$$\mathbf{E}, \mathbf{H} = \mathbf{E}(r), \mathbf{H}(r) \tag{1.11}$$

where we assume that the variation of **E** and **H** with θ and ϕ is negligible compared to the variation with r. This functional variation of the fields accurately describes configurations with sources or reflecting objects in the vicinity of the origin if point P is far from sources or scatterers. It is incorrect in the vicinity of sources or reflectors.

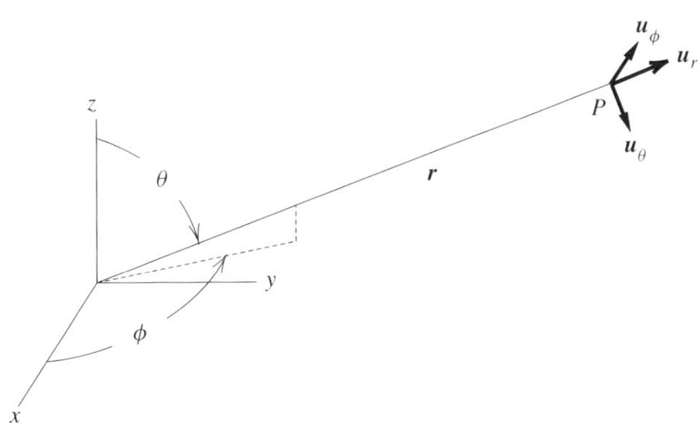

Fig. 1.1. Coordinate system for the traveling wave.

If the curl equations 1.9 and 1.10 are expanded while treating the field components as functions only of r, they become

$$\nabla \times \mathbf{E} = -\frac{1}{r}\frac{d(rE_\phi)}{dr}\mathbf{u}_\theta + \frac{1}{r}\frac{d(rE_\theta)}{dr}\mathbf{u}_\phi = -j\omega\mu_0(H_r\mathbf{u}_r + H_\theta\mathbf{u}_\theta + H_\phi\mathbf{u}_\phi)$$

$$\nabla \times \mathbf{H} = -\frac{1}{r}\frac{d(rH_\phi)}{dr}\mathbf{u}_\theta + \frac{1}{r}\frac{d(rH_\theta)}{dr}\mathbf{u}_\phi = j\omega\epsilon_0(E_r\mathbf{u}_r + E_\theta\mathbf{u}_\theta + E_\phi\mathbf{u}_\phi)$$

where $\mathbf{u}_r$, $\mathbf{u}_\theta$, and $\mathbf{u}_\phi$ are real unit vectors, shown in Fig. 1.1. Equating coefficients of like unit vectors in the first equation, differentiating the resulting equations, and substituting the coefficients of $\mathbf{u}_\theta$ and $\mathbf{u}_\phi$ from the curl of $\mathbf{H}$ gives

$$\frac{d^2(rE_{\theta,\phi})}{dr^2} + k^2 rE_{\theta,\phi} = 0$$

where $k^2 = \omega^2\mu_0\epsilon_0$. Solutions to these equations are

$$E_{\theta,\phi} = \frac{C_{\theta,\phi}}{r}e^{\pm jkr}$$

$\mathbf{H}$ can be found similarly, and its components are

$$H_\theta = \pm\frac{C_\phi/Z_0}{r}e^{\pm jkr}$$

$$H_\phi = \mp\frac{C_\theta/Z_0}{r}e^{\pm jkr}$$

where $Z_0 = \sqrt{\mu_0/\epsilon_0}$.

If we take one of the field components, say E_θ, and write its corresponding time-varying form, we obtain

$$\tilde{E}_\theta = \text{Re}\left(E_\theta e^{j\omega t}\right) = \frac{|C_\theta|}{r}\cos(\omega t \pm kr + \Phi)$$

where Φ is some phase angle. This equation represents a component of a traveling electromagnetic wave that appears to move in the increasing r direction for the negative sign in the cosine function and in the decreasing r direction for the positive sign. We reject the wave traveling toward the origin for physical reasons and retain the negative sign in the cosine wave.

When the Maxwell curl equations were expanded in spherical coordinates, with the assumption that the fields vary only with r, the curl of $\mathbf{E}$ had no radial component. It follows from (1.9) that $\mathbf{H}$ does not have a radial component. Likewise, since the curl of $\mathbf{H}$ has no radial component, $\mathbf{E}$ does not have a radial component.

Now, we can write the field vectors whose components vary only with radial distance r. The solutions represent a specialized electromagnetic field, but one of great importance in remote sensing. The fields are

$$\mathbf{E} = \frac{1}{r}(C_\theta \mathbf{u}_\theta + C_\phi \mathbf{u}_\phi)e^{-jkr} \tag{1.12}$$

$$\mathbf{H} = \frac{1}{r}\frac{1}{Z_0}(-C_\phi \mathbf{u}_\theta + C_\theta \mathbf{u}_\phi)e^{-jkr} \tag{1.13}$$

It is apparent from these equations that $\mathbf{E}$ and $\mathbf{H}$ are perpendicular to $\mathbf{u}_r$, the direction of wave travel. Further, the scalar product of $\mathbf{E}$ and $\mathbf{H}$, neglecting the phase variation with r, is zero, showing that $\mathbf{E}$ and $\mathbf{H}$ are perpendicular to each other. The phenomenon described by these equations is a spherical wave traveling outward from a coordinate origin.

The assumption (1.11) is valid only in a local sense; that is, in the vicinity of a selected point, P. At a different point, the field coefficients will differ from the values at P and the direction of wave travel will be different. At P, radial distance r is large compared to the dimensions of the region in which we require our assumptions about the behavior of the fields to be valid. Then a surface of constant r is almost a plane surface. It is common to describe the electromagnetic wave as a *plane wave*, one with amplitudes and phases constant over a plane, rather than over a spherical surface, and to construct at P a rectangular coordinate system with an axis pointing in the direction of wave travel. If the wave travels in the z direction, the fields can be written as

$$\mathbf{E} = (E_x \mathbf{u}_x + E_y \mathbf{u}_y)e^{-jkz} \tag{1.14}$$

$$\mathbf{H} = \frac{1}{Z_0}(-E_y \mathbf{u}_x + E_x \mathbf{u}_y)e^{-jkz} \tag{1.15}$$

where $\mathbf{u}_x$ and $\mathbf{u}_y$ are unit vectors. In the newly constructed coordinate system, the phase referenced to the original origin may be discarded, and the coefficient amplitudes vary so slowly with large r that the variation is neglected. Note that the fields of (1.14) and (1.15) satisfy the free-space wave equation to be discussed in Chapter 2 and have significance other than as an approximation to (1.12) and (1.13).

Two other wave descriptors are commonly given. From $\cos(\omega t - kr)$, we note that at a constant position r_0, the phase of the wave changes by 2π radians when time changes by $2\pi/\omega$. The time interval

$$T = \frac{2\pi}{\omega} = \frac{1}{f}$$

is the *wave period*. At constant time t_0, the phase changes by 2π radians when the radial distance changes by $2\pi/k$. The increment in r corresponding to this

2π phase change is the *wavelength*. It is

$$\lambda = \Delta r = \frac{2\pi}{k}$$

If one envisions the wave in space at a constant time, the wavelength is the length of one complete sine wave cycle. In a lossless region with constants those of a vacuum, or approximately those of air,

$$\lambda = \frac{2\pi}{2\pi f \sqrt{\mu_0 \epsilon_0}} = \frac{c}{f}$$

where c is the velocity of light in a vacuum.

1.3. POWER DENSITY

An electromagnetic field stores energy. If the field varies with time, the energy storage is dynamic and there is a relationship between the rate of change of the stored energy and the flow of energy. *Poynting's theorem* states that the rate of energy flow across surface S is given by

$$\iint_S (\tilde{\mathbf{E}} \times \tilde{\mathbf{H}}) \cdot \mathbf{n}\, da$$

where $\mathbf{n}$ is a surface normal vector. A time average of this integral is of interest. It represents the power density in W/m^2 in an electromagnetic wave. It is straightforward to show that

$$\overline{\tilde{\mathbf{E}} \times \tilde{\mathbf{H}}} = \frac{1}{2}\operatorname{Re}(\mathbf{E} \times \mathbf{H}^*)$$

where the overline denotes the time average.

We define a *complex Poynting vector* by

$$\mathcal{P}_c = \frac{1}{2}\mathbf{E} \times \mathbf{H}^* \tag{1.16}$$

The real power flow is given by

$$\mathcal{P} = \operatorname{Re}(\mathcal{P}_c) \tag{1.17}$$

The Poynting vector gives both the direction of wave travel and the power density.

From the fields of a plane wave, (1.14) and (1.15), the complex Poynting vector of the wave is found to be

$$\mathcal{P}_c = \frac{1}{2}\mathbf{E} \times \mathbf{H}^* = \frac{|E_x|^2 + |E_y|^2}{2Z_0^*}\mathbf{u}_z = \frac{1}{2Z_0^*}|\mathbf{E}|^2\mathbf{u}_z \tag{1.18}$$

1.4. THE POLARIZATION ELLIPSE

The tip of the electric field intensity vector of a single-frequency wave traces an ellipse at a fixed position in space as time increases (Mott, 1992, p. 117). Such a wave is said to be *elliptically polarized.* This property is shown here for a plane wave. A plane wave traveling in the z direction has two complex components and may be written as

$$\mathbf{E} = (E_x \mathbf{u}_x + E_y \mathbf{u}_y)e^{-jkz} = (\mathbf{u}_x|E_x|e^{j\Phi_x} + \mathbf{u}_y|E_y|e^{j\Phi_y})e^{-jkz}$$

The corresponding time-varying field is

$$\tilde{\mathbf{E}} = \mathbf{u}_x|E_x|\cos(\beta + \Phi_x) + \mathbf{u}_y|E_y|\cos(\beta + \Phi_y) \tag{1.19}$$

where

$$\beta = \omega t - kz \tag{1.20}$$

The components can be combined to give

$$\frac{\tilde{E}_x^2}{|E_x|^2} - 2\frac{\tilde{E}_x}{|E_x|}\frac{\tilde{E}_y}{|E_y|}\cos\Phi + \frac{\tilde{E}_y^2}{|E_y|^2} = \sin^2\Phi$$

where $\Phi = \Phi_y - \Phi_x$. This is the equation of an ellipse whose major axis is tilted at angle τ to the $\tilde{E}_x$ axis. It is shown in Fig. 1.2. With increasing time at fixed position z, the tip of the electric field vector traces the ellipse. Tilt angle τ is defined over the range $-\pi/2 \leq \tau \leq \pi/2$.

Define rotated coordinates $\tilde{E}_\xi$ and $\tilde{E}_\eta$ to coincide with the ellipse axes. The fields $\tilde{E}_\xi$ and $\tilde{E}_\eta$ are related to $\tilde{E}_x$ and $\tilde{E}_y$ by

$$\begin{bmatrix} \tilde{E}_\xi \\ \tilde{E}_\eta \end{bmatrix} = \begin{bmatrix} \cos\tau & \sin\tau \\ -\sin\tau & \cos\tau \end{bmatrix} \begin{bmatrix} \tilde{E}_x \\ \tilde{E}_y \end{bmatrix} \tag{1.21}$$

The fields can be written as

$$\tilde{E}_\xi = m\cos(\beta + \Phi_0) \tag{1.22}$$

$$\tilde{E}_\eta = n\cos(\beta + \Phi_0 \pm \pi/2) = \pm n\sin(\beta + \Phi_0) \tag{1.23}$$

where Φ_0 is a phase angle that need not be determined, and m and n are positive real. If we require that $m \geq n$, m is the semimajor axis of the ellipse and n the semiminor.

If the positive sign is used before $\pi/2$ in (1.23), the electric vector rotates with one sense as time increases; if the negative sign is used, the rotation has the opposite sense.

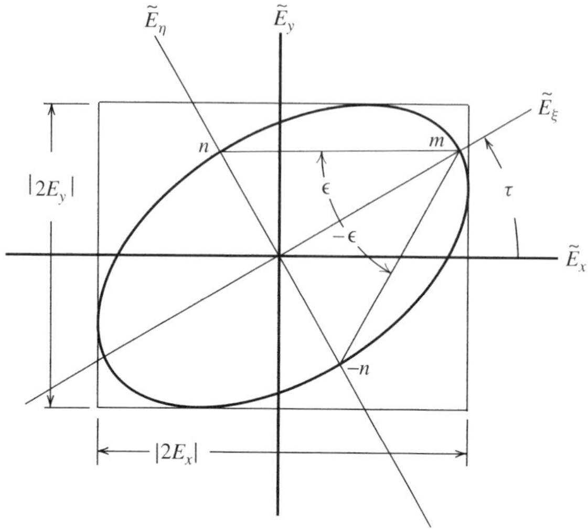

Fig. 1.2. Polarization ellipse.

We equate (1.21) and (1.22)–(1.23) and use (1.19) for $\tilde{E}_x$ and $\tilde{E}_y$. This gives

$$m\cos(\beta+\Phi_0) = |E_x|\cos(\beta+\Phi_x)\cos\tau + |E_y|\cos(\beta+\Phi_y)\sin\tau$$
$$\pm n\sin(\beta+\Phi_0) = -|E_x|\cos(\beta+\Phi_x)\sin\tau + |E_y|\cos(\beta+\Phi_y)\cos\tau$$

If the coefficients of $\cos\beta$ are equated and also the coefficients of $\sin\beta$, the following relationships are obtained:

$$m^2 + n^2 = |E_x|^2 + |E_y|^2 \tag{1.24}$$

$$\pm mn = -|E_x||E_y|\sin\Phi \tag{1.25}$$

$$\pm\frac{n}{m} = \frac{|E_x|\sin\Phi_x\sin\tau - |E_y|\sin\Phi_y\cos\tau}{|E_x|\cos\Phi_x\cos\tau + |E_y|\cos\Phi_y\sin\tau}$$
$$= \frac{-|E_x|\cos\Phi_x\sin\tau + |E_y|\cos\Phi_y\cos\tau}{|E_x|\sin\Phi_x\cos\tau + |E_y|\sin\Phi_y\sin\tau} \tag{1.26}$$

Cross-multiplying and collecting terms in the last equation of this set gives

$$\left(|E_x|^2 - |E_y|^2\right)\sin 2\tau = 2|E_x||E_y|\cos 2\tau\cos\Phi$$

Tilt angle τ may be found from

$$\tan 2\tau = \frac{2|E_x||E_y|}{|E_x|^2 - |E_y|^2}\cos\Phi \tag{1.27}$$

The ellipse shape can be specified by the axial ratio m/n or by the ellipticity angle ϵ shown in Fig. 1.2. Both positive and negative values of ϵ, with the same magnitude, are shown. It is desirable for a graphic representation of wave polarization to use the negative value of ϵ if positive signs are used in (1.25) and (1.26). We therefore define

$$\tan\epsilon = \mp\frac{n}{m} \qquad -\frac{\pi}{4} \leq \epsilon \leq \frac{\pi}{4}$$

If this equation is combined with (1.24) and (1.25), the result is

$$\sin 2\epsilon = \frac{2|E_x||E_y|}{|E_x|^2 + |E_y|^2}\sin\Phi \tag{1.28}$$

The time-varying angle of $\tilde{\mathbf{E}}$, measured from the x toward the y axis, is

$$\Psi = \tan^{-1}\frac{\tilde{E}_y}{\tilde{E}_x} = \tan^{-1}\frac{|E_y|\cos(\beta + \Phi_y)}{|E_x|\cos(\beta + \Phi_x)} \tag{1.29}$$

If the derivative of this angle with respect to β is examined, it will be seen that

$$\frac{\partial\Psi}{\partial\beta}\begin{cases} < 0, & 0 < \Phi < \pi \\ > 0, & \pi < \Phi < 2\pi \end{cases}$$

If we look in the direction of wave propagation, a positive value of $\partial\Psi/\partial\beta$ corresponds to clockwise rotation of $\tilde{\mathbf{E}}$ as β (or time) increases. By definition, this is right-handed rotation of the vector. A negative value of the derivative corresponds to counterclockwise or left-handed rotation. It can be seen from (1.28) and the ranges of Φ that the corresponding ellipticity angle ranges are

$$\epsilon\begin{cases} < 0, & \text{right-handed rotation} \\ > 0, & \text{left-handed rotation} \end{cases}$$

If the defining equation for the ellipticity angle is used in (1.22) and (1.23), the field components can be written as

$$\tilde{E}_\xi = m\cos(\beta + \Phi_0)$$
$$\tilde{E}_\eta = -m\tan\epsilon\cos(\beta + \Phi_0 - \pi/2)$$

If the variation with time and distance is suppressed, the field is

$$\mathbf{E}(\xi, \eta) = m\begin{bmatrix} 1 \\ j\tan\epsilon \end{bmatrix} e^{j\Phi_0} = \frac{m}{\cos\epsilon}\begin{bmatrix} \cos\epsilon \\ j\sin\epsilon \end{bmatrix} e^{j\Phi_0}$$

The electric field intensities in the two coordinate systems are related by

$$\mathbf{E}(x, y) = \begin{bmatrix} \cos\tau & -\sin\tau \\ \sin\tau & \cos\tau \end{bmatrix} \mathbf{E}(\xi, \eta)$$

Combining the last two equations allows the electric field of a single-frequency plane wave to be written in terms of tilt and ellipticity angles by

$$\mathbf{E}(x, y) = \frac{m}{\cos\epsilon} \begin{bmatrix} \cos\tau & -\sin\tau \\ \sin\tau & \cos\tau \end{bmatrix} \begin{bmatrix} \cos\epsilon \\ j\sin\epsilon \end{bmatrix} e^{j\Phi_0} \tag{1.30}$$

The common phase term is normally of little significance and can be omitted.

Note: In the preceding three equations and in similar equations throughout the text, the symbols in parentheses denote vector components and not a functional relationship.

Linear and Circular Polarization

In the special cases of $|E_x| = 0$, or $|E_y| = 0$, or $\Phi = 0$, the polarization ellipse degenerates to a straight line, and the wave is *linearly polarized*. The axial ratio is infinite, the tilt angle can be found by the standard equations, and rotation sense is meaningless.

If $|E_x| = |E_y|$ and $\Phi = \pi/2$, the axial ratio is equal to 1, the polarization ellipse degenerates to a circle, and the wave is *circularly polarized*, right circular if $\Phi = -\pi/2$ and left circular if $\Phi = \pi/2$.

Rotation of $\tilde{\text{E}}$ with Distance

The rotation rates of the electric field vector with time t and distance z, from (1.20), are

$$\frac{\partial\Psi}{\partial t} = \omega\frac{\partial\Psi}{\partial\beta} \qquad \frac{\partial\Psi}{\partial z} = -k\frac{\partial\Psi}{\partial\beta}$$

where Ψ is the angle of the time-varying electric field vector measured from the x-axis. We see from these equations that a wave which appears to rotate in a clockwise sense as we look in the direction of wave travel at some fixed position as t increases, corresponding to our definition of a right-handed wave, appears to rotate counterclockwise with an increase in z at a fixed time. A right-handed circular wave drawn in space at some constant time looks like a left-handed screw. Conversely, a left-handed circular wave appears in space to be a right-handed screw. With increasing time, the left-handed screw representing a right-handed wave rotates in a clockwise direction as we look in the direction of wave motion.

It can be seen from (1.20) and (1.29) that the distance between two points of the wave having parallel field vectors at constant time is

$$\Delta z = \frac{2\pi}{k} = \lambda$$

The wave appears to rotate once in space in a distance of one wavelength.

1.5. POLARIZATION VECTOR AND POLARIZATION RATIO

A description of an elliptically polarized wave in terms of tilt angle, axial ratio, and rotation sense leads to a physical understanding of the wave, but a more convenient mathematical description of the wave is needed. The time-invariant electric field itself contains all information about the wave polarization, and so does $\mathbf{p}$, the *polarization vector* or the *polarization state* of the wave. This is the electric field normalized by its magnitude. Another useful descriptor is the *complex polarization ratio, P*. The electric field can be written as

$$\mathbf{E} = E_x \mathbf{u}_x + E_y \mathbf{u}_y = E_x \left(\mathbf{u}_x + P\mathbf{u}_y\right) e^{-jkz}$$

where P is the polarization ratio,

$$P = \frac{E_y}{E_x} = \frac{|E_y|}{|E_x|} e^{j\Phi} \tag{1.31}$$

The value of E_x does not affect the wave polarization, and it can be neglected unless power density or received power is required. Values of the polarization ratio are ∞, 0, j, $-j$ for linear vertical (y-directed), linear horizontal (x-directed), and left- and right-circular waves, respectively.

1.6. CIRCULAR WAVE COMPONENTS

To this point, a plane electromagnetic wave has been considered the sum of two linearly polarized plane waves perpendicular to each other and to the direction of wave travel. It may also be considered the sum of left- and right-circular plane waves, and the common use of such a description justifies the formulation.

The orthonormal vectors

$$\mathbf{u}_L = \frac{1}{\sqrt{2}}(\mathbf{u}_x + j\mathbf{u}_y) = \frac{1}{\sqrt{2}}\mathbf{u}_x + \frac{1}{\sqrt{2}}e^{j\pi/2}\mathbf{u}_y \tag{1.32}$$

$$\mathbf{u}_R = \frac{1}{\sqrt{2}}(\mathbf{u}_x - j\mathbf{u}_y) = \frac{1}{\sqrt{2}}\mathbf{u}_x + \frac{1}{\sqrt{2}}e^{-j\pi/2}\mathbf{u}_y \tag{1.33}$$

together with $\mathbf{u}_z$, which is orthogonal to both, are a vector triplet that defines a coordinate system that we will call a circular-coordinate system. A time-invariant electric field expressed in terms of an xyz rectangular system can be converted to the circular-coordinate system using (1.32) and (1.33),

$$\mathbf{E} = E_x\mathbf{u}_x + E_y\mathbf{u}_y = E_L\mathbf{u}_L + E_R\mathbf{u}_R = |E_L|e^{j\Theta_L}\mathbf{u}_L + |E_R|e^{j\Theta_R}\mathbf{u}_R$$

The field components are related by

$$\begin{bmatrix} E_L \\ E_R \end{bmatrix} = \frac{1}{\sqrt{2}}\begin{bmatrix} 1 & -j \\ 1 & j \end{bmatrix}\begin{bmatrix} E_x \\ E_y \end{bmatrix} \tag{1.34}$$

Let us examine the real time-varying field components associated with $\mathbf{u}_L$ and $\mathbf{u}_R$. They are

$$\begin{aligned}
\mathrm{Re}\left(E_L\mathbf{u}_L e^{j(\omega t - kz)}\right) &= \frac{|E_L|}{\sqrt{2}}\mathbf{u}_x\cos(\omega t - kz + \Theta_L) \\
&\quad + \frac{|E_L|}{\sqrt{2}}\mathbf{u}_y\cos(\omega t - kz + \Theta_L + \pi/2) \\
\mathrm{Re}\left(E_R\mathbf{u}_R e^{j(\omega t - kz)}\right) &= \frac{|E_R|}{\sqrt{2}}\mathbf{u}_x\cos(\omega t - kz + \Theta_R) \\
&\quad + \frac{|E_R|}{\sqrt{2}}\mathbf{u}_y\cos(\omega t - kz + \Theta_R - \pi/2)
\end{aligned}$$

These can be recognized as real time-varying vectors of constant amplitude rotating, respectively, in a left- and right-handed sense. The vector $\mathbf{u}_L$ represents a left-circular wave and $\mathbf{u}_R$ a right-circular, and E_L and E_R are circular wave components. The linear polarization ratio of a wave was defined previously as the ratio of linear components of the wave. We define the *circular polarization ratio* as the ratio of circular wave components,

$$Q = \frac{E_R}{E_L} = \frac{|E_R|}{|E_L|}e^{j\Theta} \tag{1.35}$$

where $\Theta = \Theta_R - \Theta_L$.

1.7. CHANGE OF POLARIZATION BASIS

A plane wave traveling in the z direction can be considered the sum of two orthogonally polarized waves, neither linear nor circular, but elliptical. Let

$$\mathbf{E} = \mathbf{E}_1 + \mathbf{E}_2 = E_1\mathbf{u}_1 + E_2\mathbf{u}_2 \tag{1.36}$$

where $\mathbf{u}_1$ and $\mathbf{u}_2$ are complex orthonormal unit vectors.

The Euclidean inner product of two vectors, treated as column matrices, is

$$\langle \mathbf{a}, \mathbf{b} \rangle = \sum_{1}^{N} a_n b_n^* = \mathbf{a}^T \mathbf{b}^* = \mathbf{b}^\dagger \mathbf{a}$$

where T indicates the transpose, $*$ the conjugate, and $\dagger$ the combined transpose and conjugate operation, or Hermitian adjoint. An equivalent notation is the scalar or dot product of vectors, $\langle \mathbf{a}, \mathbf{b} \rangle = \mathbf{a} \cdot \mathbf{b}^*$.

The polarization characteristics of a component vector of (1.36), say $\mathbf{E}_1$, can be specified in xyz coordinates. To do so, we write $\mathbf{u}_1$ in terms of a linear polarization ratio P_1 in the xyz system,

$$\mathbf{u}_1 = u_{1x}\mathbf{u}_x + u_{1y}\mathbf{u}_y = u_{1x}(\mathbf{u}_x + P_1\mathbf{u}_y)$$

The phase of the x-component of $\mathbf{E}_1$ can be included in the multiplier E_1 and u_{1x} taken as real. Doing so and noting that $\mathbf{u}_1$ has unit length leads to

$$\mathbf{u}_1 = \frac{\mathbf{u}_x + P_1\mathbf{u}_y}{\left(1 + |P_1|^2\right)^{1/2}}$$

In the same way, $\mathbf{u}_2$ is found to be

$$\mathbf{u}_2 = \frac{\mathbf{u}_x + P_2\mathbf{u}_y}{\left(1 + |P_2|^2\right)^{1/2}}$$

where P_2 is the linear polarization ratio of the second elliptical component of the wave. The orthogonality condition $\langle \mathbf{u}_1, \mathbf{u}_2 \rangle = 0$ leads to $P_2 = -1/P_1^*$.

It can readily be seen that $\mathbf{u}_L$ and $\mathbf{u}_R$ of the previous section are specializations, with $P_1 = j$, of more general orthogonal vectors.

Change of Basis

Conversion of the field from linear-component form to a general orthonormal basis can be done by first writing

$$\mathbf{E} = E_1\mathbf{u}_1 + E_2\mathbf{u}_2 = E_x\mathbf{u}_x + E_y\mathbf{u}_y$$

The field components are related by

$$\begin{bmatrix} E_1 \\ E_2 \end{bmatrix} = \begin{bmatrix} \langle \mathbf{u}_x, \mathbf{u}_1 \rangle & \langle \mathbf{u}_y, \mathbf{u}_1 \rangle \\ \langle \mathbf{u}_x, \mathbf{u}_2 \rangle & \langle \mathbf{u}_y, \mathbf{u}_2 \rangle \end{bmatrix} \begin{bmatrix} E_x \\ E_y \end{bmatrix}$$

or

$$\mathbf{E}_{\mathbf{u}_1,\mathbf{u}_2} = \mathbf{U}\mathbf{E}_{\mathbf{u}_x,\mathbf{u}_y}$$

If $\mathbf{u}_1$ and $\mathbf{u}_2$ are expressed in terms of the linear polarization ratio in x and y coordinates, the matrix $\mathbf{U}$ that transforms $\mathbf{E}$ from the $\mathbf{u}_x, \mathbf{u}_y$ polarization basis to the $\mathbf{u}_1, \mathbf{u}_2$ basis becomes

$$\mathbf{U} = \begin{bmatrix} \langle \mathbf{u}_x, \mathbf{u}_1 \rangle & \langle \mathbf{u}_y, \mathbf{u}_1 \rangle \\ \langle \mathbf{u}_x, \mathbf{u}_2 \rangle & \langle \mathbf{u}_y, \mathbf{u}_2 \rangle \end{bmatrix} = \frac{1}{\left(1 + |P_1|^2\right)^{1/2}} \begin{bmatrix} 1 & P_1^* \\ |P_1| & -|P_1|/P_1 \end{bmatrix} \tag{1.37}$$

It is readily seen that $\mathbf{U}$ is unitary.

Coordinate Rotation

Of special interest is the transformation caused by a coordinate rotation. For this, $\mathbf{u}_1$ is a real unit vector rotated by angle θ (in the direction $x \to y$) from the x axis of the old coordinate system, and $\mathbf{U}$ has a simple form which we denote by

$$\mathbf{R} = \begin{bmatrix} \cos\theta & \sin\theta \\ -\sin\theta & \cos\theta \end{bmatrix}$$

1.8. ELLIPSE CHARACTERISTICS IN TERMS OF *P* AND *Q*

If the magnitudes of x and y components of the electric field intensity are equal, the tilt angle of the polarization ellipse cannot be found from (1.27). It can be found if the circular polarization ratio of the wave is used.

The circular polarization ratio of a single-frequency plane wave is

$$Q = |Q|e^{j\Theta} = \frac{E_R}{E_L} = \frac{E_x + jE_y}{E_x - jE_y} = \frac{1 + jP}{1 - jP} \tag{1.38}$$

The electric field vector can be written as

$$\mathbf{E} = |E_L|e^{j\Theta_L}(\mathbf{u}_L + Q\mathbf{u}_R)$$

The time-varying field vector is then

$$\frac{\tilde{\mathbf{E}}}{|E_L|} = \mathrm{Re}\left[e^{j\gamma}(\mathbf{u}_L + Q\mathbf{u}_R)\right] \tag{1.39}$$

where $\gamma = \omega t - kz + \Theta_L$. The unit vectors in the expression for the time-varying electric field are orthonormal and conjugate. If these characteristics are used, the square of the field intensity becomes

$$\frac{\tilde{\mathbf{E}}^2}{|E_L|^2} = \frac{1}{2}\left[1 + |Q|^2 + 2|Q|\cos(2\gamma + \Theta)\right]$$

The field intensity magnitude has maximum and minimum values $(1 \pm |Q|)/\sqrt{2}$ at $2\gamma + \Theta = 0, 2\pi, \ldots$ and $2\gamma + \Theta = \pi, 3\pi, \ldots$, respectively. It follows that the axial ratio of the polarization ellipse is

$$AR = \left| \frac{1 + |Q|}{1 - |Q|} \right| \tag{1.40}$$

The time-varying electric field where it has maximum magnitude is found by substituting $-\Theta/2$ for γ in (1.39). This field can be transformed to rectangular form by substitution; it is

$$\frac{\tilde{\mathbf{E}}}{|E_L|} = \frac{1}{\sqrt{2}}(1 + |Q|) \left[\cos\left(\frac{\Theta}{2}\right) \mathbf{u}_x + \sin\left(\frac{\Theta}{2}\right) \mathbf{u}_y \right]$$

This is the field having the greatest magnitude, and its angle measured from the x axis is the tilt angle of the polarization ellipse. It is

$$\tau = \tan^{-1} \frac{\tilde{E}_y}{\tilde{E}_x} = \frac{\Theta}{2}, \frac{\Theta}{2} \pm \pi$$

The axial ratio of the polarization ellipse in terms of P is, from (1.38) and (1.40),

$$AR = \left| \frac{|1 + jP| + |1 - jP|}{|1 + jP| - |1 - jP|} \right|$$

The tilt angle can be found from

$$e^{j2\tau} = \frac{Q}{|Q|} = \frac{(1 + jP)/(1 - jP)}{|(1 + jP)/(1 - jP)|}$$

A more convenient form, readily derived from this, is

$$\tan 2\tau = \frac{\mathrm{Im}(Q)}{\mathrm{Re}(Q)} = \frac{2\mathrm{Re}(P)}{1 - |P|^2}$$

1.9. COHERENCY AND STOKES VECTORS

Alternative ways of describing waves that span a range of frequencies are necessary, and here we introduce the coherency vector and the Stokes vector. The rationale for their use will be deferred, and in this section only their forms will be considered. In addition to being useful for multifrequency waves, the Stokes vector suggests a means, the *Poincaré sphere*, by which the polarization of a single-frequency wave can be represented graphically.

The Coherency Vector

Consider a single-frequency plane wave traveling in the z direction of a right-handed coordinate system. The *coherency vector* of the wave is defined as

$$\mathbf{J} = \mathbf{E} \otimes \mathbf{E}^*$$

where $\otimes$ denotes the Kronecker product or direct product, defined for two-element vectors by

$$\mathbf{A} \otimes \mathbf{B} = \begin{bmatrix} A_1\mathbf{B} \\ A_2\mathbf{B} \end{bmatrix} = \begin{bmatrix} A_1B_1 \\ A_1B_2 \\ A_2B_1 \\ A_2B_2 \end{bmatrix} \tag{1.41}$$

Note: Born and Wolf (1965, p. 544) give the elements of the coherency vector in a 2×2 matrix called by them the *coherency matrix*. The vector form is more convenient for our purposes.

The Stokes Vector

The elements of the coherency vector are complex, and it is sometimes desirable to describe the wave by real quantities. In his studies of nonmonochromatic light, Stokes introduced four parameters to characterize the amplitude and polarization of a wave. The parameters can be arranged in a vector form, the *Stokes vector* **G**. It is a transform of the coherency vector,

$$\mathbf{G} = \mathbf{QJ} \tag{1.42}$$

where

$$\mathbf{Q} = \begin{bmatrix} 1 & 0 & 0 & 1 \\ 1 & 0 & 0 & -1 \\ 0 & 1 & 1 & 0 \\ 0 & j & -j & 0 \end{bmatrix} \tag{1.43}$$

The elements of the Stokes vector of a monochromatic wave can be found by carrying out the multiplication in (1.42). They are

$$G_0 = |E_x|^2 + |E_y|^2 \tag{1.44}$$

$$G_1 = |E_x|^2 - |E_y|^2 \tag{1.45}$$

$$G_2 = 2|E_x||E_y| \cos \Phi \tag{1.46}$$

$$G_3 = 2|E_x||E_y| \sin \Phi \tag{1.47}$$

where Φ is the phase difference between the y and x components of the wave. We follow tradition and common usage in subscripting the first element as 0. This is not done for other vectors in this text.

The Stokes vector is sufficient to describe both amplitude and polarization of the wave. Parameter G_0 gives the amplitude directly, while $|E_x|$ and $|E_y|$ can be found from G_0 and G_1. The phase Φ can be determined from either G_2 or G_3. Only three of the equations are independent for the monochromatic case, since

$$G_0^2 = G_1^2 + G_2^2 + G_3^2$$

From (1.28) and the definitions of the Stokes parameters,

$$G_3 = G_0 \sin 2\epsilon$$

where ϵ is the ellipticity angle of the polarization ellipse. From (1.27) and the definitions of the Stokes parameters,

$$\tan 2\tau = \frac{G_2}{G_1} \tag{1.48}$$

If this equation is combined with the two previous ones, the result is

$$G_1 = G_0 \cos 2\epsilon \cos 2\tau$$

Substitution of this equation into (1.48) gives

$$G_2 = G_0 \cos 2\epsilon \sin 2\tau$$

1.10. THE POINCARÉ SPHERE

From the expression of G_1, G_2, and G_3 in terms of tilt and ellipticity angles, it is apparent that they can be considered the Cartesian coordinates of a point on a sphere of radius G_0. Angles 2ϵ and 2τ are latitude and azimuth angles measured to the point. This interpretation was introduced by Poincaré, and the sphere is called the Poincaré sphere. It is shown, with the Stokes parameters and tilt and ellipticity angles, in Fig. 1.3. A single-frequency wave can be described by a point on the Poincaré sphere. To every state of polarization there corresponds one point on the sphere and vice versa.

Special Points on the Poincaré Sphere

For a left-circular wave, $|E_x| = |E_y|$, $\Phi = \pi/2$, $G_1 = G_2 = 0$, and $G_3 = G_0$. The point representing left-circular polarization is the "north" pole ($+z$-axis) of

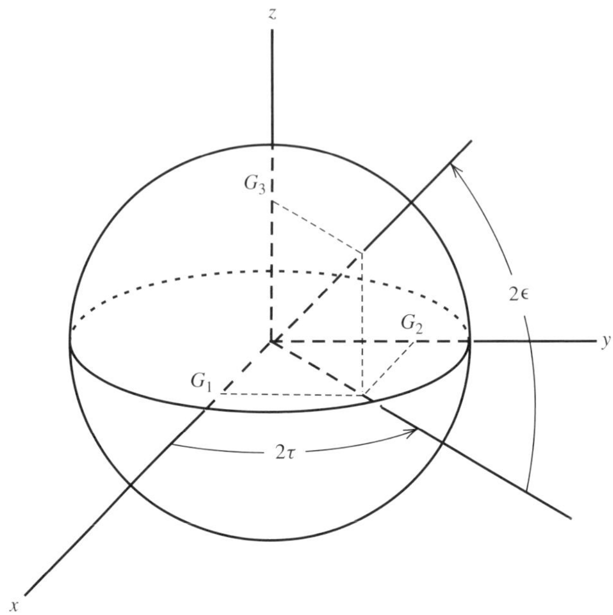

Fig. 1.3. The Poincaré sphere.

the Poincaré sphere. For a right-circular wave, $|E_x| = |E_y|$, $\Phi = -\pi/2$, $G_1 = G_2 = 0$, and $G_3 = -G_0$. This is the south pole of the sphere.

For a left-elliptic wave, $0 < \Phi < \pi$. It follows from (1.47) that $G_3 > 0$. All points for left-elliptic polarizations are plotted on the upper hemisphere. For right-elliptic polarizations, $\pi < \Phi < 2\pi$ and $G_3 < 0$. Right-elliptic polarization points are in the lower hemisphere.

For linear polarizations, if $|E_x|$ and $|E_y|$ are nonzero, $\Phi = 0$ or π, and $G_3 = 0$. All linear polarization points are at the equator. For linear vertical polarization, the Poincaré sphere point is at the $-x$-axis intersection with the sphere; for linear horizontal, it is at the $+x$-axis intersection. The $+y$-axis intersection corresponds to linear polarization with a tilt angle of $\pi/4$, and the $-y$-axis intersection to linear polarization with a tilt angle of $-\pi/4$.

Mapping the Poincaré Sphere onto a Plane

The Poincaré sphere can be mapped onto a plane using methods developed for creating plane maps of the earth. Plane maps of the Poincaré sphere distort the relationships between points representing different polarizations, just as plane maps of the earth distort the relationships between geographic features, but maps of the Poincaré sphere are nonetheless useful. Mott (1992, p. 159ff) discusses several maps, gives the mapping equations, and shows the resulting maps in detail.

REFERENCES

M. Born and E. Wolf, *Principles of Optics*, 3rd ed., Pergamon, New York, 1965.

R. S. Elliott, *Antenna Theory and Design*, Prentice-Hall, Englewood Cliffs, NJ, 1981.

H. Mott, *Antennas for Radar and Communications: A Polarimetric Approach*, Wiley-Interscience, New York, 1992.

PROBLEMS

1.1. (a) Derive the equation of conservation of charge from the Maxwell equations. (b) Use the Maxwell curl equations and the equation of conservation of charge to derive the Maxwell divergence equations.

1.2. Assume that electric charge cannot be created or destroyed and that electric current consists of the movement of charge. Express mathematically the relationship between current leaving a finite volume and the charge within the volume. From this relationship, derive the differential equation that expresses the law of conservation of charge.

1.3. Show that **E** and **H** of (1.12) and (1.13) satisfy the time-invariant Maxwell equations.

1.4. An electromagnetic wave in air has magnetic field

$$\tilde{\mathbf{H}} = 2\mathbf{u}_x \cos(\omega t - kz) - \mathbf{u}_y \cos(\omega t - kz + \pi/8)$$

(a) Find the corresponding electric field $\tilde{\mathbf{E}}$. (b) Find the polarization ratio, tilt angle, ellipticity angle, and rotation sense of the wave.

1.5. Find the coherency and Stokes vectors of the wave of Problem 1.4.

1.6. Find the time-invariant circular electric field components of the wave of Problem 1.4. Find the circular polarization ratio.

1.7. Write the time-invariant circular electric field components of Problem 1.4 as the sum of two elliptical waves, $\mathbf{E}_1 + \mathbf{E}_2$, where $\mathbf{E}_1$ has linear polarization ratio $P_1 = e^{j\pi/6}$.

1.8. A right-handed wave has tilt angle 30° and axial ratio 3. Find polarization ratio P.

1.9. The Maxwell equations are often presented as integrals. Write the integrals that correspond to the differential forms of the equations.

1.10. What are the SI units of the wave coherency vector? The Stokes vector?

1.11. Does the transformation of the magnetic vector **H** from the $\mathbf{u}_x, \mathbf{u}_y$ basis to the $\mathbf{u}_1, \mathbf{u}_2$ basis obey the same equation, with matrix **U** of (1.37), as that of the electric field? Prove your answer.

CHAPTER 2

ANTENNAS

In radar remote sensing, an electromagnetic wave of suitable frequency and appropriate polarization is launched by a transmitting antenna, and the reflected wave is received by a receiving antenna, which may be the same antenna used to transmit the wave.

2.1. ELEMENTS OF THE ANTENNA SYSTEM

The antenna system of a radar can be broken into simpler subsystems whose analysis and design are readily understood and carried out. Consider two antennas as shown in Fig. 2.1: a transmitting antenna connected to a generator and a receiving antenna connected to a receiver. The transmitting antenna appears to be a load impedance connected to the generator through a transmission line or waveguide. The impedance can be found and the power accepted from the generator determined. Part of the power accepted is radiated and part is dissipated as heat, and the separate parts can be found. The transmitting antenna does not radiate equally in all directions, and the directional characteristics can be determined.

To the receiver, the receiving antenna appears to be a voltage or current source, with the source value determined by the incident power density, the direction from which the incident wave arrives, and the polarization properties of the antenna and incident wave. The receiver, in turn, appears to be a load impedance on the receiving antenna. The receiving antenna has some internal impedance, which can be determined. When the source value and the receiving-antenna internal impedance are determined, power to the receiver load can be found.

Remote Sensing with Polarimetric Radar, by Harold Mott

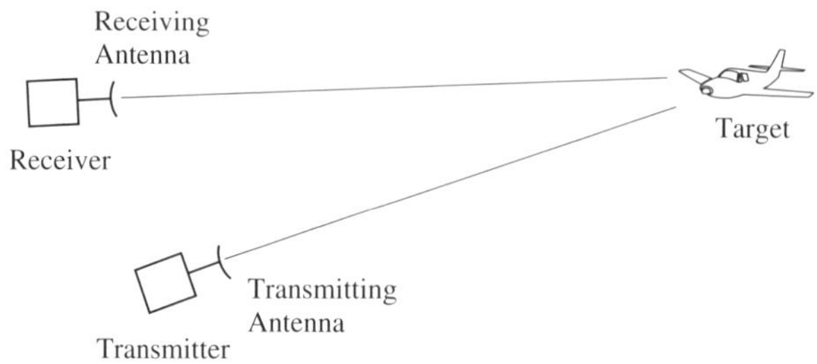

Fig. 2.1. Transmitter, receiver, and target.

The path from transmitting antenna to receiving antenna involves a reflection from a target or *scatterer*. The reflected (scattered) wave provides all the information that we can obtain about many targets.

These factors depend on frequency, and antennas consequently have a finite bandwidth determined by impedance, radiation pattern, and polarization pattern. This bandwidth can be measured or calculated.

Some concepts of antenna theory do not explicitly take into account the polarimetric properties of the antenna. These concepts are discussed in the first part of this chapter. Antenna impedance, for example, falls into this category. Directivity and gain may be expressed in a nonpolarimetric form, but implicit in their use is an assumption that a polarization match exists between two antennas. Polarization matching and other antenna polarization properties are discussed later in the chapter. At this point, it is noted only that if one antenna transmits a wave toward another, the second antenna receives maximum power if the antennas are polarization matched.

In this chapter, the principles of antenna analysis and design are presented at a level that allows antenna function in a radar system to be understood. More detail is given by Mott (1992).

2.2. THE VECTOR POTENTIALS

For linear relationships between **E** and **D** and between **H** and **B**, the time-invariant Maxwell equations can be separated into equations containing only electric sources and those containing only magnetic sources, with the total fields formed by superposing solutions for the two cases.

Electric Sources

The time-invariant Maxwell equations, with electric sources only, are

$$\nabla \times \mathbf{E}_J = -j\omega \mathbf{B}_J = -j\omega\mu \mathbf{H}_J \tag{2.1}$$

$$\nabla \times \mathbf{H}_J = \mathbf{J} + j\omega \mathbf{D}_J = \mathbf{J} + j\omega\epsilon \mathbf{E}_J \tag{2.2}$$

$$\nabla \cdot \mathbf{D}_J = \epsilon \nabla \cdot \mathbf{E}_J = \rho \tag{2.3}$$

$$\nabla \cdot \mathbf{B}_J = \nabla \cdot \mathbf{H}_J = 0 \tag{2.4}$$

where subscript J identifies these vectors as the partial fields produced by electric current density $\mathbf{J}$ and electric charge density ρ.

Since the divergence of $\mathbf{B}_J$ is zero, it can be represented as the curl of a *magnetic vector potential* $\mathbf{A}$,

$$\mathbf{B}_J = \mu \mathbf{H}_J = \nabla \times \mathbf{A} \tag{2.5}$$

Substituting this equation into (2.1) leads to

$$\nabla \times (\mathbf{E}_J + j\omega \mathbf{A}) = 0$$

The curl of the gradient of a scalar function is identically zero, so we may set

$$\mathbf{E}_J + j\omega \mathbf{A} = -\nabla \Phi_J \tag{2.6}$$

where Φ_J is the *electric scalar potential.*

The curl of (2.5) substituted into (2.2) gives

$$\nabla \times \nabla \times \mathbf{A} = \mu(\mathbf{J} + j\omega\epsilon \mathbf{E}_J)$$

Substituting (2.6) into the right side of this equation and using a vector identity gives

$$\nabla(\nabla \cdot \mathbf{A}) - \nabla^2 \mathbf{A} = \mu \mathbf{J} - j\omega\epsilon \nabla \Phi_J + \omega^2 \mu\epsilon \mathbf{A} \tag{2.7}$$

We are free to choose the divergence of the magnetic vector potential, since only the curl is specified by (2.5), and choose it as

$$\nabla \cdot \mathbf{A} = -j\omega\mu\epsilon \Phi_J \tag{2.8}$$

Then, (2.7) becomes

$$\nabla^2 \mathbf{A} + k^2 \mathbf{A} = -\mu \mathbf{J} \tag{2.9}$$

where $k^2 = \omega^2 \mu\epsilon$. The potentials $\mathbf{A}$ and Φ_J are related by (2.8), so it is unnecessary to find Φ_J. The magnetic flux density $\mathbf{B}_J$ may be found from the vector potential $\mathbf{A}$. Then $\mathbf{E}_J$ may be found at points away from the sources with (2.6) and (2.8),

$$\mathbf{E}_J = -j\omega \mathbf{A} - \frac{j}{\omega\mu\epsilon} \nabla(\nabla \cdot \mathbf{A})$$

Magnetic Sources

If the time-invariant Maxwell equations are specialized for use with magnetic sources only, potentials can be formed from the resulting equations by analogy with the process used with electric sources. *Electric vector potential* **F** is defined, with arbitrary negative sign, as

$$\mathbf{D}_M = \epsilon \mathbf{E}_M = -\nabla \times \mathbf{F}$$

where subscript M identifies the vectors as being produced by magnetic current density **M** and magnetic charge density ρ_M.

F satisfies

$$\nabla^2 \mathbf{F} + k^2 \mathbf{F} = -\epsilon \mathbf{M}$$

and $\mathbf{H}_M$ is found from

$$\mathbf{H}_M = -j\omega \mathbf{F} - \frac{j}{\omega\mu\epsilon} \nabla(\nabla \cdot \mathbf{F})$$

Superposition

The solutions for electric and magnetic sources can be superposed to give

$$\mathbf{E} = -j\omega \mathbf{A} - \frac{j}{\omega\mu\epsilon} \nabla(\nabla \cdot \mathbf{A}) - \frac{1}{\epsilon} \nabla \times \mathbf{F}$$

$$\mathbf{H} = -j\omega \mathbf{F} - \frac{j}{\omega\mu\epsilon} \nabla(\nabla \cdot \mathbf{F}) + \frac{1}{\mu} \nabla \times \mathbf{A}$$

2.3. SOLUTIONS FOR THE VECTOR POTENTIALS

If an infinitesimal z-directed electric current element is located at the origin of a spherical coordinate system, (2.9) becomes

$$\nabla^2 A_z + k^2 A_z = -\mu J_z \tag{2.10}$$

J_z is zero everywhere except at the origin, and the source has infinitesimal length. A_z is therefore spherically symmetric and satisfies

$$\frac{1}{r^2} \frac{d}{dr} \left(r^2 \frac{dA_z}{dr} \right) + k^2 A_z = 0$$

This equation has two independent solutions, $e^{\pm jkr}/r$, which represent inward- and outward-traveling spherical waves. We choose the outward-traveling wave,

$$A_z = \frac{C}{r} e^{-jkr} \tag{2.11}$$

As $k \to 0$, (2.10) reduces to Poisson's equation, with solution,

$$A_z = \frac{\mu}{4\pi} \iiint \frac{J_z}{r}\, dv'$$

If we replace $J_z\, dv'$ by $I\, dz'$ and integrate, A_z becomes $\mu I L / 4\pi r$, where L is the source length. If this value is compared to (2.11) with k zero, constant C can be found, and the magnetic vector potential of the infinitesimal current element is

$$A_z = \frac{\mu I L}{4\pi r} e^{-jkr} \tag{2.12}$$

If the infinitesimal element is located at vector distance $\mathbf{r}'$ from the origin and oriented along a line parallel to unit vector $\mathbf{u}$, the potential is

$$\mathbf{A}(\mathbf{r}) = \mathbf{u} \frac{\mu I L}{4\pi |\mathbf{r} - \mathbf{r}'|} e^{-jk|\mathbf{r}-\mathbf{r}'|}$$

If an electric current is distributed on a thin wire, this solution can be generalized to

$$\mathbf{A}(\mathbf{r}) = \frac{\mu}{4\pi} \int \frac{I(\mathbf{r}')}{|\mathbf{r} - \mathbf{r}'|} e^{-jk|\mathbf{r}-\mathbf{r}'|}\, d\mathbf{L}' \tag{2.13}$$

The integral can be written in terms of current density $\mathbf{J}$ or surface current density $\mathbf{J}_s$ if $I\, d\mathbf{L}'$ is replaced by $\mathbf{J}\, dv'$ or $\mathbf{J}_s\, da'$ and the integration performed over a volume or surface.

By analogy, an integral for the electric vector potential $\mathbf{F}$ is

$$\mathbf{F}(\mathbf{r}) = \frac{\epsilon}{4\pi} \int \frac{K(\mathbf{r}')}{|\mathbf{r} - \mathbf{r}'|} e^{-jk|\mathbf{r}-\mathbf{r}'|}\, d\mathbf{L}'$$

where K is a magnetic current.

2.4. FAR-ZONE FIELDS

Integration to find the vector potentials is difficult to carry out in the general case, and approximations are desirable in the distance term of the integrals, which can be expanded as

$$|\mathbf{r} - \mathbf{r}'| = \left[r^2 - 2\mathbf{r} \cdot \mathbf{r}' + r'^2\right]^{1/2} = r - \mathbf{u}_r \cdot \mathbf{r}' + \frac{1}{2r}\left[r'^2 - (\mathbf{u}_r \cdot \mathbf{r}')^2\right] + \dots$$

$\mathbf{u}_r = \mathbf{r}/r$, and terms in r^{-2}, r^{-3}, and so on, have been dropped.

Our primary concern in remote sensing is in the fields far from sources and scattering objects. At great distances from the sources, the r^{-1} terms of the binomial expansion can be dropped in amplitude and phase, and the equation for $\mathbf{A}(\mathbf{r})$ becomes

$$\mathbf{A}(\mathbf{r}) = \frac{\mu}{4\pi r} e^{-jkr} \iiint \mathbf{J}(\mathbf{r}') e^{jk\mathbf{u}_r \cdot \mathbf{r}'}\, dv'$$

The distance at which this approximation can be used depends on $\mathbf{J}$ and $\mathbf{M}$ and is somewhat arbitrary. A commonly used boundary for an antenna with greatest linear dimension L is $r = 2L^2/\lambda$.

From (2.5) and the volume-integral equivalent to (2.13),

$$\begin{aligned}\mathbf{H}_J &= \frac{1}{4\pi} \iiint \nabla \times \left(\frac{\mathbf{J}(\mathbf{r}')}{|\mathbf{r} - \mathbf{r}'|} e^{-jk|\mathbf{r}-\mathbf{r}'|} \right) dv' \\ &= \frac{-1}{4\pi} \iiint \mathbf{J}(\mathbf{r}') \times \nabla \left(\frac{e^{-jk|\mathbf{r}-\mathbf{r}'|}}{|\mathbf{r} - \mathbf{r}'|} \right) dv'\end{aligned}$$

The electric field intensity is

$$\mathbf{E}_J = \frac{1}{j\omega\epsilon} \nabla \times \mathbf{H}_J = \frac{-1}{j4\pi\omega\epsilon} \iiint \nabla \times \left[\mathbf{J} \times \nabla \left(\frac{e^{-jk|\mathbf{r}-\mathbf{r}'|}}{|\mathbf{r} - \mathbf{r}'|} \right) \right] dv'$$

If a vector identity is used, the electric field becomes

$$\mathbf{E}_J = \frac{-1}{j4\pi\omega\epsilon} \iiint \left[\mathbf{J}\nabla^2 \left(\frac{e^{-jk|\mathbf{r}-\mathbf{r}'|}}{|\mathbf{r} - \mathbf{r}'|} \right) - (\mathbf{J} \cdot \nabla)\nabla \left(\frac{e^{-jk|\mathbf{r}-\mathbf{r}'|}}{|\mathbf{r} - \mathbf{r}'|} \right) \right] dv'$$

The function of $\mathbf{r}$ and $\mathbf{r}'$ in parentheses of this integral is a solution of the scalar Helmholtz equation without sources,

$$\left(\nabla^2 + k^2\right) \frac{e^{-jk|\mathbf{r}-\mathbf{r}'|}}{|\mathbf{r} - \mathbf{r}'|} = 0$$

If this is used, $\mathbf{E}_J$ becomes

$$\mathbf{E}_J = \frac{1}{j4\pi\omega\epsilon} \iiint \left[\mathbf{J}k^2 \left(\frac{e^{-jk|\mathbf{r}-\mathbf{r}'|}}{|\mathbf{r} - \mathbf{r}'|} \right) + (\mathbf{J} \cdot \nabla)\nabla \left(\frac{e^{-jk|\mathbf{r}-\mathbf{r}'|}}{|\mathbf{r} - \mathbf{r}'|} \right) \right] dv' \quad (2.14)$$

If this procedure is repeated for the electric vector potential $\mathbf{F}$ with magnetic source $\mathbf{M}$, we obtain

$$\mathbf{E}_M = \frac{-1}{\epsilon} \nabla \times \mathbf{F} = \frac{1}{4\pi} \iiint \mathbf{M}(\mathbf{r}') \times \nabla \left(\frac{e^{-jk|\mathbf{r}-\mathbf{r}'|}}{|\mathbf{r} - \mathbf{r}'|} \right) dv'$$

and

$$\mathbf{H}_M = \frac{-1}{j\omega\mu}\nabla \times \mathbf{E}_M$$
$$= \frac{1}{j4\pi\omega\mu}\iiint\left[\mathbf{M}k^2\left(\frac{e^{-jk|\mathbf{r}-\mathbf{r}'|}}{|\mathbf{r}-\mathbf{r}'|}\right) + (\mathbf{M}\cdot\nabla)\nabla\left(\frac{e^{-jk|\mathbf{r}-\mathbf{r}'|}}{|\mathbf{r}-\mathbf{r}'|}\right)\right]dv'$$

We use the far-field approximation,

$$\left(\frac{e^{-jk|\mathbf{r}-\mathbf{r}'|}}{|\mathbf{r}-\mathbf{r}'|}\right) = \frac{e^{-jk(r-\mathbf{u_r}\cdot\mathbf{r}')}}{r} \tag{2.15}$$

If only terms of order $1/r$ are retained, the gradient appearing in the equations for the fields is

$$\nabla\left(\frac{e^{-jk(r-\mathbf{u}_r\cdot\mathbf{r}')}}{r}\right) \approx \frac{-jk}{r}e^{-jkr}e^{jk\mathbf{u}_r\cdot\mathbf{r}'}\mathbf{u}_r \tag{2.16}$$

If this is used in the equation for $\mathbf{H}_J$, it becomes

$$\mathbf{H}_J = \frac{jk}{4\pi r}e^{-jkr}\iiint \mathbf{J}(\mathbf{r}') \times \mathbf{u}_r e^{jk\mathbf{u}_r\cdot\mathbf{r}'}\,dv'$$
$$= \frac{jk}{4\pi r}e^{-jkr}\iiint (J_\phi\mathbf{u}_\theta - J_\theta\mathbf{u}_\phi)e^{jk\mathbf{u}_r\cdot\mathbf{r}'}\,dv'$$

The corresponding value of the electric field for an electric source distribution can be found by substituting (2.15) and (2.16) into (2.14). The result is

$$\mathbf{E}_J = \frac{-jkZ_0}{4\pi r}e^{-jkr}\iiint (J_\theta\mathbf{u}_\theta + J_\phi\mathbf{u}_\phi)e^{jk\mathbf{u}_r\cdot\mathbf{r}'}\,dv'$$

The fields of a magnetic source distribution are found in the same manner and are

$$\mathbf{E}_M = -\frac{jk}{4\pi r}e^{-jkr}\iiint (M_\phi\mathbf{u}_\theta - M_\theta\mathbf{u}_\phi)e^{jk\mathbf{u}_r\cdot\mathbf{r}'}\,dv'$$
$$\mathbf{H}_M = -\frac{jk}{4\pi Z_0 r}e^{-jkr}\iiint (M_\theta\mathbf{u}_\theta + M_\phi\mathbf{u}_\phi)e^{jk\mathbf{u}_r\cdot\mathbf{r}'}\,dv'$$

where Z_0 is the intrinsic impedance of the medium of interest.

In summary, the far fields are

Electric Sources:

$$E_r = H_r = 0 \qquad E_\theta = -j\omega A_\theta \qquad E_\phi = -j\omega A_\phi$$

Magnetic Sources:

$$H_r = E_r = 0 \qquad H_\theta = -j\omega F_\theta \qquad H_\phi = -j\omega F_\phi$$

Either or Both Sources:

$$E_\theta = Z_0 H_\phi \qquad E_\phi = -jZ_0 H_\theta$$

The **E** and **H** fields are perpendicular to each other and to **r** in the far zone. This verifies the conclusion drawn from the example of Section 1.2.

2.5. RADIATION PATTERN

A transmitting antenna does not radiate power isotropically; that is, equally in all directions, nor is the polarization of the wave independent of direction. The radiation pattern of an antenna illustrates the directional properties of the antenna.

Radiation Intensity

The *radiation intensity* of a wave radiated by an antenna in a given direction is the power radiated per unit solid angle in that direction. A three-dimensional figure can be created by a closed bundle of contiguous rays intersecting at a common point. If a sphere of radius r is constructed with center at the ray intersection, the rays subtend area A on the sphere surface. The ratio $\Omega = A/r^2$ is independent of the sphere radius and defines the solid angle formed at the intersection by the rays. On a more general surface, as in Fig. 2.2, the projection of the surface area element onto a sphere centered at the ray intersection is $\mathbf{u}_r \cdot \mathbf{n}\, da$, where $\mathbf{n}$ is the unit normal vector to the surface and $\mathbf{u}_r$ is the unit vector in the direction of the vector from ray intersection to the surface element. Then a solid-angle element is

$$d\Omega = \frac{da}{r^2}\mathbf{u}_r \cdot \mathbf{n}$$

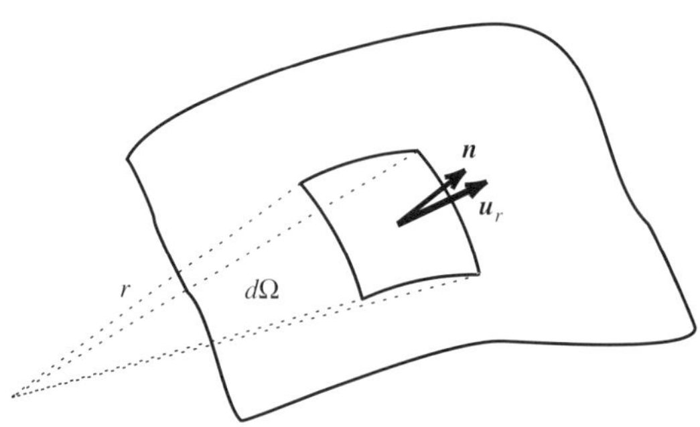

Fig. 2.2. Elementary solid angle.

Radiated power W and radiation intensity U are related by

$$W = \iint \frac{U(\theta, \phi)}{r^2} \mathbf{u}_r \cdot \mathbf{n}\, da$$

The radiated power is also

$$W = \iint \mathcal{P}(r, \theta, \phi) \mathbf{u}_r \cdot \mathbf{n}\, da$$

where $\mathcal{P}$ is the Poynting vector power density. From a comparison of the integrals,

$$U(\theta, \phi) = r^2 \mathcal{P}(r, \theta, \phi) \tag{2.17}$$

Radiation Pattern

The three-dimensional *radiation pattern* of an antenna shows some property of the antenna's electromagnetic field in the far-field region as a function of polar and azimuth angles measured at the antenna. Radiation properties that can be presented include power flux density (magnitude of the Poynting vector), radiation intensity, field strength, and received power to a polarization-matched antenna. Power flux density is most often used.

Figure 2.3 shows a section of a radiation pattern. The plot is of power density in the yz plane as a function of angle measured from the z-axis of a coordinate system at the antenna. The main beam and minor pattern lobes, or sidelobes, are shown. Also shown are angles for which the power density in the main lobe is one-half its greatest value. The angle between those limits is the *half-power beamwidth* of the antenna. In the xz plane, the pattern section might have a main beam with the same width as the one shown. Such a beam is sometimes referred to as a "pencil beam", particularly if it is narrow. The main beam section in the

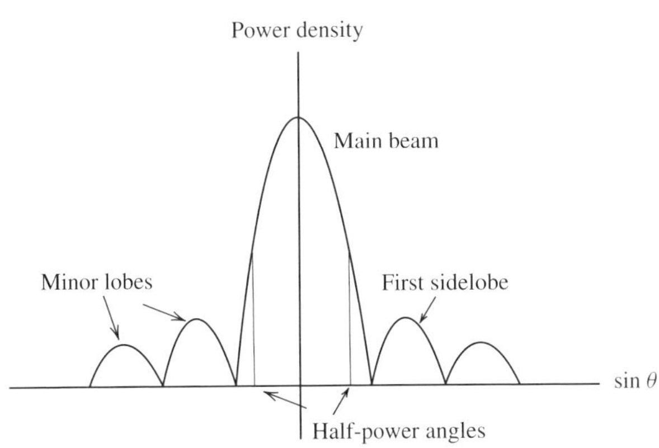

Fig. 2.3. Antenna pattern section.

xz plane may, on the other hand, be broader or narrower than that in the yz plane. This has been called a "fan beam". Finally, the pattern section in the xz plane, unlike the one shown, may not be symmetric about the z-axis.

2.6. GAIN AND DIRECTIVITY

The property that causes radiation intensity from an antenna to be greater in some directions than in others is the *directivity* of the antenna. It is the ratio of the radiation intensity in a given direction to the radiation intensity averaged over all directions.

The Infinitesimal Current Element

The concept of directivity can be illustrated readily by radiation from an infinitesimal current element. If the spherical components of the magnetic vector potential produced by a z-directed current element, (2.12), are used to find the field components, they will be found to be

$$E_\theta = -j\omega A_\theta = \frac{j\mu I L}{4\pi r} \sin\theta e^{-jkr} \tag{2.18}$$

$$H_\phi = \frac{E_\theta}{Z_0} = \frac{j\omega\mu I L}{4\pi Z_0 r} \sin\theta e^{-jkr} \tag{2.19}$$

The time-average Poynting vector found by using these field components is

$$\mathcal{P} = \frac{1}{2}\mathrm{Re}(\mathbf{E} \times \mathbf{H}^*) = \frac{Z_0|I|^2 L^2}{8\lambda^2 r^2} \sin^2\theta \mathbf{u}_r$$

and the radiation intensity is

$$U(\theta, \phi) = \frac{Z_0|I|^2 L^2}{8\lambda^2} \sin^2\theta$$

Radiated power is determined by integrating over the surface of a sphere of large radius,

$$W_{\mathrm{rad}} = \iint U(\theta, \phi)\, d\Omega = \frac{Z_0|I|^2 L^2}{8\lambda^2} \int_0^{2\pi} \int_0^{\pi} \sin^3\theta \, d\theta \, d\phi = \frac{\pi Z_0|I|^2 L^2}{3\lambda^2} \tag{2.20}$$

The average radiation intensity U_{av} is this power divided by 4π, and the directivity is

$$D(\theta, \phi) = \frac{U(\theta, \phi)}{U_{\mathrm{av}}} = \frac{3}{2} \sin^2\theta \tag{2.21}$$

We will make use of this result in Section 2.7.

Radiation Resistance

Radiated power is lost to the transmitting system, and to the generator the loss is indistinguishable from heat loss in a resistance. We therefore define an equivalent resistance called the *radiation resistance* of the antenna. It is the power radiated by the antenna divided by one-half of the square of the root mean square (rms) current at a specified point.

A widely used antenna is a circular cylindrical center-fed dipole, shown in Fig. 2.4. If it is made of a wire whose radius is much smaller than a wavelength and much smaller than the dipole length, the current, to a good approximation, is sinusoidal (Kraus, 1988, p. 369).

$$I(z') = I_m \sin\left[k\left(\frac{1}{2}L - |z'|\right)\right] \qquad -\frac{1}{2}L \le z' \le \frac{1}{2}L \tag{2.22}$$

At the antenna feed point,

$$I_{\text{in}} = I(0) = I_m \sin(kL/2)$$

We can define the radiation resistance of the dipole at the point of maximum current,

$$R_{\text{rad}} = \frac{2W_{\text{rad}}}{|I_m|^2}$$

or with reference to the feed point,

$$R_{\text{in}} = \frac{2W_{\text{rad}}}{|I_{\text{in}}|^2}$$

Both definitions are used in the literature.

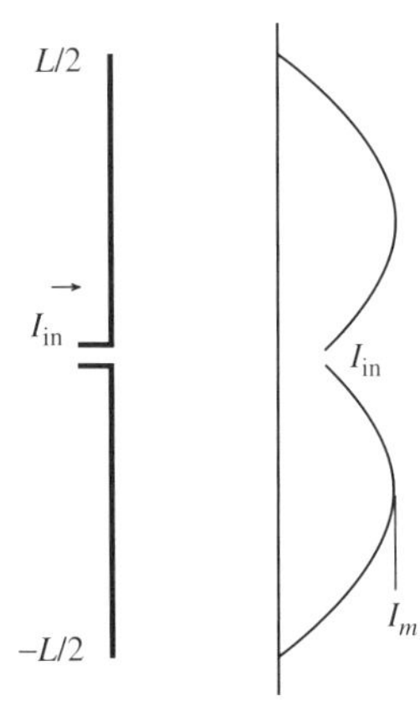

Fig. 2.4. Dipole and its current distribution.

Obtaining the radiation resistance of a center-fed dipole requires extensive calculations (Balanis, 1982, p. 120). A much simpler problem is that of an infinitesimal current source, which is given here as an example. From (2.20) and the definition of radiation resistance, it is

$$R_{\text{rad}} = \frac{2\pi Z_0}{3}\left(\frac{L}{\lambda}\right)^2 \tag{2.23}$$

Antenna Losses and Radiation Efficiency

The transmitter delivers power to the antenna, part of which is reflected because of an impedance mismatch and part of which is accepted by the antenna. Some of the power accepted is radiated and some is lost as heat in conductors and dielectrics. The losses occur because of the finite conductivity of the antenna and lossy dielectrics near the antenna. The determination of losses requires a knowledge of the fields in the vicinity of the antenna. A simple example is that of finding losses in a center-fed circular cylindrical antenna made of a wire with conductivity σ and carrying a known current, $I(z')$.

High-frequency currents in a conductor flow very near the surface. We may treat an axial high-frequency current in a wire of radius a as though it flows with constant density to a small depth δ, where δ is the *skin depth*, given by (Harrington, 1961, p. 53)

$$\delta = \frac{1}{\sqrt{\pi f \mu \sigma}}$$

The high-frequency resistance per unit length of the wire is approximately

$$R_{\text{hf}} = \frac{1}{2\pi a \delta \sigma}$$

Power loss for an antenna of length L is

$$W_{\text{loss}} = \frac{1}{2}\int_{-L/2}^{L/2} |I(z')|^2 R_{hf}\, dz'$$

Loss resistance for the dipole antenna, referred to the input, is

$$R_{\text{loss}} = \frac{2W_{\text{loss}}}{|I_{\text{in}}|^2} = \frac{1}{|I_{\text{in}}|^2}\int_{-L/2}^{L/2} |I(z')|^2 R_{\text{hf}}\, dz'$$

The loss resistance for other antennas can be much more difficult to obtain than for this simple case.

Radiation efficiency is defined as the ratio of the power radiated by the antenna to the power accepted. It is

$$e = \frac{W_{\text{rad}}}{W_{\text{rad}} + W_{\text{loss}}} = \frac{W_{\text{rad}}}{W_{\text{acc}}}$$

If all of the losses are attributed to a loss resistance, referred to the same point as the radiation resistance, the efficiency is

$$e = \frac{R_{\rm rad}}{R_{\rm rad} + R_{\rm loss}}$$

Gain

Antenna gain is the ratio of the radiation intensity in a given direction to the radiation intensity that would exist if the power accepted by the antenna were radiated equally in all directions. It is

$$G(\theta, \phi) = \frac{U(\theta, \phi)}{W_{\rm acc}/4\pi}$$

Gain does not include losses arising from impedance and polarization mismatches. It does include heat losses. If the direction is not specified, the direction of maximum radiation intensity is implied.

The relation between radiated and accepted power can be used to relate gain and directivity,

$$G(\theta, \phi) = \frac{U(\theta, \phi)}{[1/(4\pi)]W_{\rm rad}/e} = eD(\theta, \phi)$$

Input Impedance

The antenna is seen by a generator as an impedance at the end of a connecting transmission line. The impedance consists of loss resistance $R_{\rm loss}$, radiation resistance R_r, and antenna reactance X_a. These parameters can be used in an equivalent circuit with the generator open-circuit voltage and impedance to find power accepted by the antenna, power radiated, and power lost as heat.

It is reasonable to consider feeding a dipole antenna with a two-wire transmission line having transverse electric and magnetic (TEM) fields, but other antennas may be fed with a less clearly defined feed system. The concept of antenna impedance is dependent on defining a driving point, or input port, for the antenna. Silver (1949, p. 37) points out that the current distribution in the line must be that characteristic of a transmission line up to the assigned driving point if impedance is to be an unambiguous concept. At high frequencies, interaction between the radiating system and the line may disturb the line currents back over a considerable distance; there is then no definite transition between transmission line currents and antenna currents. The concept of "antenna impedance" is ambiguous in such a case. Some antennas are fed by waveguides, which do not propagate the TEM mode. If the waveguide propagates a single mode, a "mode impedance" of the waveguide can be defined, and the antenna impedance can be expressed in terms of this mode impedance. As with the TEM line, the validity of the impedance concept depends on our ability to define an antenna driving point with only a single waveguide mode on one or both sides of this driving point. Obtaining the input impedance or admittance is not easy, even for a simple antenna, and idealizations of the antenna and its feed system must be used. These idealizations may affect the impedance significantly.

A general method of determining input impedance is not applicable to all antenna types, although for all antennas a knowledge of the electric and magnetic fields near the antenna is necessary. This is unlike the situation for finding radiation resistance, which can be found from the far fields in many cases. We will note briefly here the procedure used for a simple antenna, a cylindrical center-fed dipole whose current distribution is given by (2.22), with the current flowing in a thin layer at the antenna surface. The dipole is fed by a current across an infinitesimal gap at the dipole center. The magnetic vector potential can be found from (2.13), and from it the electric and magnetic fields at the surface. The fields are used to find the input resistance and reactance. The process results in equations that must be evaluated numerically, even for such a simple, idealized problem (Balanis, 1982, p. 293).

2.7. THE RECEIVING ANTENNA

We have discussed an antenna primarily as a transmitter. In this section, its operation as a receiver is considered.

Reciprocity

Consider two sets of sources, $\mathbf{J}^1$, $\mathbf{M}^1$ and $\mathbf{J}^2$, $\mathbf{M}^2$ in a linear isotropic medium. The Maxwell curl equations are

$$\nabla \times \mathbf{H}^i = \mathbf{J}^i + j\omega\epsilon\mathbf{E}^i \qquad i = 1, 2 \tag{2.24}$$

$$\nabla \times \mathbf{E}^i = -\mathbf{M}^i - j\omega\mu\mathbf{H}^i \qquad i = 1, 2 \tag{2.25}$$

where $\mathbf{E}^1$, $\mathbf{H}^1$ and $\mathbf{E}^2$, $\mathbf{H}^2$ are the fields produced by sources 1 and 2, respectively. Dot multiplying the first of these equations, with $i = 1$, by $\mathbf{E}^2$ and the second equation, with $i = 2$, by $\mathbf{H}^1$, adding, using a vector identity, and repeating, except with the multiplication of the first equation, with $i = 2$, by $\mathbf{E}^1$ and the second equation, with $i = 1$, by $\mathbf{H}^2$, leads to

$$-\nabla \cdot (\mathbf{E}^1 \times \mathbf{H}^2 - \mathbf{E}^2 \times \mathbf{H}^1) = \mathbf{E}^1 \cdot \mathbf{J}^2 - \mathbf{E}^2 \cdot \mathbf{J}^1 + \mathbf{H}^2 \cdot \mathbf{M}^1 - \mathbf{H}^1 \cdot \mathbf{M}^2$$

Integrating over a volume and using the divergence theorem on the left side yields

$$-\iint_{\text{closed}} (\mathbf{E}^1 \times \mathbf{H}^2 - \mathbf{E}^2 \times \mathbf{H}^1) \cdot d\mathbf{a}$$

$$= \iiint (\mathbf{E}^1 \cdot \mathbf{J}^2 - \mathbf{E}^2 \cdot \mathbf{J}^1 + \mathbf{H}^2 \cdot \mathbf{M}^1 - \mathbf{H}^1 \cdot \mathbf{M}^2)\, dv \tag{2.26}$$

This equation represents the *Lorentz reciprocity theorem*. In a source-free region it reduces to

$$\iint_{\text{closed}} (\mathbf{E}^1 \times \mathbf{H}^2 - \mathbf{E}^2 \times \mathbf{H}^1) \cdot d\mathbf{a} = 0$$

Impedance

The fields far from sources and material objects are related by

$$E_\theta = Z_0 H_\phi \qquad E_\phi = -Z_0 H_\theta$$

and the surface integral of (2.26) becomes, with integration over an infinitely large sphere,

$$-Z_0 \iint_{\text{closed}} (H_\theta^1 H_\theta^2 + H_\phi^1 H_\phi^2 - H_\theta^2 H_\theta^1 - H_\phi^2 H_\phi^1)\, da = 0$$

Then the Lorentz reciprocity theorem becomes

$$\iiint (\mathbf{E}^1 \cdot \mathbf{J}^2 - \mathbf{H}^1 \cdot \mathbf{M}^2)\, dv = \iiint (\mathbf{E}^2 \cdot \mathbf{J}^1 - \mathbf{H}^2 \cdot \mathbf{M}^1)\, dv \tag{2.27}$$

The integrals in this equation are called *reaction* (Rumsey, 1954). The reaction of field 1 on source 2 is

$$\langle 1, 2 \rangle = \iiint (\mathbf{E}^1 \cdot \mathbf{J}^2 - \mathbf{H}^1 \cdot \mathbf{M}^2)\, dv$$

In this notation, the reciprocity theorem in the form (2.27) is $\langle 1, 2 \rangle = \langle 2, 1 \rangle$.

For a current source I_2 with $\mathbf{M}^2 = 0$, the reaction is

$$\langle 1, 2 \rangle = \iiint \mathbf{E}^1 \cdot \mathbf{J}^2\, dv = \int \mathbf{E}^1 \cdot I_2\, d\mathbf{L} = I_2 \int \mathbf{E}^1 \cdot d\mathbf{L} = -V_{21} I_2 \tag{2.28}$$

where V_{21} is the voltage across source 2 due to the fields produced by source 1. Similarly, the voltage across source 1 due to fields produced by source 2 is $\langle 2, 1 \rangle = -V_{12} I_1$.

The reaction theorem is applicable to antennas in a transmit-receive configuration. It is also applicable to the linear two-port network shown in Fig. 2.5. The antenna configuration may therefore be considered a two-port network with voltages and currents related by

$$\begin{bmatrix} V_1 \\ V_2 \end{bmatrix} = \begin{bmatrix} Z_{11} & Z_{12} \\ Z_{21} & Z_{22} \end{bmatrix} \begin{bmatrix} I_1 \\ I_2 \end{bmatrix} \tag{2.29}$$

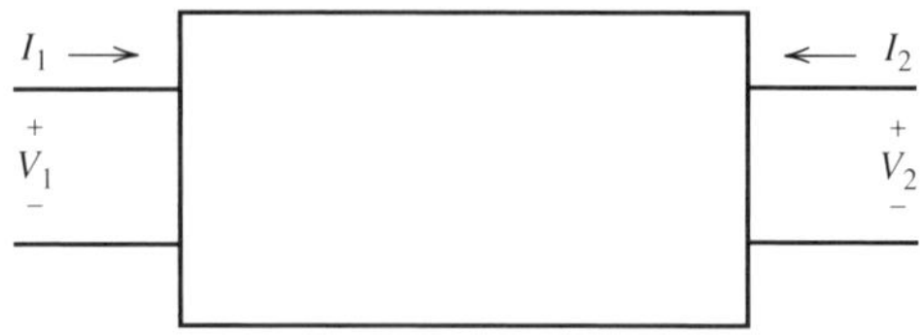

Fig. 2.5. Two-port network.

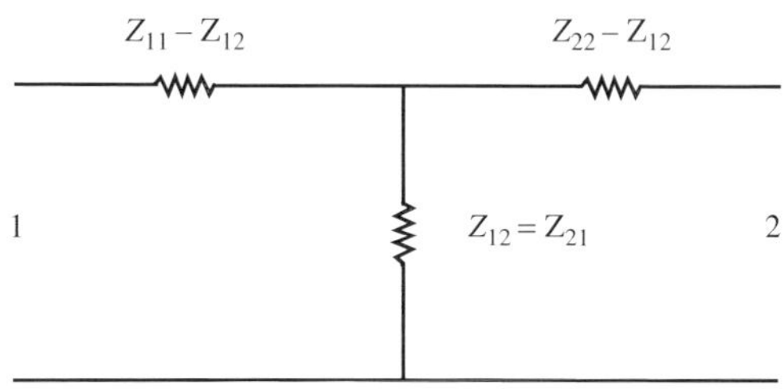

Fig. 2.6. Equivalent circuit of two antennas.

where the matrix is the impedance matrix. Each voltage in this equation may be written as the sum of two partial voltages,

$$V_1 = V_{11} + V_{12} = Z_{11}I_1 + Z_{12}I_2$$
$$V_2 = V_{21} + V_{22} = Z_{21}I_1 + Z_{22}I_2$$

If current sources are applied to ports 1 and 2, the partial voltage at port 2 due to the source at port 1 is $V_{21} = Z_{21}I_1$. This partial voltage is that of (2.28). Combining the equations gives

$$Z_{21} = \frac{V_{21}}{I_1} = -\frac{\langle 1, 2\rangle}{I_1 I_2}$$

If a current source at port 2 and partial voltage at port 1 are considered, it will be found that

$$Z_{12} = -\frac{\langle 2, 1\rangle}{I_1 I_2}$$

Since the reactions are equal, $Z_{12} = Z_{21}$. The two-port network can be antennas in a transmit-receive configuration with the equivalent circuit of Fig. 2.6. It is important to note that (2.29) and the equivalent circuit of Fig. 2.6 hold no matter which antenna transmits. If the antennas are widely separated, Z_{12} will be small, and an approximate equivalent circuit for the two antennas, with one of them, say 1, transmitting and the other receiving is shown in Fig. 2.7 (Ramo et al., 1984, p. 655).

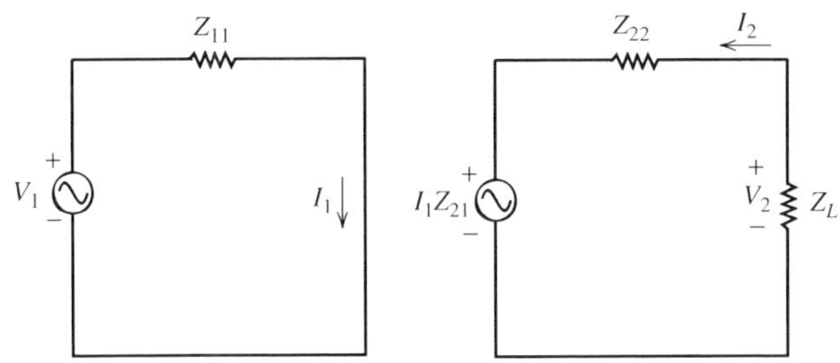

Fig. 2.7. Approximate equivalent circuit.

We may consider Z_{11} the input impedance of the transmitting antenna and neglect the effect of the receiving antenna on the transmitting antenna; this is an excellent approximation for widely separated antennas. The impedance of the receiving antenna in the approximate equivalent circuit, Z_{22}, would be the input impedance of antenna 2 if it were transmitting. We can call Z_{11} and Z_{22} the self-impedances of the antennas, and the approximate equivalent circuit shows that the self-impedance of an antenna is the same with the antenna transmitting and receiving. An interesting aspect of this equality is that it must hold for lossless antennas and for antennas with losses. If the self-impedance consists of a radiation resistance in series with a loss resistance, the total resistance is the same with the antenna transmitting and receiving. Also, the radiation resistance alone is the same with the antenna transmitting and receiving (the lossless case). It follows that loss resistance and antenna efficiency are the same with the antenna transmitting and receiving.

This discussion was given in terms of antenna impedances, but it would apply equally well if the antennas were described by their admittances. The two forms can be included in a general conclusion that the impedance (or admittance) and efficiency of an antenna are the same when the antenna is receiving and transmitting.

Receiving Pattern

An antenna receives a wave in a manner that is directional. It has a receiving pattern as well as a radiation pattern. The receiving pattern is a representation of the received power (or voltage) as a function of polar and azimuth angles when a polarization-matched plane wave is incident on the antenna. Consider two positions for antenna 2 of the transmit-receive antenna configuration of Fig. 2.8. Antenna 2 is moved along the surface of a sphere centered at antenna 1. Its orientation with respect to a line drawn between the two antennas remains the same. If the sphere radius is large and if the absolute phase of the signal at antenna 2 is not important, the exact location of the sphere center is unimportant if it is near antenna 1. The antennas are far apart, and a wave transmitted by one is effectively a plane wave at the other.

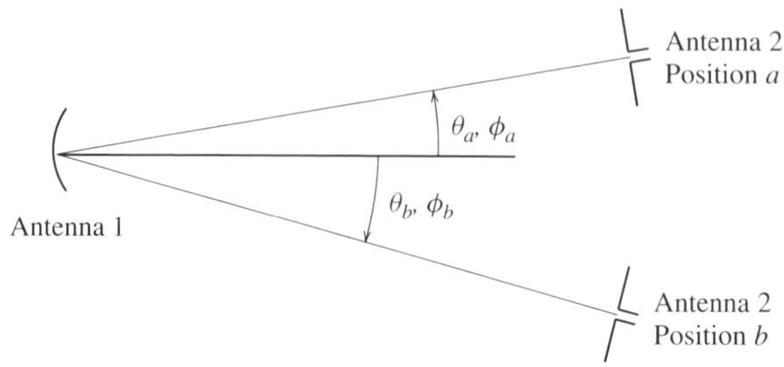

Fig. 2.8. Measurement of antenna patterns.

Let antenna 1 transmit and use the equivalent circuit of Fig. 2.7. Note that the mutual impedance term Z_{21} is a function of θ and ϕ. The ratio of powers received in positions a and b is

$$\frac{W_{2b}}{W_{2a}} = \frac{|Z_{21}(\theta_b, \phi_b)|^2}{|Z_{21}(\theta_a, \phi_a)|^2}$$

If position a is a reference position, this equation describes the relative radiation pattern of antenna 1.

In addition to the measurement of power to the load of antenna 2, at each position of antenna 2 the generator is connected to antenna 2 and the load to antenna 1; load power is then measured. An equivalent circuit similar to Fig. 2.7, with generator $Z_{12}I_2$, leads to a ratio of load powers in positions a and b,

$$\frac{W_{1b}}{W_{1a}} = \frac{|Z_{12}(\theta_b, \phi_b)|^2}{|Z_{12}(\theta_a, \phi_a)|^2}$$

This equation is the relative receiving pattern of antenna 1.

It was shown earlier that $Z_{12}(\theta_i, \phi_i) = Z_{21}(\theta_i, \phi_i)$. We conclude that the relative radiation and receiving patterns of an antenna are equal if the patterns represent received power.

Effective Area

The *effective area* presented by a receiving antenna to a polarization-matched plane wave incident from a given direction is the ratio of available power at the terminals of the antenna to the power density of the wave. "Available power" is the power that would be supplied to an impedance-matched load on the antenna terminals. The effective area of an antenna is normally a more useful concept than the transmitter current and mutual impedance because it is independent of the transmitter parameters and distance between the antennas. For aperture antennas it appears to be a natural characteristic. For wire antennas, the effective area does not correspond to a physical area of the antenna; nevertheless, it is a dimensionally correct and useful way to describe even a wire antenna.

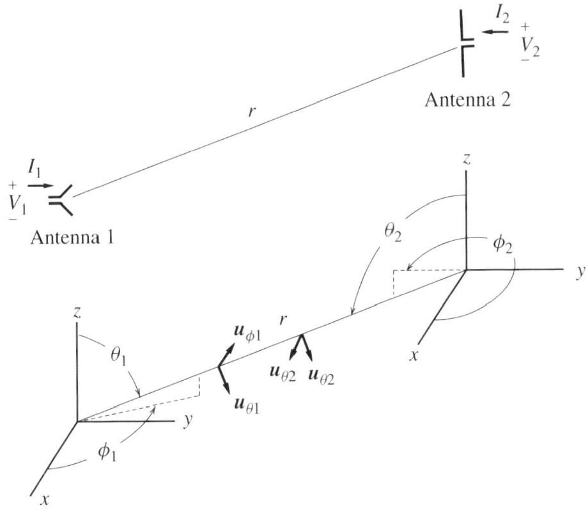

Fig. 2.9. Transmission and reception.

Consider two antennas in a transmit-receive configuration, as shown in Fig. 2.9. The antennas are oriented arbitrarily with respect to their coordinate systems and to each other. Antenna 1 is transmitting and 2 is receiving with an impedance-matched load. If antenna 1 accepts power W_{a1} from its generator and has gain G_1, the power density at 2 is

$$\mathcal{P} = \frac{W_{a1} G_1(\theta_1, \phi_1)}{4\pi r^2}$$

Power to the impedance-matched load is

$$W_{L2} = \mathcal{P} A_{e2}(\theta_2, \phi_2)$$

where A_{e2} is the effective area of antenna 2. Combining these equations gives

$$G_1(\theta_1, \phi_1) A_{e2}(\theta_2, \phi_2) = \frac{W_{L2}}{W_{a1}}(4\pi r^2) \tag{2.30}$$

If we reverse the transmitting and receiving roles of the antennas by connecting a generator to antenna 2 and causing it to accept power W_{a2}, the power to an impedance-matched load on antenna 1 is

$$W_{L1} = \frac{W_{a2} G_2(\theta_2, \phi_2) A_{e1}(\theta_1, \phi_1)}{4\pi r^2}$$

which gives

$$G_2(\theta_2, \phi_2)A_{e1}(\theta_1, \phi_1) = \frac{W_{L1}}{W_{a2}}(4\pi r^2) \tag{2.31}$$

With antenna 1 transmitting and power W_{a1} supplied to Z_{11} in Fig. 2.7, the ratio of load power in load Z_L to the power accepted by Z_{11} is

$$\frac{W_{L2}}{W_{a1}} = \frac{|Z_{21}|^2}{4\text{Re}(Z_{11})\text{Re}(Z_{22})}$$

If the roles of transmitter and receiver are reversed, the equivalent circuit shows that

$$\frac{W_{L1}}{W_{a2}} = \frac{|Z_{12}|^2}{4\text{Re}(Z_{11})\text{Re}(Z_{22})}$$

From the equality of Z_{12} and Z_{21}, it follows that

$$\frac{W_{L2}}{W_{a1}} = \frac{W_{L1}}{W_{a2}}$$

Using this relation, a comparison of (2.30) and (2.31) results in

$$\frac{A_{e1}(\theta_1, \phi_1)}{G_1(\theta_1, \phi_1)} = \frac{A_{e2}(\theta_2, \phi_2)}{G_2(\theta_2, \phi_2)}$$

The equation holds for lossless and lossy antennas. Antenna types were not specified in the development, and if ratio A_e/G is found for one antenna, lossless or lossy, it is known for all.

The simplest antenna from which the desired ratio can be found is an infinitesimal z-directed current source, for which the directivity and radiation resistance are given by (2.21) and (2.23). If this antenna is receiving, with a wave $\mathbf{E}$ incident on it from the direction specified by angle θ, the open-circuit voltage at the antenna terminals is $V_{\text{oc}} = EL\sin\theta$, where V_{oc} and E are taken as peak values of the corresponding sinusoidal form. Then the power to a matched load is

$$W = \frac{|V_{\text{oc}}|^2}{8R_{\text{rad}}} = \frac{|E|^2L^2\sin^2\theta}{8R_{\text{rad}}}$$

The power density at the antenna is

$$\mathcal{P} = \frac{1}{2Z_0}|E|^2$$

and therefore,

$$W = \frac{Z_0\mathcal{P}L^2\sin^2\theta}{4R_{\text{rad}}}$$

This gives an effective area for the lossless infinitesimal antenna,

$$A_e = \frac{W}{\mathcal{P}} = \frac{3\lambda^2 \sin^2 \theta}{8\pi}$$

The ratio of effective area to directivity is $\lambda^2/4\pi$. This ratio was obtained for a lossless example, but we saw earlier that it holds for the lossy case also. As a general rule, therefore, the effective area and gain of an antenna are related by

$$\frac{A_e(\theta, \phi)}{G(\theta, \phi)} = \frac{\lambda^2}{4\pi}$$

2.8. TRANSMISSION BETWEEN ANTENNAS

If the power accepted by antenna 1 in Fig. 2.9 were W_{at}, and if the antenna radiated isotropically, the power density at 2 would be $W_{at}/4\pi r^2$. Since it does not radiate isotropically but has gain G_t, the power density at 2 is $W_{at}G_t(\theta_t, \phi_t)/4\pi r^2$. The power W_r in the load on the receiving antenna is

$$W_r = \frac{W_{at}G_t(\theta_t, \phi_t)A_{er}(\theta_r, \phi_r)}{4\pi r^2} \tag{2.32}$$

with subscript r indicating the receiving antenna. This equation is known as the *Friis equation*. If the receiving antenna is not terminated by a matched load, (2.32) must be multiplied by an impedance match factor, or efficiency, to account for the mismatch loss. If the receiving antenna is represented by the series combination of R_a, including both radiation and loss resistances, and X_a, the antenna reactance, and if the load impedance is $R_L + jX_L$, it is easy to show that the impedance match factor is

$$M_z = \frac{4R_a R_L}{(R_a + R_L)^2 + (X_a + X_L)^2}$$

If the antennas are not polarization matched, (2.32) must be multiplied by a polarization efficiency.

2.9. ANTENNA ARRAYS

A configuration of identical radiating elements is an *antenna array*. The elements are fed by currents or voltages whose amplitudes and phases can be varied. Arrays and synthetic arrays are widely used in radar and are of interest in remote sensing. An array to be carried by an aircraft or spacecraft often has a more convenient shape and size than a single antenna with the same directivity. In fact, an array can be constructed to conform to the surface of an aircraft, and offers minimal resistance to aircraft motion. The main lobe of the array radiation pattern can be moved at will, or *scanned*, by electronic phase shifters for the feed currents. This beam motion is more rapid and flexible than a mechanical motion of the antenna. In this section, we consider special cases of a general array, specifically planar and linear arrays, which illustrate array principles.

Planar Arrays

Figure 2.10 shows the elements of a planar array on a rectangular grid in the xy plane, with element separations d_x and d_y in the x and y directions. The field produced by the array is

$$\mathbf{E}(\mathbf{r}) = \frac{1}{\sqrt{4\pi}} \sum_{0}^{M-1} \sum_{0}^{N-1} \frac{I_{mn}\mathbf{f}(\theta_{mn}, \phi_{mn})}{r_{mn}} e^{-jkr_{mn}}$$

where I_{mn} is the feed current in the mn element. In this equation, $\mathbf{f}(\theta_{mn}, \phi_{mn})$ accounts for the radiative properties of one element of the array and is called the *element pattern*. We take it to be independent of its position in the array and also

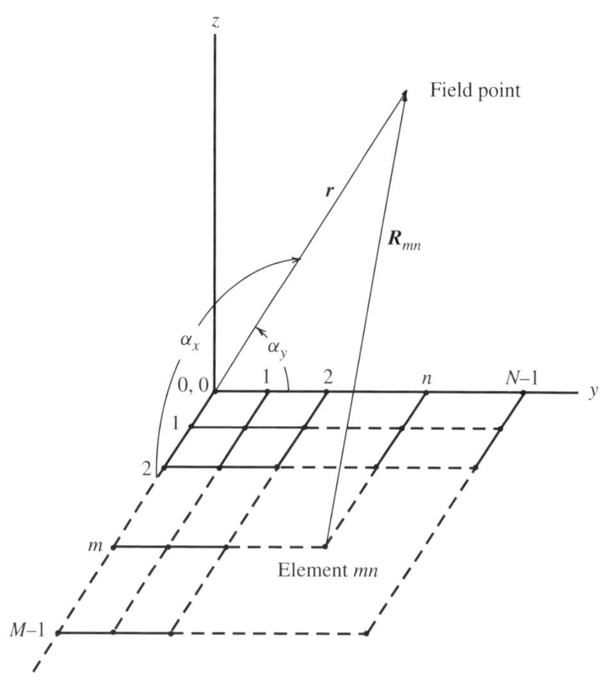

Fig. 2.10. Planar array.

take r_{mn} in the amplitude as constant. The radiated field then is

$$\mathbf{E}(\mathbf{r}) = \frac{\mathbf{f}(\theta, \phi)\, e^{-jkr}}{\sqrt{4\pi}\, r} \sum_{0}^{M-1} \sum_{0}^{N-1} I_{mn} e^{-jk(r_{mn}-r)} \tag{2.33}$$

The term preceding the double summation is the field that would be radiated by an array element located at the origin. It contains all information about the polarization of the wave from the array. The double sum in (2.33),

$$F(\theta, \phi) = \sum_{0}^{M-1} \sum_{0}^{N-1} I_{mn} e^{-jk(r_{mn}-r)}$$

is called the *array factor* and without r in the phase term would be the scalar field produced by an array of isotropic sources. We can now write the radiated field as

$$\mathbf{E}(\mathbf{r}) = \frac{1}{\sqrt{4\pi}\, r} \mathbf{f}(\theta, \phi) F(\theta, \phi) e^{-jkr} \tag{2.34}$$

The distance r_{mn} in the phase term can be approximated by

$$r_{mn} = \left[(x - md_x)^2 + \left(y - nd_y \right)^2 + z^2 \right]^{1/2} \approx r - md_x \cos\alpha_x - nd_y \cos\alpha_y$$

where α_x and α_y are angles measured from the x and y axes, respectively, to the line from the array to field point. The cosines in the equation are projections of a unit length along $\mathbf{r}$ onto the axes, so

$$\cos\alpha_x = \sin\theta \cos\phi$$
$$\cos\alpha_y = \sin\theta \sin\phi$$

where θ and ϕ are the polar and azimuth angles measured to the field point from the origin.

It is common practice to operate a planar array, as a transmitter, with feed-current phase advance δ_x between adjacent rows and feed-current phase advance δ_y between adjacent columns. When an array is used to receive a wave, the signals from the array elements are combined by phase shifting them before addition.

When an array is fed in the manner indicated, its radiated field is given by (2.34) with

$$F(\theta, \phi) = \sum_{0}^{M-1} \sum_{0}^{N-1} |I_{mn}| e^{j(m\Psi_x + n\Psi_y)} \tag{2.35}$$

where

$$\Psi_x = kd_x \cos\alpha_x + \delta_x = kd_x \sin\theta \cos\phi + \delta_x$$

$$\Psi_y = kd_y \cos\alpha_y + \delta_y = kd_y \sin\theta \sin\phi + \delta_y$$

Consideration of a uniform planar array, which is an array with all feed currents having the same amplitude, can lead to a better understanding of the radiation pattern structure. The series of (2.35) can be summed, for a uniform array, to be

$$F(\theta,\phi) = \frac{\sin(M\Psi_x/2)\sin(N\Psi_y/2)}{M\sin(\Psi_x/2)N\sin(\Psi_y/2)} \tag{2.36}$$

F is normalized to have a maximum value of one, and the phase of F is referenced to the array center.

The array pattern has maxima at angles for which $\sin(\Psi_x/2) = \sin(\Psi_y/2) = 0$. It can be seen from (2.36) that the array pattern has multiple maxima for sufficiently large values of kd_x and kd_y. The additional, and normally undesired, maxima are called *grating lobes* by analogy to optical scattering from a diffraction grating. The effect of undesired grating lobes can be reduced by using array elements with a small beamwidth.

The array pattern also has sidelobes with local maxima at angles for which $|\sin(M\Psi_x/2)| = |\sin(N\Psi_y/2)| = 1$.

Linear Arrays

If the planar array of Fig. 2.10 has only one row, say that on the y axis, the array factor (2.35) becomes, if we drop the unnecessary subscript y,

$$F(\theta,\phi) = \sum_{0}^{N-1} |I_n| e^{jn\Psi}$$

where

$$\Psi = kd\cos\alpha + \delta$$

The resulting array is a linear array. A uniform linear array has all current amplitudes the same and a constant phase difference between adjacent elements. Its normalized array factor, with phase referenced to the array center, is

$$F = \frac{\sin(N\Psi/2)}{N\sin(\Psi/2)}$$

This array factor shows that if the element factor **f** is not considered, the radiated field is maximum in directions for which $\Psi = 0, \pm\pi, \pm 2\pi \ldots$. Moreover, F has local magnitude maxima in directions for which $\sin(N\Psi/2) = \pm 1$. The array pattern has a lobed structure in a plane containing the line of the array and is rotationally symmetric about the array line. With choices of N, d, and δ, the pattern can be made to vary over a wide range.

A linear array with a moderate number of low-gain, closely spaced elements is of interest in radar remote sensing. The radiation pattern, determined in a plane containing the array, will be similar to Fig. 2.3. If the feed-current phase difference is zero, the array radiates primarily in a direction transverse to the line of the array, in the broadside direction. The array pattern whose section is shown is rotationally symmetric, but the directional properties of the elements will cause the overall pattern to lose rotational symmetry.

In many radar applications, the sidelobe amplitude of the pattern is unacceptably large. It can be reduced by choosing an array element with directional properties and by *tapering* the array currents; that is, by supplying the elements with unequal currents, with currents nearer the array center having greater magnitudes than those nearer the array ends. Tapering also increases the width of the main radiation lobe.

Another antenna configuration used in remote sensing of the earth is the *synthetic array* formed by using an antenna, or array, at multiple locations as it moves along a known path and combining the received signals. For purposes of an overall antenna pattern, this can be considered a linear array whose element spacing d is large compared to a wavelength. Figure 2.11 shows a portion of the array pattern of such an array. The principal pattern maxima, or grating lobes, occur at intervals of 2π in angle Ψ, and there are many such maxima in the region $0 \leq \alpha \leq \pi$. Between the grating lobes in the figure are local maxima at values of Ψ for which

$$N\Psi/2 = \frac{N\pi d}{\lambda}\cos\alpha + \frac{\delta}{2} = m\frac{\pi}{2} \qquad m = \pm 3, \pm 5, \ldots \qquad 0 \leq \alpha \leq \pi$$

with relative magnitudes

$$F = \frac{1}{N\sin\left(\frac{m}{N}\frac{\pi}{2}\right)}$$

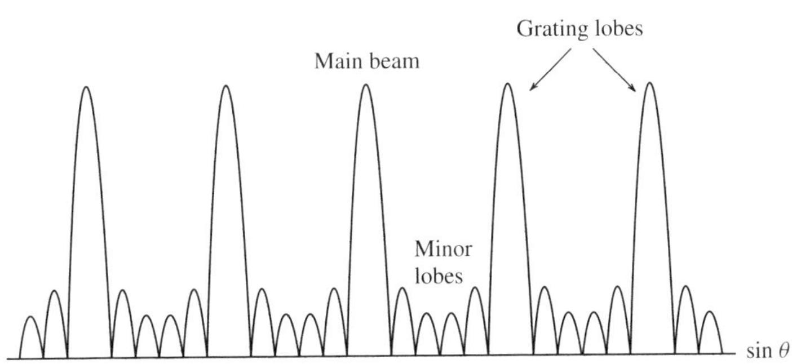

Fig. 2.11. Linear array factor with grating and minor lobes.

Another array of interest is that of a *radar interferometer*. It is an array of two elements (which may themselves be arrays) separated by many wavelengths. Its array pattern is similar to that of Fig. 2.11 with many grating lobes in the range $0 \leq \alpha \leq \pi$.

Beamwidth

With a linear array, choose $\delta = 0$ for convenience and consider the pattern lobe perpendicular to the line of the array, with beam maximum at $\Psi = kd\cos\alpha = 0$. Throughout this lobe, if the array spacing is sufficiently large to yield a narrow pattern lobe, Ψ is small and the array factor can be approximated by

$$F = \frac{\sin(N\Psi/2)}{N\Psi/2}$$

The half-power beamwidth corresponding to this array factor is

$$\frac{0.8856}{Nd/\lambda}$$

The product Nd approximates the array length L for N moderately large. Then, to a good approximation, the beamwidth of a linear array of length L is λ/L. It was noted earlier that the beamwidth of an antenna is inversely proportional to the antenna dimension. This development of the beamwidth of a linear array supports that statement.

If the main beam is scanned by altering phase angle δ, the beamwidth increases as the main beam is scanned away from the perpendicular (Mott, 1992, p. 73).

The Effect of Scanning on Element and Array Patterns

This consideration of arrays has been idealized in that we assumed an element pattern unaffected by the feed-current phase changes used to scan the array beams and that the feed-current phases can be selected without affecting the feed current amplitudes. In real arrays, there are coupling effects between the antenna elements that change when the beam position changes. The input impedance of an array element is affected by the overall field configuration at the array surface, and elements near the end or edge of the array are affected differently from those nearer the array center. Consequently, some array elements have feed-current amplitudes different from the expected amplitudes, and the array pattern is altered. In addition, the surface field effects may alter the current distribution on an element of the array, and the element pattern may not be the same for all elements. The design of antenna arrays, taking coupling effects into account, is a complex task.

2.10. EFFECTIVE LENGTH OF AN ANTENNA

The electric field in the far zone of a short dipole antenna of length L directed along the z-axis, given by (2.18), can be generalized to give the transmitted field of any antenna (Sinclair, 1950),

$$\mathbf{E}^t(r, \theta, \phi) = \frac{jZ_0 I}{2\lambda r} e^{-jkr} \mathbf{h}(\theta, \phi) \tag{2.37}$$

Current I is an input current at an arbitrary pair of terminals, and $\mathbf{h}(\theta, \phi)$ is the *effective length* of the antenna. Impedance Z_0 is the intrinsic impedance of free space, k the free space propagation constant, and λ the wavelength. The effective length does not necessarily correspond to a physical length of the antenna, but it is dimensionally a length for all antennas. For the short dipole of (2.18), $\mathbf{h} = \mathbf{u}_\theta h_\theta = \mathbf{u}_\theta L \sin\theta$. Current I is that at an arbitrary pair of terminals, and $\mathbf{h}(\theta, \phi)$ depends on the choice of terminal pair. If $\mathbf{E}^t$ is to describe an elliptically-polarized field, $\mathbf{h}$ is complex. With a proper choice of coordinate system, $\mathbf{E}^t$ and $\mathbf{h}$ will have only two components.

Let antenna 1 in Fig. 2.9 be general and antenna 2 be a short dipole on the z-axis of its coordinate system. The general antenna, fed by a current source of one ampere, transmits a wave toward the short dipole. If antenna 1 has effective length $\mathbf{h}$, its field at the dipole is given by (2.37) and the open-circuit voltage across the dipole terminals, with the polarity shown, is

$$V_2 = \mathbf{E}^{tT}\mathbf{L} = \frac{jZ_0}{2\lambda r} e^{-jkr} \mathbf{h}^T(\theta_1, \phi_1)\mathbf{L}(\theta_1, \phi_1)$$

where $\mathbf{L}$ is the effective length of the dipole.

Suppose the dipole, fed by a one-ampere current source, transmits and the general antenna, with open terminals, receives. The field produced at the general antenna is given by

$$\begin{bmatrix} E^i_{\theta_2} \\ E^i_{\phi_2} \end{bmatrix} = \frac{jZ_0}{2\lambda r} e^{-jkr} \begin{bmatrix} L_{\theta_2} \\ L_{\phi_2} \end{bmatrix} \tag{2.38}$$

If the coordinate systems of Fig. 2.9 are identical except for their origins, $\mathbf{u}_{\theta_1} = \mathbf{u}_{\theta_2}$ and $\mathbf{u}_{\phi_1} = -\mathbf{u}_{\phi_2}$. It follows that

$$\begin{bmatrix} E^i_{\theta_2} \\ E^i_{\phi_2} \end{bmatrix} = \begin{bmatrix} 1 & 0 \\ 0 & -1 \end{bmatrix} \begin{bmatrix} E^i_{\theta_1} \\ E^i_{\phi_1} \end{bmatrix}$$

$$\begin{bmatrix} L_{\theta_2} \\ L_{\phi_2} \end{bmatrix} = \begin{bmatrix} 1 & 0 \\ 0 & -1 \end{bmatrix} \begin{bmatrix} L_{\theta_1} \\ L_{\phi_1} \end{bmatrix}$$

If these equations are combined with (2.38), the incident field at the general antenna is

$$\mathbf{E}^i(\theta_1, \phi_1) = \frac{jZ_0}{2\lambda r} e^{-jkr} \mathbf{L}(\theta_1, \phi_1) \tag{2.39}$$

The open-circuit voltage induced in the general antenna is V_1. By the principle of reciprocity, if two antennas are fed by equal current sources, the open-circuit voltage produced across the terminals of antenna 1 by the current source feeding antenna 2 is equal to the open-circuit voltage produced across the terminals of antenna 2 by the current source feeding antenna 1. Then,

$$V_1 = V_2 = \frac{jZ_0}{2\lambda r} e^{-jkr} \mathbf{h}^T(\theta_1, \phi_1)\mathbf{L}(\theta_1, \phi_1)$$

If this expression is compared to (2.39), the received voltage can be written in terms of the incident field,

$$V_1 = \mathbf{h}^T(\theta, \phi)\mathbf{E}^i(\theta, \phi) \tag{2.40}$$

Subscript 1 for the angles is omitted from this equation because it is not needed. The equation is valid if both $\mathbf{E}^i$ and $\mathbf{h}$ are specified in a coordinate system with an axis pointing outward from the antenna. If $\mathbf{E}^i$ is given in a right-handed system with an axis in the direction of wave travel, V_1 becomes

$$V_1 = \mathbf{h}^T(\theta_1, \phi_1) \begin{bmatrix} 1 & 0 \\ 0 & -1 \end{bmatrix} \mathbf{E}^i(\theta_2, \phi_2) = \mathbf{h}^T(\theta_1, \phi_1)\mathbf{diag}(1, \text{-}1)\mathbf{E}^i(\theta_2, \phi_2)$$

In specifying **h** for an antenna, a terminal pair at which input current is to be measured must be chosen. Then V_1 is the open-circuit voltage measured across those terminals.

2.11. RECEPTION OF COMPLETELY POLARIZED WAVES

We will primarily use rectangular coordinate systems in succeeding developments, and it is desirable to reexamine the coordinate systems and the resulting equation for the received voltage.

Electric field **E** of a wave incident on an antenna can be written with two components in properly chosen coordinates because it is transverse to the direction of wave travel. Antenna effective length **h**, in coordinates with an axis oppositely directed to the direction of travel of the incident wave, may have three components, but only those transverse to the direction of the incoming wave are effective in producing a voltage at the antenna terminals.

Figure 2.12 shows coordinate systems applicable to a receiving antenna with an incoming wave. Let the antenna be at the origin of the xyz system with its effective length given in that system. The incident wave is described in $\xi\eta\zeta$

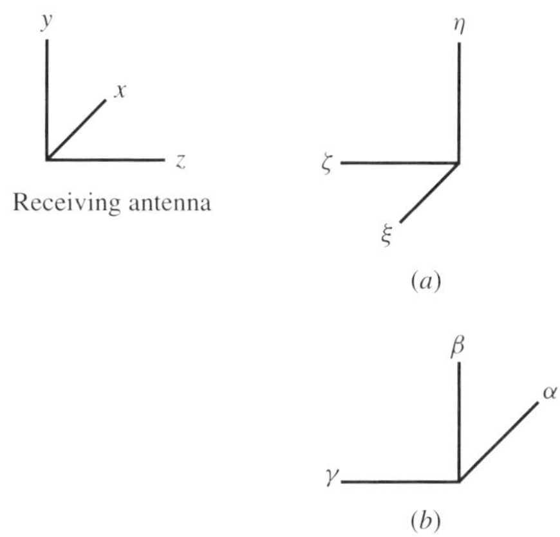

Fig. 2.12. Coordinate systems.

coordinates. We have chosen arbitrarily to align the z and ζ axes, pointing in opposite directions, and the y- and η-axes.

A z component of $\mathbf{h}$ has no effect, so we ignore it and write

$$\mathbf{h}(x, y) = \begin{bmatrix} h_x \\ h_y \end{bmatrix} \qquad \mathbf{E}(\xi, \eta) = \begin{bmatrix} E_\xi \\ E_\eta \end{bmatrix}$$

The electric field intensity vector in this form is sometimes called the Jones vector of the wave. Note again that the parenthetical symbols signify antenna and field components.

The equation for the received voltage with $\mathbf{E}$ and $\mathbf{h}$ expressed in the same coordinate system is

$$V = \mathbf{h}^T \mathbf{E}$$

In this work, $\mathbf{h}$ is always expressed in a right-handed coordinate system with an axis directed outward from the antenna; we therefore convert $\mathbf{E}$ to xyz coordinates,

$$\mathbf{E}(x, y) = \mathbf{diag}(-1, 1)\mathbf{E}(\xi, \eta)$$

The received voltage is

$$V = \mathbf{h}^T(x, y)\mathbf{E}(x, y) = \mathbf{h}^T(x, y)\mathbf{diag}(-1, 1)\mathbf{E}(\xi, \eta) \tag{2.41}$$

Left-Handed Coordinates

The electromagnetic wave scattered by a target is often given in a coordinate system that does not conform to the requirements outlined above. The simplest form to illustrate this case is shown in Fig. 2.12, for which we assume that a radar transmitting antenna is colocated with the receiving antenna.

If the wave from the transmitter incident on the target has components E_x^i and E_y^i, it is a common practice to use the reflected or scattered field components as $E_\alpha^s = E_x^s$ and $E_\beta^s = E_y^s$. The received voltage is

$$V = \mathbf{h}^T(x, y)\mathbf{E}^s(\alpha, \beta) = \mathbf{h}^T(x, y)\mathbf{E}^s(x, y)$$

The $\alpha\beta\gamma$ coordinate system of Fig. 2.12 is left handed. With our choice of the α and β axes to coincide with the x- and y-axes, we must use a left-handed coordinate system for the scattered wave or consider that the scattered wave travels in a direction opposite that of the third coordinate axis. With either choice, the wave descriptions that were developed previously for describing the wave, such as polarization ratio, ellipse tilt angle, axial ratio, and rotation sense, are no longer valid. A left-handed system is used in scattering problems primarily to find received voltage and power and is little used for a description of the wave itself.

Circular Components

Circular wave components are of sufficient interest in radar to warrant development of an equation for received voltage using them. A conversion previously developed between rectangular and circular wave components requires that the wave be in a right-handed system with z axis in the direction of wave travel. This is the $\xi\eta\zeta$ system of Fig. 2.12. Inversion of (1.34) yields

$$\mathbf{E}(\xi, \eta) = \begin{bmatrix} E_\xi \\ E_\eta \end{bmatrix} = \frac{1}{\sqrt{2}} \begin{bmatrix} 1 & 1 \\ j & -j \end{bmatrix} \begin{bmatrix} E_L \\ E_R \end{bmatrix} \tag{2.42}$$

and

$$\mathbf{h}(x, y) = \begin{bmatrix} h_x \\ h_y \end{bmatrix} = \frac{1}{\sqrt{2}} \begin{bmatrix} 1 & 1 \\ j & -j \end{bmatrix} \begin{bmatrix} h_L \\ h_R \end{bmatrix}$$

Bear in mind that E_L and E_R are associated with $\xi\eta\zeta$ coordinates and h_L and h_R with xyz.

With appropriate substitutions, the received voltage becomes

$$V = \mathbf{h}^T(x, y)\mathbf{diag}(-1, 1)\mathbf{E}(\xi, \eta) = -\begin{bmatrix} h_L & h_R \end{bmatrix} \begin{bmatrix} E_L \\ E_R \end{bmatrix} \tag{2.43}$$

General Orthonormal Basis

The incident wave can be transformed to an orthonormal basis $\mathbf{u}_1, \mathbf{u}_2$ by

$$\mathbf{E}(1, 2) = \mathbf{U}E(\xi, \eta) \tag{2.44}$$

where **U** is the unitary matrix, adapted with notational changes from (1.37),

$$\mathbf{U} = \begin{bmatrix} \langle \mathbf{u}_\xi, \mathbf{u}_1 \rangle & \langle \mathbf{u}_\eta, \mathbf{u}_1 \rangle \\ \langle \mathbf{u}_\xi, \mathbf{u}_2 \rangle & \langle \mathbf{u}_\eta, \mathbf{u}_2 \rangle \end{bmatrix}$$

The effective length of the receiving antenna can be transformed to basis $\mathbf{u}_3, \mathbf{u}_4$ by

$$\mathbf{h}(3, 4) = \mathbf{U}_h \mathbf{h}(x, y) \tag{2.45}$$

where

$$\mathbf{U}_h = \begin{bmatrix} \langle \mathbf{u}_x, \mathbf{u}_3 \rangle & \langle \mathbf{u}_y, \mathbf{u}_3 \rangle \\ \langle \mathbf{u}_x, \mathbf{u}_4 \rangle & \langle \mathbf{u}_y, \mathbf{u}_4 \rangle \end{bmatrix}$$

We require $\mathbf{u}_3$ and $\mathbf{u}_4$ to be related to $\mathbf{u}_x$ and $\mathbf{u}_y$ in the same way that $\mathbf{u}_1$ and $\mathbf{u}_2$ are related to $\mathbf{u}_\xi$ and $\mathbf{u}_\eta$. The transforming matrices **U** and $\mathbf{U}_h$ are therefore equal, and we can find the received voltage by substituting the transforms (2.44) and (2.45) into (2.41) and setting $\mathbf{U}_h = \mathbf{U}$. We also use the unitary nature of **U**, omit the functional notation, and obtain

$$V = \mathbf{h}^T \mathbf{U}^* \mathbf{diag}(-1, 1) \mathbf{U}^{-1} \mathbf{E}$$

2.12. GAIN, EFFECTIVE AREA, AND RADIATION RESISTANCE

From (1.17), (1.18), and (2.17), using Z_0 as real, the intensity of a radiated wave is

$$U(\theta, \phi) = r^2 \mathcal{P}(r, \theta, \phi) = \frac{1}{2Z_0} |\mathbf{E}(\theta, \phi)|^2$$

Antenna directivity is then

$$D(\theta, \phi) = \frac{|\mathbf{E}(\theta, \phi)|^2}{\frac{1}{4\pi} \iint_{4\pi} |\mathbf{E}(\theta, \phi)|^2 \, d\Omega} = \frac{|\mathbf{h}(\theta, \phi)|^2}{\frac{1}{4\pi} \iint_{4\pi} |\mathbf{h}(\theta, \phi)|^2 \, d\Omega}$$

Gain is the antenna efficiency e times the directivity, and gain and effective area are related by $\lambda^2/4\pi$, so that

$$A_e(\theta, \phi) = \frac{\lambda^2 e |\mathbf{h}(\theta, \phi)|^2}{\iint_{4\pi} |\mathbf{h}(\theta, \phi)|^2 \, d\Omega}$$

In this equation, θ and ϕ refer to the direction from which the wave comes to strike the receiving antenna.

Radiation resistance is the ratio of radiated power to the square of the rms current at arbitrarily chosen terminals. Then,

$$R_{\text{rad}} = \frac{2}{I^2} \frac{r^2}{2Z_0} \iint_{4\pi} |\mathbf{E}(\theta, \phi)|^2 \, d\Omega$$

where I is the peak value of a sine wave of current. If we use (2.37), the last equation becomes

$$R_{\text{rad}} = \frac{Z_0}{4\lambda^2} \iint_{4\pi} |\mathbf{h}(\theta, \phi)|^2 \, d\Omega$$

2.13. MAXIMUM RECEIVED POWER

Received power can be altered by selecting the effective length of the receiving antenna. If the problems of impedance mismatch are neglected, the power received by the antenna is proportional to the square of the magnitude of the open-circuit voltage. Received power is

$$W = \frac{VV^*}{8R_a} = \frac{1}{8R_a} |\mathbf{h}^T \mathbf{E}^i|^2$$

where R_a is the antenna resistance and $\mathbf{E}^i$ and $\mathbf{h}$ are expressed in the same coordinates.

Write the power as an inner product and use the Cauchy-Schwarz inequality (Horn and Johnson, 1990, p. 261),

$$|\langle \mathbf{h}, \mathbf{E}^{i*} \rangle|^2 \leq \langle \mathbf{h}, \mathbf{h} \rangle \langle \mathbf{E}^{i*}, \mathbf{E}^{i*} \rangle$$

Equality occurs only if $\mathbf{h}$ and $\mathbf{E}^{i*}$ are linearly related. For maximum received power, $\mathbf{h} = c\mathbf{E}^{i*}$, where c is some constant, real or complex, and the vectors are expressed in the same coordinate system. The maximum received power is then

$$W_m = \frac{1}{8R_a} |\mathbf{h}|^2 |\mathbf{E}^i|^2$$

2.14. POLARIZATION EFFICIENCY

The ratio of power received by an antenna of length $\mathbf{h}$ to that received under the most favorable circumstances from an incident wave is

$$\rho = \frac{|\mathbf{h}^T \mathbf{diag}(-1, 1) \mathbf{E}^i|^2}{|\mathbf{E}^i|^2 |\mathbf{h}|^2}$$

where $\mathbf{h}$ and $\mathbf{E}^i$ are in the right-handed coordinate systems of Fig. 2.12. If they are expressed in the same coordinate system, the diagonal matrix of this equation is omitted. The parameter ρ is the *polarization efficiency* or *polarization match factor* of the antenna and incident-wave combination. Take the receiving antenna length as $\mathbf{h}_1$, described in xyz coordinates at the origin of the coordinate system of Fig. 2.12. Consider that incident wave $\mathbf{E}_2$ was transmitted by an antenna with effective length $\mathbf{h}_2$ at the origin of the $\xi\eta\zeta$ system and described in those coordinates. Then ρ is

$$\rho = \frac{|\mathbf{h}_1^T \mathbf{diag}(-1, 1)\mathbf{h}_2|^2}{|\mathbf{h}_1|^2 |\mathbf{h}_2|^2} \tag{2.46}$$

The polarization ratio of an antenna is defined as the polarization ratio of the wave that it transmits. We can therefore write

$$\mathbf{h}_i = h_{ix}(\mathbf{u}_x + P_i \mathbf{u}_y) \qquad i = 1, 2$$

If these forms for the effective lengths are substituted into (2.46), the polarization efficiency becomes

$$\rho = \frac{(1 - P_1 P_2)(1 - P_1^* P_2^*)}{(1 + P_1 P_1^*)(1 + P_2 P_2^*)} \tag{2.47}$$

Special Cases

The antenna polarizations for the special cases giving efficiencies of one and zero are of interest. They are described here.

Polarization-Matched Antennas. If two polarization-matched antennas are in a transmit–receive configuration, the polarization efficiency, (2.47), is equal to one. A solution to the resulting equation is $P_2 = -P_1^*$. It can be shown that the axial ratios and rotation senses of the two antennas are equal, and the tilt angle of one is the negative of the tilt angle of the other (Mott, 1992, p. 196). It can also be shown that the rotation senses of the waves they transmit are the same. Then if the antennas were to transmit a wave toward each other simultaneously, the two waves would appear to rotate in opposite directions at a point in space at which they "meet".

Cross-Polarized Antennas. Two antennas in a transmit–receive configuration that are so polarized that no signal is received are said to be *cross-polarized*. For this situation, match factor (2.47) is zero, and $P_1 = 1/P_2$. The axial ratios of the antennas are equal and rotation senses of the polarization ellipses of the antennas are opposite, so that if both antennas transmitted simultaneously their field vectors would appear to rotate in the same direction (Mott, 1992, p. 196). The major axis of one polarization ellipse coincides with the minor axis of the other.

Orthogonal Wave Components. Consider orthogonal elliptically-polarized plane waves with fields $\mathbf{E}_a = E_a \mathbf{u}_a$ and $\mathbf{E}_b = E_b \mathbf{u}_b$. Vectors $\mathbf{u}_a$ and $\mathbf{u}_b$ are orthonormal, and the polarization ratios of the waves are respectively P_a and $P_b = -1/P_a^*$.

Let $\mathbf{E}_a$ be incident on a receiving antenna whose polarization ratio P_r maximizes received power. The polarization ratio value, from the preceding discussion in this section, is $P_r = -P_a^*$. Next, let the orthogonal wave $\mathbf{E}_b$ be incident on the same antenna. If, in the equation for polarization efficiency, (2.47), we set $P_1 = P_r = -P_a^*$ and $P_2 = P_b = -1/P_a^*$, polarization efficiency $\rho = 0$. We see that if a receiving antenna has a polarization that causes it to receive a wave without loss, it cannot receive an orthogonally polarized wave at all.

2.15. THE MODIFIED FRIIS TRANSMISSION EQUATION

The power received in a transmitting-receiving antenna configuration is given by the Friis equation, (2.32), modified to account for polarization mismatch,

$$W_r = \frac{W_{at} G_t A_{er} \rho}{4\pi r^2} = \frac{W_{at} G_t A_{er}}{4\pi r^2} \frac{|\mathbf{h}_r^T \mathbf{diag}(-1, 1)\mathbf{h}_t|^2}{|\mathbf{h}_r|^2 |\mathbf{h}_t|^2}$$

with W_{at} the transmitted power, G_t the gain of the transmitting antenna, A_{er} the effective area of the receiving antenna, and r the distance between them.

From Section 2.12, $R_t G_t = \pi Z_0 |\mathbf{h}_t|^2/\lambda^2$ and $A_{er} R_r = Z_0 |\mathbf{h}_r|^2/4$. If these equations are combined with the previous one, another form of the Friis equation is obtained,

$$W_r = \frac{W_{at} Z_0^2}{16\lambda^2 R_t R_r r^2} |\mathbf{h}_r^T \mathbf{diag}(-1, 1)\mathbf{h}_t|^2$$

If $\mathbf{h}_r$ and $\mathbf{h}_t$ are described in the same coordinates, the diagonal matrix is omitted from this equation.

2.16. ALIGNMENT OF ANTENNAS

In a transmit–receive configuration the radiated field components and the effective length components of the receiving antenna may be known in separate coordinate systems having axes that are not aligned. The effective lengths and field components must be transformed to other coordinate systems before determining the polarization efficiency of the antenna pair.

Figure 2.13 shows the coordinate systems to be considered. The system without superscripts or subscripts is a reference system. The a system at x_1, y_1, z_1 (position 1) is appropriate to the transmitting antenna, with the radiated fields known in that system. The b system is rotated so that its z axis points toward the receiving antenna. Likewise, the c system at position 2 is the natural one for the

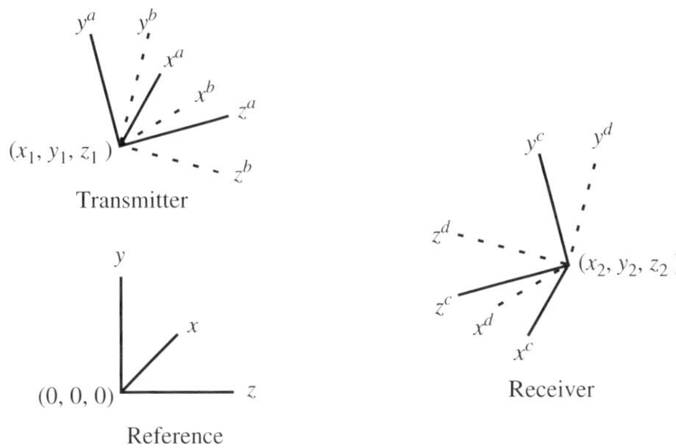

Fig. 2.13. Misaligned transmitting and receiving antennas.

receiving antenna, the one in which its radiated field is known. The d system is rotated so that its z axis points to the transmitting antenna. The z axes of systems b and d are antiparallel.

Coordinate systems b and d have a degree of arbitrariness. It is required only that their z axes be directed toward the other antenna. Their x and y axes may be chosen at will, subject to the right-handed orthogonality of each.

We use **E** to represent the field of the transmitting antenna and **h** for the effective length of the receiving antenna. A letter superscript refers to the coordinate system in which a quantity is measured. The locations of transmitter and receiver are

$$\mathbf{X}_i = \begin{bmatrix} x_i & y_i & z_i \end{bmatrix}^T \qquad i = 1, 2$$

The transformation by rotations of a point from coordinate system p to system q, having the same origin, is carried out by the Euler-angle matrix,

$$\mathbf{U}_E = \begin{bmatrix} \cos\beta\cos\gamma & \cos\beta\sin\gamma & -\sin\beta \\ \sin\alpha\sin\beta\cos\gamma - \cos\alpha\sin\gamma & \sin\alpha\sin\beta\sin\gamma + \cos\alpha\cos\gamma & \sin\alpha\cos\beta \\ \cos\alpha\sin\beta\cos\gamma + \sin\alpha\sin\gamma & \cos\alpha\sin\beta\sin\gamma - \sin\alpha\cos\gamma & \cos\alpha\cos\beta \end{bmatrix}$$

The angles α, β, γ are measured from an axis in the old system (p) toward the corresponding axis in the new (q). The rotations are taken in order:

1. γ around the z-axis in the direction $x \to y$.
2. β around the y-axis in the direction $z \to x$.
3. α around the x-axis in the direction $y \to z$.

For two systems having the same origin, the location of point $\mathbf{X}_q$ in the new system is related to its location $\mathbf{X}_p$ in the old system by

$$\mathbf{X}_q = \mathbf{U}_E \mathbf{X}_p$$

Transformation of vector functions is carried out by the same matrix; thus

$$\mathbf{F}_q = \mathbf{U}_E \mathbf{F}_p$$

The following Euler-angle matrices are used for the coordinate systems of Fig. 2.13:

A from the ground/reference system (translated to x_1, y_1, z_1) to system a.
B from the ground/reference system (translated to x_1, y_1, z_1) to system b.
C from the ground/reference system (translated to x_2, y_2, z_2) to system c.
D from the ground/reference system (translated to x_2, y_2, z_2) to system d.

In many cases, the geometry is simpler than this general case. The transmitter, for example, may also be the reference system and z^a may already point to the receiver, making two transformations unnecessary.

The positions and orientations of the transmitting and receiving antennas with respect to the reference system are known; that is, locations $\mathbf{X}_1$ and $\mathbf{X}_2$ and matrices **A** and **C** are known. From the considerations given above, we can create the Euler-angle matrices **B** and **D**. Further, we know the fields E_θ^a and E_ϕ^a of the transmitting antenna in its natural coordinate system and the effective length components h_θ^c and h_ϕ^c of the receiving antenna in its natural coordinate system.

Below is a process for transforming the antenna effective lengths and field components to allow the received power to be found. It is not the only feasible process, but it is simple and efficient.

Step 1. Translate the reference system to point 1 (at x_1, y_1, z_1). Find the position $\mathbf{X}_2'$ of point 2 (at x_2, y_2, z_2) with respect to point 1.

Step 2. Use the Euler angle matrix **A** to find point 2 in the natural system (system a) of the transmitter.

$$\mathbf{X}_2^a = \mathbf{A} X_2'$$

Determine the colatitude and azimuth angles of point 2 in system a.

Step 3. From the known properties of the transmitter, find E_θ^a and E_ϕ^a at point 2. The absolute values of the fields must be found if power levels throughout are needed. This requires knowledge of transmitted power and the distance from transmitter to point 2. If relative values are sufficient, this distance and the transmitter power may be neglected.

Step 4. Convert E_θ^a and E_ϕ^a at point 2 to rectangular form.

$$E_x^a = E_\theta^a \cos\theta \cos\phi - E_\phi^a \sin\phi$$
$$E_y^a = E_\theta^a \cos\theta \sin\phi + E_\phi^a \cos\phi$$
$$E_z^a = -E_\theta^a \sin\theta$$

where θ and ϕ are the known values at the point 2 found in step 2. The subscripts refer to axes in the a system.

Step 5. Transform the field components at position 2 to system c, going to the reference system as an intermediate step and then to system c using the known matrix **C**,

$$\mathbf{E} = \mathbf{A}^T \mathbf{E}^a$$
$$\mathbf{E}^c = \mathbf{C}E = \mathbf{C}A^T \mathbf{E}^a$$

Step 6. Find the receiver open-circuit voltage from $\mathbf{E}^c$ and $\mathbf{h}_2^c$.

REFERENCES

C. A. Balanis, *Antenna Theory: Analysis and Design*, Harper & Row, New York, 1982.

R. F. Harrington, *Time-Harmonic Electromagnetic Fields*, McGraw-Hill, New York, 1961.

R. A. Horn and C. R. Johnson, *Matrix Analysis*, Cambridge University Press, Cambridge, UK, 1990.

J. D. Kraus, *Antennas*, 2nd ed., McGraw-Hill, New York, 1988.

H. Mott, *Antennas for Radar and Communications: A Polarimetric Approach*, Wiley-Interscience, New York, 1992.

S. Ramo, J. R. Whinnery, and T. Van Duzer, *Fields and Waves in Communication Electronics*,Wiley, New York, 1984.

V. H. Rumsey, “The Reaction Concept in Electromagnetic Theory”, *Phys. Rev., Ser. 2*, **94**, (6), 1483–1491, (June 15, 1954).

S. Silver, *Microwave Antenna Theory and Design*, Boston Technical Lithographers, Lexington, MA, 1963. (MIT Radiation Lab. Series, Vol. 12, McGraw-Hill, New York, 1949.)

G. Sinclair, “The Transmission and Reception of Elliptically Polarized Waves”, *Proc. IRE*, **38**, (2), 148–151, (February, 1950).

PROBLEMS

2.1. A radar transmitting antenna has a gain of 30 dB in the direction of a target and transmits pulses with peak power 1 MW. Find the power density at the target 50 km from the radar.

2.2. The time-average Poynting vector of a wave radiated by an antenna is

$$\mathcal{P}(\theta, \phi) = \frac{C}{r^2} \sin^2\theta \, \sin^2\phi \qquad 0 \le \theta \le \pi,\ 0 \le \phi \le 2\pi$$

Find the directivity $D(\theta, \phi)$ of the antenna. If the design frequency is 1 GHz and the radiation efficiency is 0.96, find the effective receiving area of the antenna.

2.3. A linear array has 12 isotropic elements spaced one-half wavelength. All elements are fed in phase. Find the antenna beamwidth in a plane containing the array. Find the angular location of the first sidelobe and its amplitude compared to the main lobe.

2.4. If the isotropic elements of the array of Problem 2.3 are replaced by short dipole elements lying in the line of the array, find the beamwidth.

2.5. An antenna with gain 30 dB at 10 GHz is pointed at the sun through a lossy radome and lossy atmosphere. The spectral flux density of the sun's rays is 10^{-17}W/m^2 Hz. Assume that 92% of the power is transmitted through the radome and that atmospheric loss is 0.5 dB. Find the equivalent antenna temperature.

2.6. A 12×12 array in the xy plane has isotropic elements spaced 0.75 wavelengths apart in each row and column. The phase advance of the feed currents between rows and columns is 30°. Find the polar and aximuth angles of the main antenna beam. Find the location of the sidelobe closest to the main beam. Find the location of the grating lobe nearest the main beam.

2.7. To obtain information about an antenna, it is used as a transmitter with a feed current of 5 A. A receiving antenna with an effective area of 0.5 m^2, polarization matched to the incoming wave and impedance-matched to its receiver, is placed 500 m from the transmitter. The transmitting antenna is rotated so that the receiver sees the transmitter over a full range of polar and azimuth angles. The received power is found to be given by

$$W = 600 \sin^2\theta \, \sin^2\phi \qquad \mu\text{W} \qquad 0 \le \theta \le \pi,\ 0 \le \phi \le \pi$$

and zero outside this range. Find the directivity of the transmitting antenna. Find its radiation resistance.

CHAPTER 3

COHERENTLY SCATTERING TARGETS

In this chapter, we consider mathematical forms for describing scattering by targets whose scattered waves are completely polarized.

3.1. RADAR TARGETS

When a radar wave strikes a target, part of the incident energy is reflected, or scattered. If the incident wave is monochromatic, if the target is unchanging, and if the radar-target aspect angle is constant, the scattered wave will also be monochromatic and completely polarized. Such a target has been called a *single target*, a *point target*, or a *deterministic target*. A more descriptive name is *coherently scattering target*, and that designation is used here. If the target's scattering properties change with time, as would be the case if the radar were examining ocean waves or wind-stirred tree foliage, the scattered wave is modified by the target motion and covers a band of frequencies; it is partially coherent and partially polarized. The target may have more than one scattering center (a point at which the incident wave can be considered to be reflected). If the radar transmits and receives multiple pulses, each from a different location, as is common with synthetic aperture radars, a different target is effectively seen with each pulse. The received pulses are combined, and the result is equivalent to a time-varying target that incoherently scatters and depolarizes the wave. A target belonging to either category has been called a *distributed target*, but we will normally refer to it as an *incoherently scattering* or *depolarizing* target. In the following discussion, we consider only coherently scattering targets.

Remote Sensing with Polarimetric Radar, by Harold Mott

Coordinate Systems

Figure 3.1 shows the coordinate systems to be used in this discussion. Wave travel is in the $\pm z$ directions of the coordinates, and all antennas are described with z-axes pointing away from the antenna in the direction of wave propagation of interest, or, in the case of a receiving antenna, in the direction from which an incident wave arrives. For a transmitting antenna, the z-axis need not be pointing in the direction of the antenna beam maximum; for a receiving antenna, the z-axis need not be pointing in the direction for which the antenna would develop maximum received voltage from an incoming wave. The antenna behavior is accounted for by use of its effective length as a function of direction. The wave components are taken, without any restriction on their utility, in the x and y directions. It is common to give wave components as horizontal (H) and vertical (V), but we avoid that. What is meant by horizontal and vertical must be carefully defined, and a convenient symbol for the propagation axis is not universally used. When one speaks of a right-handed xyz coordinate system, no confusion is likely to result.

A transmitting antenna radiates a wave in the z_1 direction, with x_1 and y_1 components. This radiated wave is incident on a target and is scattered at all angles. The scattered wave can be described by the right-handed coordinates $x_2y_2z_2$, with z_2 pointing away from the target in the direction of a receiving antenna. To describe the receiving antenna, right-handed coordinates $x_3y_3z_3$, with z_3 pointing from the receiving antenna to the target, are used. The $x_2y_2z_2$ coordinate system is sometimes called "wave-oriented" and the $x_3y_3z_3$ system an "antenna-oriented" coordinate system. The scattered wave can also be described in the left-handed system $x_4y_4z_4$ and its components given as x_4 and y_4, which coincide with x_3 and y_3.

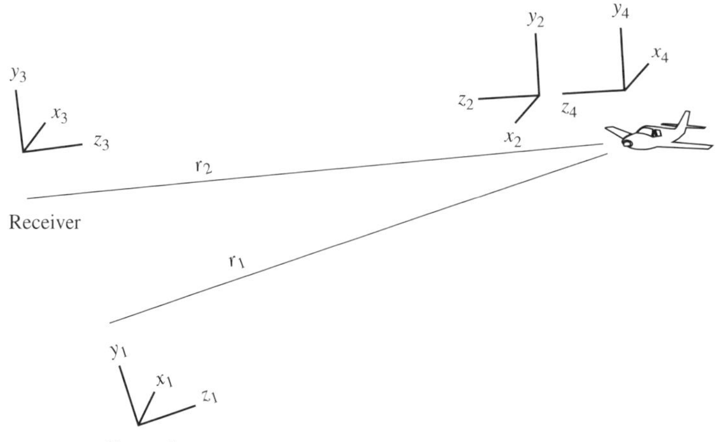

Fig. 3.1. Coordinate systems for scattering.

A *plane of scattering* is defined by propagation vectors pointing, respectively, in the directions of incident and scattered waves. Since the target may scatter waves over a range of angles, it is more precise to say that the propagation vectors point, respectively, from transmitter to target and target to receiver. If the angle between the propagation vectors is 0 or π, a scattering plane is undefined and may be chosen arbitrarily. In Fig. 3.1, z_1 and z_2 determine the plane of scattering. The coordinate systems were chosen for this discussion so that the y- and z-axes of all three coordinate systems lie in this plane. Both z_1 and z_3 point at the target, and z_2 points toward the receiver from the target. These choices are in accord with the convention adopted in Section 2.13 to describe a wave incident on a receiving antenna.

If the angle between the z_1-axis and the z_2-axis is between $\pi/2$ and π, the scattering from the target is called *backward scattering*; if the angle is between 0 and $\pi/2$, it is called *forward scattering*. If the angle is π and the receiver is located in the same direction from the target as the transmitter, the scattering process is called *backscattering*.

The Scattered Wave

The scattered wave from a target is dependent on the target and on the incident wave. The incident and scattered waves are described by two-element vectors, and the relationship between them is a 2×2 matrix, which depends only on the target. We call this matrix the "scattering matrix," a general name which includes the Sinclair and Jones matrices introduced in this chapter, and for some purposes the 4×4 target matrices introduced in Chapter 7.

3.2. THE JONES MATRIX

If the incident wave is in $x_1y_1z_1$ coordinates and the scattered wave in $x_2y_2z_2$, the waves are related by

$$\begin{bmatrix} E^s_{x2} \\ E^s_{y2} \end{bmatrix} = \frac{1}{\sqrt{4\pi}\, r} \begin{bmatrix} T_{x2x1} & T_{x2y1} \\ T_{y2x1} & T_{y2y1} \end{bmatrix} \begin{bmatrix} E^i_{x1} \\ E^i_{y1} \end{bmatrix} e^{-jkr}$$

where r is the distance between the point at which the scattered field is measured and an arbitrary reference plane at the target where the matrix elements are determined.

If the coordinate systems used for describing the waves are kept in mind, the numerical subscripts can be dropped and the field relationship written as

$$\begin{bmatrix} E^s_{x} \\ E^s_{y} \end{bmatrix} = \frac{1}{\sqrt{4\pi}\, r} \begin{bmatrix} T_{xx} & T_{xy} \\ T_{yx} & T_{yy} \end{bmatrix} \begin{bmatrix} E^i_{x} \\ E^i_{y} \end{bmatrix} e^{-jkr} \qquad (3.1)$$

or as

$$\mathbf{E}^s = \frac{1}{\sqrt{4\pi} r} \mathbf{T}\mathbf{E}^i e^{-jkr} \tag{3.2}$$

Matrix **T** is the *Jones Matrix*. The equation is valid either for forward or backward scattering.

The incident field is related to the transmitting antenna effective length by

$$\mathbf{E}^i(x_1, y_1) = \frac{jZ_0 I}{2\lambda r_1} \mathbf{h}_t(x_1, y_1) e^{-jkr_1} \tag{3.3}$$

where Z_0 is the characteristic impedance of the medium, I is the transmitting antenna current at some chosen port, λ is the wavelength, and $\mathbf{h}_t$ is the effective length of the transmitting antenna. The field and effective length are written with the components in parentheses as a reminder that coordinates $x_1 y_1 z_1$ are used, and the parenthetical terms are not functional arguments.

Received Voltage and Power

If we adapt (2.41) to find the voltage induced in the receiving antenna of length $\mathbf{h}_r$, we get

$$\begin{aligned} V &= \mathbf{h}_r^T(x_3, y_3)\,\mathbf{diag}(-1, 1)\mathbf{E}^s(x_2, y_2) \\ &= \frac{jZ_0 I}{2\sqrt{4\pi}\lambda r_1 r_2} \mathbf{h}_r^T(x_3, y_3)\,\mathbf{diag}(-1, 1)\mathbf{T}\mathbf{h}_t(x_1, y_1) e^{-jk(r_1 + r_2)} \end{aligned}$$

The parenthetical coordinate notation for the effective lengths can be dropped if we keep in mind that both antennas are described in right-handed coordinates with the z-axis pointing from antenna to target. Power to a matched load is

$$W = \frac{VV^*}{8R_a} = \frac{Z_0^2 I^2}{128\pi R_a \lambda^2 r_1^2 r_2^2} |\mathbf{h}_r^T \mathbf{diag}(-1, 1)\mathbf{T}\mathbf{h}_t|^2$$

3.3. THE SINCLAIR MATRIX

If the incident wave is in $x_1 y_1 z_1$ coordinates and the scattered wave has x_3 and y_3 components, the relationship between scattered and incident waves is

$$\begin{bmatrix} E_{x3}^s \\ E_{y3}^s \end{bmatrix} = \frac{1}{\sqrt{4\pi} r} \begin{bmatrix} S_{x3x1} & S_{x3y1} \\ S_{y3x1} & S_{y3y1} \end{bmatrix} \begin{bmatrix} E_{x1}^i \\ E_{y1}^i \end{bmatrix} e^{-jkr}$$

If we keep the coordinates in mind, we can drop the numerical subscripts and write

$$\begin{bmatrix} E_x^s \\ E_y^s \end{bmatrix} = \frac{1}{\sqrt{4\pi} r} \begin{bmatrix} S_{xx} & S_{xy} \\ S_{yx} & S_{yy} \end{bmatrix} \begin{bmatrix} E_x^i \\ E_y^i \end{bmatrix} e^{-jkr}$$

or

$$\mathbf{E}^s = \frac{1}{\sqrt{4\pi}\,r}\mathbf{S}\mathbf{E}^i e^{-jkr} \tag{3.4}$$

$\mathbf{S}$ is the *Sinclair Matrix*.

The scattered field components in this formulation are written in what would be a left-handed system if the wave were described in the normal manner with a coordinate axis in the direction of wave propagation. If the Sinclair matrix is used, care must be exercised to account for the left-handed system.

Use of the Sinclair matrix, like that of the Jones matrix, is valid for backward or forward scattering. The matrix is particularly useful for backscattering, for which coordinates $x_1y_1z_1$ and $x_3y_3z_3$ coincide except for a translation along the common z-axis.

If all sources and matter are of finite extent and if there are no magnetic sources, the Lorentz reciprocity theorem becomes

$$\iiint (\mathbf{E}_1^T\mathbf{J}_2 - \mathbf{E}_2^T\mathbf{J}_1)\,dv = 0$$

where $\mathbf{E}_1$ and $\mathbf{E}_2$ are the fields produced by sources $\mathbf{J}_1$ and $\mathbf{J}_2$, respectively. The theorem is valid for the scattering configuration of Fig. 3.1 if we use antenna 1 as the transmitter and antenna 2 as the receiver, or *vice versa*. The presence of a scatterer does not invalidate the theorem.

Let the antennas of Fig. 3.1 have infinitesimal extent. Then the integral of the current density over the antenna volume is proportional to the effective length of the antenna, and the Lorentz reciprocity theorem becomes

$$\mathbf{E}_1^T\mathbf{h}_2 = \mathbf{E}_2^T\mathbf{h}_1 \tag{3.5}$$

$\mathbf{E}_1$ in this equation is the field produced at position 2 by the antenna of length $\mathbf{h}_1$. $\mathbf{E}_2$ is the field produced at position 1 by the antenna of length $\mathbf{h}_2$. $\mathbf{E}_1$ and $\mathbf{h}_2$ are described in the same coordinate system, and so are $\mathbf{E}_2$ and $\mathbf{h}_1$.

Let the antenna at position 1 transmit. Then, neglecting constants in the definition of the Sinclair matrix, the electric field at position 2 is

$$\mathbf{E}_1(x_3, y_3) = \mathbf{S}_1(x_1, y_1, x_3, y_3)\mathbf{h}_1(x_1, y_1)$$

where the parenthetical notation indicates the vector components and is not a functional relationship.

Let the antenna at position 2 transmit. The field it produces at position 1 is

$$\mathbf{E}_2(x_1, y_1) = \mathbf{S}_2(x_3, y_3, x_1, y_1)\mathbf{h}_2(x_3, y_3)$$

From (3.5) and the equations for the fields,

$$(\mathbf{S}_1\mathbf{h}_1)^T\mathbf{h}_2 = (\mathbf{S}_2\mathbf{h}_2)^T\mathbf{h}_1$$

Transposing terms appropriately gives $\mathbf{h}_2^T\mathbf{S}_1\mathbf{h}_1 = \mathbf{h}_2^T\mathbf{S}_2^T\mathbf{h}_1$, and it follows that $\mathbf{S}_2 = \mathbf{S}_1^T$. For backscattering with colocated transmitter and receiver, there can be only one Sinclair matrix. Then $\mathbf{S}_2 = \mathbf{S}_1 = \mathbf{S}_1^T$. The Sinclair matrix is symmetric for backscattering.

Received Voltage and Power

If we use the Sinclair matrix and repeat the steps followed in determining received voltage and power in terms of the Jones matrix, they are found to be

$$V = \frac{jZ_0 I}{2\sqrt{4\pi}\lambda r_1 r_2}\mathbf{h}_r^T(x_3, y_3)\mathbf{S}\mathbf{h}_t(x_1, y_1)e^{-jk(r_1+r_2)} \tag{3.6}$$

and

$$W = \frac{Z_0^2 I^2}{128\pi R_a \lambda^2 r_1^2 r_2^2}|\mathbf{h}_r^T\mathbf{S}\mathbf{h}_t|^2 \tag{3.7}$$

If $\mathbf{S}$ is unknown, it can be found with the help of (3.6) by measuring V for selected values of $\mathbf{h}_r$ and $\mathbf{h}_t$.

3.4. MATRICES WITH RELATIVE PHASE

It is not possible in general to measure the transmitted wave at the "position" of the target. If the target is a point, or if it lies completely in a plane perpendicular to the transmitter-target line of sight, its position is unambiguous; it does not generally satisfy one of these requirements, and an arbitrarily located reference plane transverse to the transmitter-target line of sight and near to or intersecting the target must be chosen. The transmitted field in this plane is the incident wave $\mathbf{E}^i$, and the plane is the reference plane for the incident wave. Similarly, a plane must be chosen near the target and transverse to the line of sight from receiver to target. This is the reference plane for the scattered wave. When the reference planes are chosen, we can write (3.4) as

$$\mathbf{E}^s = \frac{1}{\sqrt{4\pi}r_0}\mathbf{S}\mathbf{E}^i e^{-jkr_0}$$

In this equation, r_0 is measured from the scattering reference plane to the receiving antenna phase center, and $\mathbf{E}^i$ is measured at the incident-wave reference plane.

If both reference planes are located within a distance from the target that is much smaller than transmitter-target and receiver-target distance, the element magnitudes in $\mathbf{S}$ are unaffected by the location of the reference planes. The phase angles of the elements of $\mathbf{S}$ are affected by the placement of the reference planes and are arbitrary. They are equally affected, however, and their relative phase angles are independent of the choice of reference planes. The received power from the scattered wave and the scattered wave polarization are not affected by a common phase term of all elements of **S**.

The Jones and Sinclair matrices with arbitrarily selected reference planes are known as the Jones and Sinclair matrices with relative phase. Since the reference planes for incident and scattered waves are essentially arbitrary, except for point targets, we will neglect the description "relative" when discussing the Sinclair and Jones matrices.

It can be seen that the Sinclair matrix has seven independent real parameters, four magnitudes and three phases. For backscattering, **S** is symmetric and has only five independent real parameters.

3.5. FSA–BSA CONVENTIONS

It is convenient to use the single coordinate system made possible by the Sinclair matrix if the radar uses colocated transmitting and receiving antennas. More generally, any backward scattering configuration using the Sinclair matrix has been referred to as the Backscatter Alignment, BSA, (van Zyl et al., 1987; Lüneburg, 1995). Conversely, forward scattering is often of interest in optics, and it is convenient to use the Jones matrix. This configuration is called the Forward Scatter Alignment, FSA. Both sets of coordinates and both scattering matrices, Jones and Sinclair, are valid for all single-frequency scattering, although with different degrees of convenience. The names BSA and FSA imply limitations; rather than use them, we give explicitly the coordinate systems being used.

3.6. RELATIONSHIP BETWEEN JONES AND SINCLAIR MATRICES

In Fig. 3.1, the scattered field components are related by

$$\mathbf{E}^s(x_3, y_3) = \begin{bmatrix} -1 & 0 \\ 0 & 1 \end{bmatrix} \mathbf{E}^s(x_2, y_2) = \mathbf{diag}(-1, 1)\mathbf{E}^s(x_2, y_2)$$

If the Sinclair and Jones matrices are used, we have

$$\frac{1}{\sqrt{4\pi}r} \begin{bmatrix} S_{x3x1} & S_{x3y1} \\ S_{y3x1} & S_{y3y1} \end{bmatrix} \begin{bmatrix} E^i_{x1} \\ E^i_{y1} \end{bmatrix} e^{-jkr}$$
$$= \frac{1}{\sqrt{4\pi}r}\mathbf{diag}(-1, 1) \begin{bmatrix} T_{x2x1} & T_{x2y1} \\ T_{y2x1} & T_{y2y1} \end{bmatrix} \begin{bmatrix} E^i_{x1} \\ E^i_{y1} \end{bmatrix} e^{-jkr}$$

From this expression, if we omit the numerical subscripts as understood, we see that the Sinclair and Jones matrices are related by

$$\mathbf{S} = \mathbf{diag}(-1, 1)\mathbf{T} \tag{3.8}$$

3.7. SCATTERING WITH CIRCULAR WAVE COMPONENTS

The incident field at the target in Fig. 3.1 can be written in $x_1y_1z_1$ coordinates as

$$\mathbf{E}^i = \frac{jZ_0I}{2\lambda r_1}e^{-jkr_1}\mathbf{h}_t$$

where $\mathbf{h}_t$ is the effective length of the transmitting antenna, Z_0 the characteristic impedance of the medium, I a current at a reference port, and λ the wavelength.

The transformation between rectangular and circular wave components, adapted from (2.42), is

$$\mathbf{E}(x, y) = \begin{bmatrix} E_x \\ E_y \end{bmatrix} = \mathbf{U}_c \begin{bmatrix} E_L \\ E_R \end{bmatrix} = \mathbf{U}_c\mathbf{E}(L, R)$$

where $\mathbf{U}_c$ is the unitary matrix,

$$\mathbf{U}_c = \frac{1}{\sqrt{2}}\begin{bmatrix} 1 & 1 \\ j & -j \end{bmatrix} \tag{3.9}$$

The relation between incident field and transmitting antenna effective length is

$$\mathbf{E}^i(L, R) = \frac{jZ_0I}{2\lambda r_1}e^{-jkr_1}\mathbf{h}_t(L, R)$$

We can use the same transformation for the scattered field, with coordinates $x_1y_1z_1$ for the incident wave and $x_2y_2z_2$ for the scattered. The scattered wave at the receiving antenna, using the Jones matrix formulation to ensure that the scattered wave is in a right-handed system with the z axis in the direction of wave travel, was given by (3.2). The transformation to circular wave components causes (3.2) to become

$$\mathbf{U}_c \begin{bmatrix} E_L^s \\ E_R^s \end{bmatrix} = \frac{1}{\sqrt{4\pi}\,r_2}\begin{bmatrix} T_{xx} & T_{xy} \\ T_{yx} & T_{yy} \end{bmatrix}\mathbf{U}_c \begin{bmatrix} E_L^i \\ E_R^i \end{bmatrix} e^{-jkr_2}$$

where the first subscript in $\mathbf{T}$ refers to $x_2y_2z_2$ coordinates and the second to $x_1y_1z_1$. The subscript on r denotes target-receiver distance. The circular field components are transformed from the incident wave in $x_1y_1z_1$ coordinates; in the scattered wave, they are found from the field in $x_2y_2z_2$ coordinates.

The scattered field can be written as

$$\mathbf{E}^s(L, R) = \frac{1}{\sqrt{4\pi}\,r_2}\mathbf{A}\mathbf{E}^i(L, R)e^{-jkr_2}$$

where **A**, the circular-component scattering matrix, is given by the unitary similarity transform,

$$\mathbf{A} = \begin{bmatrix} A_{LL} & A_{LR} \\ A_{RL} & A_{RR} \end{bmatrix} = \mathbf{U}_c^{-1}\mathbf{T}\mathbf{U}_c \tag{3.10}$$

The received voltage in an antenna of effective length $\mathbf{h}_r$, when the scattered electric field impinges on it, is, from (2.43),

$$V = -\mathbf{h}_r(L, R)\mathbf{E}^s(L, R) = -\frac{1}{\sqrt{4\pi}\, r_2}\mathbf{h}_r^T(L, R)\mathbf{A}\mathbf{E}^i(L, R)e^{-jkr_2}$$

where the receiving antenna effective length is defined with circular components based on the $x_3y_3z_3$ coordinate system of Fig. 3.1.

If the incident field is written in terms of the length of the transmitting antenna, the received voltage becomes

$$V = -\frac{jZ_0 I}{2\lambda\sqrt{4\pi}\, r_1 r_2}\mathbf{h}_r^T(L, R)\mathbf{A}\mathbf{h}_t(L, R)e^{-jk(r_1+r_2)}$$

The received power is given by

$$W = \frac{VV^*}{8R_a} = \frac{Z_0^2 I^2}{128\pi R_a \lambda^2 (r_1 r_2)^2}|\mathbf{h}_r^T\mathbf{A}\mathbf{h}_t|^2 \tag{3.11}$$

3.8. BACKSCATTERING

For backscattering, the reciprocity theorem requires that the Sinclair matrix be symmetric. Then the scattering matrices, from (3.8), (3.9), and (3.10), become

$$\mathbf{S} = \begin{bmatrix} S_{xx} & S_{xy} \\ S_{xy} & S_{yy} \end{bmatrix}$$

$$\mathbf{T} = \begin{bmatrix} T_{xx} & T_{xy} \\ -T_{xy} & T_{yy} \end{bmatrix}$$

$$\mathbf{A} = \begin{bmatrix} A_{LL} & A_{LR} \\ A_{LR} & A_{RR} \end{bmatrix} = \frac{1}{2}\begin{bmatrix} (T_{xx} + j2T_{xy} + T_{yy}) & (T_{xx} - T_{yy}) \\ (T_{xx} - T_{yy}) & (T_{xx} - j2T_{xy} + T_{yy}) \end{bmatrix}$$

3.9. POLARIZATION RATIO OF THE SCATTERED WAVE

The polarization ratio of the reflected wave can be obtained in terms of the incident wave and the Jones matrix. We begin with (3.1), where the incident wave is in $x_1y_1z_1$ coordinates and the scattered wave in $x_2y_2z_2$. The polarization ratios for the incident and scattered waves can be written as $P^i = E^i_{y1}/E^i_{x1}$ and $P^s = E^s_{y2}/E^s_{x2}$. If they are substituted into (3.1), it becomes

$$\begin{bmatrix} 1 \\ P^s \end{bmatrix} = \frac{1}{\sqrt{4\pi}r}\frac{E^i_{x1}}{E^s_{x2}}\begin{bmatrix} T_{xx} & T_{xy} \\ T_{yx} & T_{yy} \end{bmatrix}\begin{bmatrix} 1 \\ P^i \end{bmatrix} e^{-jkr}$$

which is easily solved for P^s,

$$P^s = \frac{T_{yy}P^i + T_{yx}}{T_{xy}P^i + T_{xx}}$$

where we assume $E^s_x \neq 0$. In terms of the Sinclair matrix, this becomes

$$P^s = -\frac{S_{yy}P^i + S_{yx}}{S_{xy}P^i + S_{xx}}$$

The circular polarization ratio Q^s can be found from the circular scattering matrix and the circular polarization ratio of the incident wave. Scattered and incident fields are related by

$$\begin{bmatrix} E^s_L \\ E^s_R \end{bmatrix} = \frac{1}{\sqrt{4\pi}r}\begin{bmatrix} A_{LL} & A_{LR} \\ A_{RL} & A_{RR} \end{bmatrix}\begin{bmatrix} E^i_L \\ E^i_R \end{bmatrix} e^{-jkr} \tag{3.12}$$

Substitute Q from (1.35) into (3.12), noting that the definition of circular polarization ratio applies to both incident and reflected waves without any concern about coordinate systems. Polarization ratio Q^s can be found from the resulting equation to be

$$Q^s = \frac{A_{RR}Q^i + A_{RL}}{A_{LR}Q^i + A_{LL}}$$

3.10. CHANGE OF POLARIZATION BASIS: THE SCATTERING MATRIX

A scattering matrix can be found if the waves are expressed in components other than rectangular or circular. The incident wave is given in $x_1y_1z_1$ coordinates of Fig. 3.1. If the scattered field is expressed in $x_2y_2z_2$, it is appropriate to use the Jones matrix **T** and write

$$\mathbf{E}^s = C\mathbf{T}\mathbf{E}^i$$

where C is a constant. The incident field can be transformed to orthonormal basis $\mathbf{u}_1, \mathbf{u}_2$ by

$$\hat{\mathbf{E}}^i = \mathbf{U}\mathbf{E}^i$$

where $\mathbf{U}$, from (1.37), is the unitary matrix,

$$\mathbf{U} = \begin{bmatrix} \langle \mathbf{u}_{x1}, \mathbf{u}_1 \rangle & \langle \mathbf{u}_{y1}, \mathbf{u}_1 \rangle \\ \langle \mathbf{u}_{x1}, \mathbf{u}_2 \rangle & \langle \mathbf{u}_{y1}, \mathbf{u}_2 \rangle \end{bmatrix} \tag{3.13}$$

The $\mathbf{u}_{x1}, \mathbf{u}_{y1}$ are unit vectors in coordinate system 1 of Fig. 3.1. Vectors $\mathbf{u}_1$ and $\mathbf{u}_2$ form a set with $\mathbf{u}_{z1}$, a vector in the direction of increasing z.

The scattered-wave field can be transformed from rectangular components in system 2 of Fig. 3.1. The vector set for the scattered wave is $\mathbf{v}_1$, $\mathbf{v}_2$, and $\mathbf{u}_{z2}$. If we require that $\mathbf{v}_1$ and $\mathbf{v}_2$ bear the same relation to $\mathbf{u}_{x2}$ and $\mathbf{u}_{y2}$ as do $\mathbf{u}_1$ and $\mathbf{u}_2$ to $\mathbf{u}_{x1}$ and $\mathbf{u}_{y1}$, the transformation of the scattered wave is carried out by the matrix used to transform the incident wave,

$$\hat{\mathbf{E}}^s = \mathbf{U}\mathbf{E}^s$$

The scattered field can then be written as

$$\hat{\mathbf{E}}^s = C\mathbf{U}\mathbf{T}\mathbf{E}^i = C\mathbf{U}\mathbf{T}\mathbf{U}^{-1}\hat{\mathbf{E}}^i = C\hat{\mathbf{T}}\hat{\mathbf{E}}^i$$

where the transformed Jones matrix is

$$\hat{\mathbf{T}} = \mathbf{U}\mathbf{T}\mathbf{U}^{-1}$$

This is a unitary similarity transform. The eigenvalues of $\mathbf{T}$ and $\hat{\mathbf{T}}$ are therefore the same (Horn and Johnson, 1990, p. 45).

To transform the Sinclair matrix, the effective length of the transmitting antenna is given in $x_1y_1z_1$ coordinates and that of the receiving antenna in $x_3y_3z_3$. Since the transmitting antenna effective length is proportional to the incident electric field, it can be transformed to orthonormal basis $\mathbf{u}_1, \mathbf{u}_2$ with the matrix $\mathbf{U}$ of (3.13), so that $\hat{\mathbf{h}}_t = \mathbf{U}\mathbf{h}_t$. The receiving antenna length can be transformed to orthonormal basis $\mathbf{w}_1$ and $\mathbf{w}_2$, where $\mathbf{w}_1$, $\mathbf{w}_2$, and $\mathbf{u}_{z3}$ are a vector set. If $\mathbf{w}_1$ and $\mathbf{w}_2$ bear the same relationship to $\mathbf{u}_{x3}$ and $\mathbf{u}_{y3}$ as do $\mathbf{u}_1$ and $\mathbf{u}_2$ to $\mathbf{u}_{x1}$ and $\mathbf{u}_{y1}$, the transformation of the receiving antenna effective length is carried out by the same matrix used for the transmitting antenna, and $\hat{\mathbf{h}}_r = \mathbf{U}\mathbf{h}_r$.

Received voltage V is

$$V = C\mathbf{h}_r^T\mathbf{S}\mathbf{h}_t = C\hat{\mathbf{h}}_r^T\hat{\mathbf{S}}\hat{\mathbf{h}}_t$$

Substituting the transformed antenna lengths into this equation and using the unitary properties of $\mathbf{U}$ gives

$$\hat{\mathbf{S}} = \mathbf{U}^*\mathbf{S}\mathbf{U}^\dagger \tag{3.14}$$

This is a unitary consimilarity transform. $\hat{\mathbf{S}}$ is also said to be tee-congruent to $\mathbf{S}$ (Horn and Johnson, 1990, p. 290).

3.11. POLARIZATIONS FOR MAXIMUM AND MINIMUM POWER

In this section, we will determine the polarizations that give maximum and minimum received power from a target

Equality of Transmitting and Receiving Antennas

The Sinclair matrix for backscattering is symmetric, and the received power from a target, given by (3.7), can be written in two forms,

$$W = \frac{Z_0^2 I^2}{128\pi R_a \lambda^2 r^4} |\mathbf{h}_r^T \mathbf{S} \mathbf{h}_t|^2 = \frac{Z_0^2 I^2}{128\pi R_a \lambda^2 r^4} |\mathbf{h}_t^T \mathbf{S} \mathbf{h}_r|^2 \tag{3.15}$$

The first form as an inner product is

$$|\mathbf{h}_r^T \mathbf{S} \mathbf{h}_t|^2 = |\langle \mathbf{h}_r, \mathbf{S}^* \mathbf{h}_t^* \rangle|^2$$

By the Cauchy–Schwarz inequality (Horn and Johnson, 1990, p. 261), this has maximum value

$$|\langle \mathbf{h}_r, \mathbf{S}^* \mathbf{h}_t^* \rangle|_m^2 = \langle \mathbf{h}_r, \mathbf{h}_r \rangle \langle \mathbf{S}^* \mathbf{h}_t^*, \mathbf{S}^* \mathbf{h}_t^* \rangle$$

if

$$\mathbf{h}_r = c_1 \mathbf{S}^* \mathbf{h}_t^* \tag{3.16}$$

where c_1 is a complex constant.

We require that the effective lengths of the antennas have unit magnitude. This restricts c_1 but has no effect on this development. With this constraint, maximum received power is

$$W_m = \frac{Z_0^2 I^2}{128\pi R_a \lambda^2 r^4} \langle \mathbf{S}^* \mathbf{h}_t^*, \mathbf{S}^* \mathbf{h}_t^* \rangle = \frac{Z_0^2 I^2}{128\pi R_a \lambda^2 r^4} |\mathbf{S} \mathbf{h}_t|^2$$

If the process is repeated with the second form for W, another equation for its maximum value is

$$W_m = \frac{Z_0^2 I^2}{128\pi R_a \lambda^2 r^4} |\mathbf{S} \mathbf{h}_r|^2$$

if

$$\mathbf{h}_t = c_2 \mathbf{S}^* \mathbf{h}_r^* \tag{3.17}$$

Maximum power is the same for both cases, and

$$|\mathbf{Sh}_t|^2 = |\mathbf{Sh}_r|^2 \tag{3.18}$$

Equations 3.16–3.18 are satisfied if transmitting and receiving antenna effective lengths are equal. If we set $\mathbf{h}_t = \mathbf{h}_r = \mathbf{h}_o$, the received power can be maximized by a proper choice of $\mathbf{h}_o$. Subscript o denotes "optimum".

We include (3.16) and (3.17) in the general form

$$\mathbf{Ax} = \gamma \mathbf{x}^* \tag{3.19}$$

and require that $\mathbf{A}$ be symmetric. The scalar γ is a coneigenvalue of $\mathbf{A}$ if it meets a requirement given below, and $\mathbf{x}$ is a coneigenvector. Not all matrices have coneigenvalues (Horn and Johnson, 1990, p. 245). In the radar community, the coneigenvalue equation (3.19) is known as *Kennaugh's pseudo–eigenvalue equation*.

The phase of the coneigenvalue cannot be determined. To see this, assume that γ is a coneigenvalue of $\mathbf{A}$ and change the phase of the corresponding coneigenvector by θ. Equation 3.19 can then be rewritten as

$$\mathbf{A}(e^{j\theta}\mathbf{x}) = (e^{j2\theta}\gamma)(e^{j\theta}\mathbf{x})^*$$

It can be seen that $e^{j2\theta}\gamma$ is also a coneigenvalue of $\mathbf{A}$.

Multiply (3.19) by the conjugate of $\mathbf{A}$,

$$\mathbf{A}^*\mathbf{Ax} = \gamma \mathbf{A}^*\mathbf{x}^* = \gamma\gamma^*\mathbf{x}$$

This is the eigenvalue equation

$$(\mathbf{A}^*\mathbf{A} - \lambda \mathbf{I})\mathbf{x} = \mathbf{0} \tag{3.20}$$

where λ has replaced $|\gamma|^2$ and $\mathbf{I}$ is the identity matrix. $\mathbf{A}^*\mathbf{A}$ is Hermitian with real eigenvalues, and if $\mathbf{A}$ is symmetric, the eigenvalues are nonnegative. The nonnegative real λ is an eigenvalue of $\mathbf{A}^*\mathbf{A}$ if and only if $\sqrt{\lambda}$ is a coneigenvalue of $\mathbf{A}$ (Horn and Johnson, 1990, p. 245).

The characteristic polynomial of (3.20) is a quadratic with roots

$$\lambda_1, \lambda_2 = \frac{B}{2} \pm \frac{1}{2}\sqrt{B^2 - 4C}$$

where

$$B = |A_{11}|^2 + 2|A_{12}|^2 + |A_{22}|^2$$
$$C = |A_{11}A_{22}|^2 + |A_{12}|^4 - 2\text{Re}\left(A_{11}A_{12}^{*2}A_{22}\right)$$

If the eigenvalues of $\mathbf{A}^*\mathbf{A}$ are distinct, $\mathbf{A}$ has at least two independent coneigenvectors (Horn and Johnson, 1990, p. 245) with corresponding coneigenvalue magnitudes $|\gamma_1| = \sqrt{\lambda_1}$ and $|\gamma_2| = \sqrt{\lambda_2}$. The coneigenvectors $\mathbf{x}_1$ and $\mathbf{x}_2$ of $\mathbf{A}$ are the eigenvectors of $\mathbf{A}^*\mathbf{A}$. The eigenvalues of $\mathbf{A}^*\mathbf{A}$ are real and $B^2 \geq 4C$. If $\mathbf{A}$ is singular, $C = 0$ and an eigenvalue is zero; therefore $\mathbf{A}^*\mathbf{A}$ is singular. If $B^2 = 4C$, the roots of the characteristic polynomial are equal. Since $\mathbf{A}^*\mathbf{A}$ is Hermitian, the eigenvectors are linearly independent and orthogonal if the eigenvalues are distinct (Pease, 1965, p. 74). If some of the eigenvalues of a Hermitian matrix are not distinct; that is, if the characteristic polynomial of the matrix has a root of multiplicity r, the matrix has r linearly independent characteristic vectors that correspond to the repeated root (Wylie and Barrett, 1982, p. 724). Any linear combination of the r independent eigenvectors is also an eigenvector of the matrix (Pease, 1965, p. 74). If the roots of the characteristic polynomial corresponding to $\mathbf{A}^*\mathbf{A}$ are repeated, it can be seen that (3.20) is satisfied by any value of $\mathbf{x}$.

Let $\mathbf{A}$ be the symmetric Sinclair matrix for backscattering and recognize that coneigenvector $\mathbf{x}$ was used to represent the normalized effective length for both antennas, $\hat{\mathbf{h}}_o$. The equations of interest are

$$\mathbf{S}\hat{\mathbf{h}}_o = \gamma \hat{\mathbf{h}}_o^* \tag{3.21}$$

$$(\mathbf{S}^*\mathbf{S} - |\gamma|^2\mathbf{I})\hat{\mathbf{h}}_o = \mathbf{0} \tag{3.22}$$

If $\mathbf{S}^*\mathbf{S}$ has distinct eigenvalues, there are two solutions of (3.22) for the eigenvalues and eigenvectors, identified as $|\gamma_1|^2, |\gamma_2|^2, \hat{\mathbf{h}}_{o1}, \hat{\mathbf{h}}_{o2}$, with coneigenvalues γ_1 and γ_2. The eigenvectors are the antenna effective lengths that give locally maximum received power. One eigenvector may give greater power than the other, and these powers are referred to here as maximum and submaximum. The eigenvectors/coneigenvectors satisfy $\hat{\mathbf{h}}_{o1}^T \hat{\mathbf{h}}_{o2}^* = \hat{\mathbf{h}}_{o2}^T \hat{\mathbf{h}}_{o1}^* = 0$. If the eigenvalues are not distinct, any vector satisfies (3.22), but not (3.21).

The Diagonal Scattering Matrix

A symmetric matrix such as the Sinclair matrix for backscattering is unitarily condiagonalizable and can be put into a diagonal form by

$$\mathbf{S}_d = \mathbf{U}_d^{-1}\mathbf{S}\mathbf{U}_d^*$$

where $\mathbf{U}_d$ is unitary (Horn and Johnson, 1990, p. 245).

We use the coneigenvectors $\hat{\mathbf{h}}_{o1}$ and $\hat{\mathbf{h}}_{o2}$ of (3.21) to form

$$\mathbf{U}_d = \begin{bmatrix} \hat{\mathbf{h}}_{o1} & \hat{\mathbf{h}}_{o2} \end{bmatrix}^* = \begin{bmatrix} \hat{h}_{o1x}^* & \hat{h}_{o2x}^* \\ \hat{h}_{o1y}^* & \hat{h}_{o2y}^* \end{bmatrix}$$

This matrix is unitary. It is not unique because the phase relationship between $\hat{\mathbf{h}}_{o1}$ and $\hat{\mathbf{h}}_{o2}$ cannot be determined from (3.21) and (3.22). A phase relationship between a nonzero component of $\hat{\mathbf{h}}_{o1}$ and a nonzero component of $\hat{\mathbf{h}}_{o2}$ can be assigned arbitrarily; the specification of $\mathbf{U}_d$ is complete if this is done.

With the substitution of this value for $\mathbf{U}_d$ and the use of (3.21), the diagonal scattering matrix becomes

$$\mathbf{S}_d = \mathbf{U}_d^\dagger \mathbf{S} \left[\, \hat{\mathbf{h}}_{o1} \;\; \hat{\mathbf{h}}_{o2} \, \right] = \left[\, \gamma_1 \mathbf{U}_d^\dagger \hat{\mathbf{h}}_{o1}^* \;\; \gamma_2 \mathbf{U}_d^\dagger \hat{\mathbf{h}}_{o2}^* \, \right]$$

$$= \left[\gamma_1 \begin{bmatrix} 1 \\ 0 \end{bmatrix} \; \gamma_2 \begin{bmatrix} 0 \\ 1 \end{bmatrix} \right] = \begin{bmatrix} \gamma_1 & 0 \\ 0 & \gamma_2 \end{bmatrix}$$

In developing this form, matrices $\mathbf{S}^*\mathbf{S}$ with non-distinct eigenvalues were excluded.

To complete the definition of $\mathbf{U}_d$, a phase difference between a component of $\hat{\mathbf{h}}_{o1}$ and a component of $\hat{\mathbf{h}}_{o2}$ was arbitrarily selected. Since the coneigenvalue phase is related to the phase of the corresponding coneigenvector, a phase difference between the coneigenvalues γ_1 and γ_2 was, in effect, chosen arbitrarily.

Maximum Backscattered Power

The received voltage with optimum polarization is

$$V = \mathbf{h}_o^T \mathbf{E}^s = \frac{jZ_0 I}{\sqrt{4\pi}(2\lambda r^2)} \mathbf{h}_o^T \mathbf{S} \mathbf{h}_o$$

where the transmitting and receiving antenna effective lengths are the same, with $\mathbf{h}_o$ the effective length corresponding to the polarization state $\hat{\mathbf{h}}_o$. With the use of (3.21), V becomes

$$V = \frac{jZ_0 I \gamma}{\sqrt{4\pi}(2\lambda r^2)} e^{-j2kr} \mathbf{h}_o^T \mathbf{h}_o^*$$

Received power is

$$W = \frac{Z_0^2 I^2 |\gamma|^2 |\mathbf{h}_o^T \mathbf{h}_o^*|^2}{128\pi R_a \lambda^2 r^4}$$

Received power for backscattering, when receiving and transmitting antennas have the same effective lengths, is called the *copolarized power*. The maximum copolarized received power corresponds to the larger eigenvalue of the two solutions to (3.22), given by

$$|\gamma_1|^2, |\gamma_2|^2 = \frac{B}{2} \pm \frac{1}{2}\sqrt{B^2 - 4C} \tag{3.23}$$

where

$$B = |S_{xx}|^2 + 2|S_{xy}|^2 + |S_{yy}|^2 \tag{3.24}$$

$$C = |S_{xx}S_{yy}|^2 + |S_{xy}|^4 - 2\mathrm{Re}\left(S_{xx}S_{xy}^{*2}S_{yy}\right) \tag{3.25}$$

In these developments we excluded a zero value for $B^2 - 4C$.

The maximum copolarized power is

$$W_m = \frac{Z_0^2 I^2 |\gamma_1|^2 |\mathbf{h}_{o1}|^4}{128\pi R_a \lambda^2 r^4}$$

where $\hat{\mathbf{h}}_{o1}$ denotes the optimum effective length corresponding to the larger eigenvalue of (3.22). A submaximum power corresponds to the smaller of the eigenvalues and is

$$W_{\text{sm}} = \frac{Z_0^2 I^2 |\gamma_2|^2 |\mathbf{h}_{o2}|^4}{128\pi R_a \lambda^2 r^4}$$

It is convenient to use polarization ratios P_1 and P_2 for the optimum antenna effective lengths and the waves that would be radiated by these antennas. The relationship is

$$\hat{\mathbf{h}}_{oi} = \frac{1}{\sqrt{1+|P_i|^2}} \begin{bmatrix} 1 \\ P_i \end{bmatrix} e^{j\Phi_i} \qquad i = 1, 2 \tag{3.26}$$

where the scalar multiplier normalizes the antenna effective length and Φ_i is an arbitrary angle attesting to our lack of knowledge of the phases of the eigenvectors. Note that the polarization ratios are completely defined. If the eigenvalues of $\mathbf{S}^*\mathbf{S}$ are distinct, expansion of

$$\mathbf{S}^*\mathbf{S}\hat{\mathbf{h}}_{oi} = |\gamma_i|^2 \hat{\mathbf{h}}_{oi} \qquad i = 1, 2$$

yields two scalar equations, from either of which the antenna polarization ratios can be found,

$$P_i = \frac{\hat{h}_{oy}}{\hat{h}_{ox}} = -\frac{S_{xx}S_{xy}^* + S_{xy}S_{yy}^*}{|S_{xy}|^2 + |S_{yy}|^2 - |\gamma_i|^2} \qquad i = 1, 2 \tag{3.27}$$

and

$$P_i = -\frac{|S_{xx}|^2 + |S_{xy}|^2 - |\gamma_i|^2}{S_{xx}^* S_{xy} + S_{xy}^* S_{yy}} \qquad i = 1, 2 \tag{3.28}$$

The orthonormality of the eigenvectors leads to a relationship between the polarization ratios of the antennas yielding maximum and submaximum copolarized powers. It is

$$P_2 = -\frac{1}{P_1^*} \tag{3.29}$$

Consider the polarization efficiency of an antenna used both as transmitter and receiver in a monostatic radar with the antenna length not optimum. The received power for arbitrary polarization is given by (3.7) with $\hat{\mathbf{h}}_r = \hat{\mathbf{h}}_t = \hat{\mathbf{h}}$, where the circumflex denotes the normalized effective length. The ratio of copolarized power for arbitrary polarization to maximum copolarized power is the backscatter polarization efficiency,

$$\rho_s = \frac{|\hat{\mathbf{h}}^T \mathbf{S} \hat{\mathbf{h}}|^2}{|\gamma_1|^2}$$

Although we have discussed the optimum antenna as though it exists physically, it need not be constructed, but may be synthesized instead. The effective length of the optimum antenna is the weighted sum of the effective lengths of two orthogonally polarized antennas. The received voltage for the optimum antenna is then the weighted sum, with the same weighting factors, of the received voltages when the orthogonal antennas are used, either simultaneously or sequentially.

Minimum Backscattered Power

An antenna polarization for a radar using one antenna for transmitting and receiving that gives minimum backscattered co-polarized power can be found. The received voltage and power are zero if $\mathbf{h}^T \mathbf{S} \mathbf{h} = 0$. If antenna polarization ratio P is used, this equation becomes either

$$S_{yy} P^2 + 2 S_{xy} P + S_{xx} = 0$$

or

$$S_{xx} \left(\frac{1}{P} \right)^2 + 2 S_{xy} \left(\frac{1}{P} \right) + S_{yy} = 0$$

If $S_{yy} \neq 0$, P can be found from the first equation and if $S_{xx} \neq 0$, from the second. The first equation is used here, but the results that follow would be unaltered if the second were used. The two roots of the equation for P are

$$P_3, P_4 = \pm \frac{R \mp S_{xy}}{S_{yy}} \tag{3.30}$$

where

$$R = \sqrt{S_{xy}^2 - S_{xx} S_{yy}} \tag{3.31}$$

If the radar antenna used for transmitting and receiving has polarization ratio P_3 or P_4, the copolarized power received from the target is zero.

Copolarization and Cross-Polarization Nulls

The backscattered power from a target can be separated into a part that can be received by the illuminating antenna, the copolarized signal, and a part that can be received with an antenna that is orthogonal to the transmitting antenna, the *cross-polarized signal.* The voltage induced in a receiving antenna, given by (3.6), can be written in terms of the antenna polarization ratios, as

$$V = \frac{jZ_0 I}{\sqrt{4\pi}(2\lambda r^2)} h_{rx} h_{tx} [\,1 \quad P_r\,] \begin{bmatrix} S_{xx} & S_{xy} \\ S_{xy} & S_{yy} \end{bmatrix} \begin{bmatrix} 1 \\ P_t \end{bmatrix} e^{-j2kr}$$

For orthogonal transmitting and receiving antennas, $P_r = -1/P_t^*$. If this relationship is substituted into the preceding equation, the cross-polarized received voltage becomes

$$V_{\text{cross}} = \frac{jZ_0 I}{\sqrt{4\pi}(2\lambda r^2)} h_{rx} h_{tx} \left(S_{xx} + S_{xy} P_t - \frac{S_{xy}}{P_t^*} - \frac{S_{yy} P_t}{P_t^*} \right) e^{-j2kr}$$

Polarization ratios P_1 and P_2 give a maximum and submaximum copolarized backscattered power. They are called copolarization maximum polarization ratios or *co-pol maxima*. Substitution of either P_1 or P_2 for P_t in the equation for the cross-polarized voltage gives $V_{\text{cross}}(P_1) = V_{\text{cross}}(P_2) = 0$. At the co-pol maxima, none of the backscattered power is cross-polarized; polarization ratios P_1 and P_2 may be called cross-polarization null polarization ratios, or *cross-pol nulls* or *X-pol nulls*. The co-pol maxima and the cross-pol nulls coincide.

Antenna polarization ratios P_3 and P_4, in a monostatic radar using one antenna for transmitting and receiving, give zero copolarized power. They are called *copolarization null polarization ratios*, or *co-pol nulls*. The cross-polarized power to a colocated orthogonal receiving antenna, when the transmitting antenna has polarizations P_3 or P_4, is of interest. The cross-polarized power is

$$W_{\text{cross}} = \frac{Z_0 I^2}{128\pi R_a \lambda^2 r^4} \left| \mathbf{h}_r^T \mathbf{S} \mathbf{h}_t \right|^2$$

where $\mathbf{h}_r$ and $\mathbf{h}_t$ are orthogonal. It can be shown by expanding this equation and requiring the transmitter polarization to be either P_3 or P_4 and the receiving antenna to be orthogonal to the transmitting antenna, that the cross-polarized power is related to the maximum copolarized power, for antenna polarization P_1, by

$$\frac{W_{\text{cross}}|_{P_3}}{W_m|_{P_1}} = \frac{|\gamma_2|}{|\gamma_1|}$$

Scattering Matrix from Polarization Nulls

The copolarization null ratios P_3 and P_4 can be used to obtain information about the target scattering matrix. From (3.30) and (3.31),

$$\frac{S_{xx}}{S_{xy}} = -\frac{2P_3P_4}{P_3 + P_4} \qquad \frac{S_{yy}}{S_{xy}} = -\frac{2}{P_3 + P_4}$$

To within a multiplying factor, the target scattering matrix can be found from the co-pol nulls. The polarization ratios do not provide amplitude information, so S_{xy} cannot be found.

It has been shown that if the square of the Euclidean norm of the scattering matrix is known, the scattering matrix, with amplitudes, can be obtained from a knowledge of the co-pol null pair, P_3 and P_4, or from a cross-pol null, P_1 or P_2, and a co-pol null, P_3 or P_4 (Boerner et al., 1981).

3.12. THE POLARIZATION FORK

The Stokes parameters corresponding to polarization ratios P_1, P_2, P_3, and P_4 form an interesting pattern on the Poincaré sphere, Fig. 3.2. It is somewhat like a

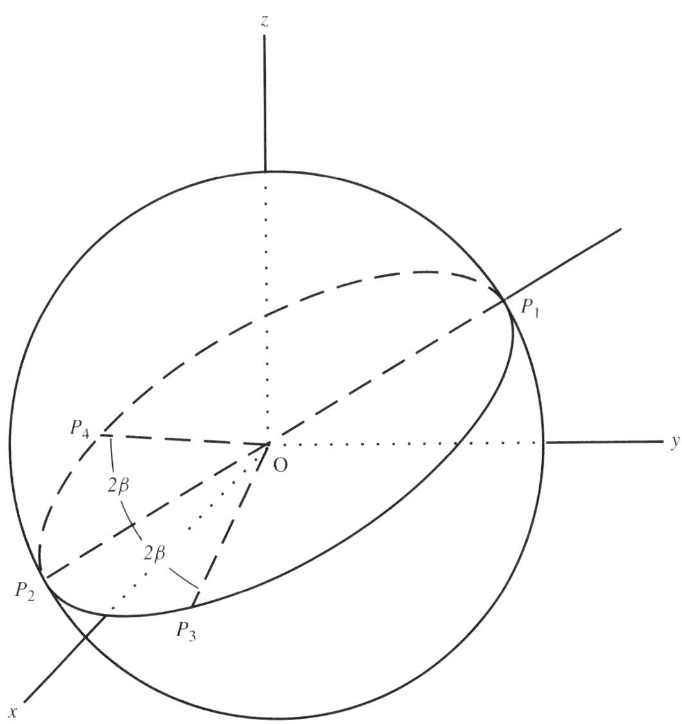

Fig. 3.2. The Huynen polarization fork.

fork with a handle and three tines, and is known as the polarization fork. It was discussed by Huynen and is sometimes called the *Huynen fork* (Huynen, 1970).

From the definitions of the Stokes parameters of a wave by (1.44)–(1.47) and the polarization ratio of a wave by (1.31), the Stokes parameters corresponding to polarization ratio P are

$$\frac{G_1}{G_0} = \frac{1 - |P|^2}{1 + |P|^2} \qquad \frac{G_2}{G_0} = \frac{2\mathrm{Re}(P)}{1 + |P|^2} \qquad \frac{G_3}{G_0} = \frac{2\mathrm{Im}(P)}{1 + |P|^2} \tag{3.32}$$

Polarization ratios P_1 and P_2 giving maximum and submaximum power are related by (3.29). The Stokes parameters corresponding to these polarization ratios, found by substituting (3.29) into (3.32), are related by

$$G_i^{(2)} = -G_i^{(1)} \qquad i = 1, 2, 3 \tag{3.33}$$

where the superscripts refer to polarization ratios P_1 and P_2. We see from this equation that the plotted Stokes parameter points corresponding to polarization ratios P_1 and P_2 lie at opposite ends of a diameter of the Poincaré sphere.

It was shown in Section 1.7 that orthogonal waves with polarization ratios P_i and P_j also obey an equation like (3.29). It follows that all orthogonal wave pairs plotted on the Poincaré sphere can be described by (3.33) and lie at antipodal points on the sphere.

Consider the location of points corresponding to P_3 and P_4. If we define a three-element vector composed of three Stokes parameters normalized to G_0,

$$\mathbf{g} = \frac{1}{G_0} [\, G_1 \;\; G_2 \;\; G_3 \,]^T$$

the angle between two rays drawn from the origin to points corresponding to polarization ratios P_a and P_b, on a Poincaré sphere of radius G_0, is given by

$$\cos\beta_{ab} = \mathbf{g}_a^T \mathbf{g}_b = \frac{G_1^a}{G_0}\frac{G_1^b}{G_0} + \frac{G_2^a}{G_0}\frac{G_2^b}{G_0} + \frac{G_3^a}{G_0}\frac{G_3^b}{G_0}$$

If a substitution of the Stokes parameters from (3.32) is made in this equation, it becomes

$$\cos\beta_{ab} = \frac{1 - |P_a|^2 - |P_b|^2 + |P_a|^2|P_b|^2 + 2P_a P_b^* + 2P_a^* P_b}{\left(1 + |P_a|^2\right)\left(1 + |P_b|^2\right)} \tag{3.34}$$

If the polarization ratio pairs P_1, P_3 and P_1, P_4 are used in this equation, it can be shown that $\cos\beta_{13} = \cos\beta_{14}$. Finally, if the appropriate polarization ratio pairs are used, it can be shown that $\beta_{34} = 2\beta_{13}$. It is apparent that the last two equations can be satisfied only if the points corresponding to P_1, P_2, P_3, and P_4 lie on a great circle of the Poincaré sphere. Further, the central angle formed by rays to the points P_1 and P_3 is equal to that between rays to P_1 and P_4. The two outer tines of the fork are symmetric about the handle and center tine.

In order to plot the polarization fork on the Poincaré sphere, only the points corresponding to P_1 and P_3 need be determined, since P_2 is at the opposite end of a sphere diameter from P_1, and P_4 is the image of P_3 about the line P_1P_2. The points can be located as Cartesian coordinates G_1, G_2, and G_3, with the Stokes parameters found from (3.32). An alternative procedure is to plot each point by using its azimuth and elevation angles. It is also readily seen that the polarization fork is determined by polarizations P_3 and P_4.

Angle β, defined as

$$\tan^2\beta = \frac{|\gamma_2|}{|\gamma_1|} \tag{3.35}$$

has a geometric significance. If polarization ratios P_1 and P_2 are found from (3.27) or (3.28) and used in (3.34), it will be seen that

$$\cos^2\beta_{23} = \frac{B - 2RR^*}{B + 2RR^*} \tag{3.36}$$

where B and R are given by (3.24) and (3.31). It may also be shown, using (3.23) and (3.35), that

$$\tan^2\beta = \frac{|B - (B^2 - 4C)^{1/2}|^{1/2}}{|B + (B^2 - 4C)^{1/2}|^{1/2}} \tag{3.37}$$

We exclude equal eigenvalues from consideration, so $B^2 > 4C$. It is straightforward, using (3.36) and (3.37), to show that

$$\beta_{23} = 2\beta \tag{3.38}$$

Angle β has been called the *characteristic angle* of a target (Huynen, 1970) and the *polarizability angle* (Holm, 1987, p. 629). Polarizability is the target characteristic that causes it to scatter an initially unpolarized wave as one with a greater degree of polarization. As an example, an unpolarized wave has x and y components that are uncorrelated and have equal power densities. If it is incident on a target that backscatters the x but not the y component, then the backscattered wave is polarized. Targets with larger β values are less able to polarize waves than those with smaller β values.

The positions of the points corresponding to the significant polarization ratios will change if the radar is rotated around the radar-target line of sight, and the orientation of the polarization fork will change, but the shape of the fork depends only on the target and is unchanged by the rotation. To show this, it is only necessary to show that β in (3.35) is unchanged by the radar rotation, since from (3.38) the polarization fork shape is determined by β.

For backscattering, we use coordinates $x_1 y_1 z_1$ of Fig. 3.1 for both transmitting and receiving antennas and omit the subscripts for convenience. Assume the Sinclair matrix of the target is known and the polarization fork has been determined in xyz coordinates. The radar antenna effective lengths are known in ξ and η components of Fig. 3.3. Neglecting unnecessary constants, the receiver voltage is given by either of the two forms,

$$V = \mathbf{h}^T(\xi, \eta)\mathbf{S}(\xi, \eta)\mathbf{h}(\xi, \eta) \tag{3.39}$$

$$V = \mathbf{h}^T(x, y)\mathbf{S}(x, y)\mathbf{h}(x, y) \tag{3.40}$$

$\mathbf{S}$ can be written with two coordinates only since the same coordinate system is used for transmitter and receiver.

Now,

$$\mathbf{h}(\xi, \eta) = \begin{bmatrix} h_\xi \\ h_\eta \end{bmatrix} = \begin{bmatrix} \cos\theta & \sin\theta \\ -\sin\theta & \cos\theta \end{bmatrix} \begin{bmatrix} h_x \\ h_y \end{bmatrix} = \mathbf{R}\mathbf{h}(x, y)$$

If this substitution is made in (3.39), it becomes

$$V = \mathbf{h}^T(x, y)\mathbf{R}^T\mathbf{S}(\xi, \eta)\mathbf{R}\mathbf{h}(x, y)$$

Comparison to (3.40) shows that

$$\mathbf{S}(x, y) = \mathbf{R}^T\mathbf{S}(\xi, \eta)\mathbf{R} = \mathbf{R}^{-1}\mathbf{S}(\xi, \eta)\mathbf{R}$$

Then,

$$\begin{aligned} \mathbf{S}^*(\xi, \eta)\mathbf{S}(\xi, \eta) &= \mathbf{R}^*\mathbf{S}^*(x, y)(\mathbf{R}^{-1})^*\mathbf{R}\mathbf{S}(x, y)\mathbf{R}^{-1} \\ &= \mathbf{R}\mathbf{S}^*(x, y)\mathbf{S}(x, y)\mathbf{R}^{-1} \end{aligned}$$

It is seen from this that $\mathbf{S}^*\mathbf{S}$ expressed in $\xi\eta$ components is similar to $\mathbf{S}^*\mathbf{S}$ expressed in xy components and has the same eigenvalues. The eigenvalues

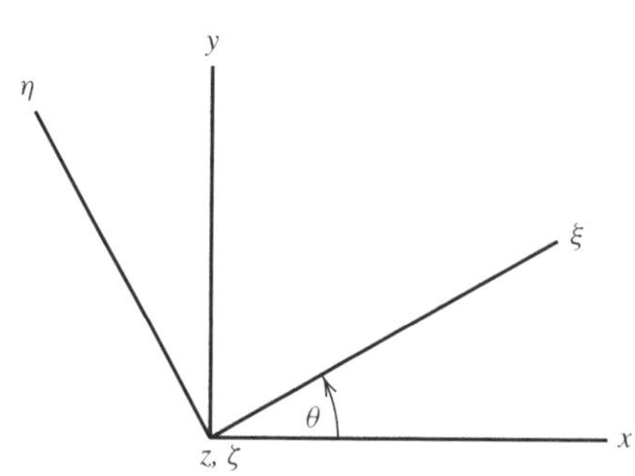

Fig. 3.3. Coordinate systems for radar rotation.

are unchanged by the radar rotation, and β is unchanged. The shape of the polarization fork is therefore unchanged by rotation of the radar or target.

We saw that **S** can be transformed to an orthonormal basis system by the unitary consimilarity transform

$$\hat{\mathbf{S}} = \mathbf{U}^*\mathbf{S}\mathbf{U}^\dagger$$

where **U** is given by (3.13). Consider the transformation of $\mathbf{S}^*\mathbf{S}$ if **S** is transformed by (3.14),

$$\hat{\mathbf{S}}^*\hat{\mathbf{S}} = \mathbf{U}\mathbf{S}^*\mathbf{U}^T\mathbf{U}^*\mathbf{S}\mathbf{U}^\dagger = \mathbf{U}\mathbf{S}^*\mathbf{S}\mathbf{U}^{-1}$$

This is a unitary similarity transform and the eigenvalues of $\mathbf{S}^*\mathbf{S}$ are preserved.

Some properties of the characteristic polarizations may be inferred from an examination of the polarization fork:

1. The cross-polarization nulls P_1 and P_2 are distinct.
2. If one cross-polarization null represents linear polarization, so does the other. If one cross-polarization null represents circular polarization, so does the other.
3. If the copolarization nulls P_3 and P_4 are identical, they coincide with P_2, which is then the polarization for zero received power.
4. If the copolarization nulls P_3 and P_4 represent orthogonal waves, $\beta = \pi/4$, and P_3 and P_4 are antipodal points on the sphere.
5. If one copolarization null, say P_3, is for left-circular polarization and the other, P_4, for right-circular, then $\beta = \pi/4$ and both cross-polarization nulls are linear.

3.13. NONALIGNED COORDINATE SYSTEMS

In Section 2.16, we considered the problem of transmission between antennas if the effective lengths were known in different coordinate systems. A similar problem occurs in scattering; the incident fields are known in one coordinate system and the target's Sinclair matrix in another. If the Sinclair matrix is known for all incident angles in a given coordinate system, the wave scattered to the receiver can be determined. We limit the discussion to backscattering with colocated transmitting and receiving antennas.

The first steps to finding the scattered wave from a target are the same as those for doing so in a two-antenna transmit–receive configuration. The notation is that of Section 2.16 and the coordinate systems are shown in Fig. 2.13.

Steps 1–5. Same as Section 2.16.

Step 6. Find the colatitude and azimuth angles of the radar at position 1 in system c. The Sinclair matrix of the target for any incidence angle is assumed known in system c. Find the scattered field at position 1, also in system c, from

$$\mathbf{E}^{cs} = \frac{1}{\sqrt{4\pi}|\mathbf{X}_1 - \mathbf{X}_2|}\mathbf{S}\mathbf{E}^c e^{-jk|\mathbf{X}_1 - \mathbf{X}_2|}$$

Step 7. If characteristics of the wave, such as polarization ratio, are needed, use the Euler angle matrices to convert the fields from system c to system d, using the ground/reference system as an intermediate step, thus

$$\mathbf{E}^d = \mathbf{D}\mathbf{C}^T\mathbf{E}^c$$

Find the polarization ratio of the scattered wave in the direction of the radar receiving antenna from this field in system d.

Step 8. If only the received voltage at the radar receiver is needed, find it by converting the scattered field to system a, by

$$\mathbf{E}^{as} = \mathbf{A}\mathbf{C}^T\mathbf{E}^{cs}$$

and then using the received-voltage equation

$$V = \mathbf{h}_1^{aT}\mathbf{E}^{as}$$

where the value of the receiving antenna effective length is determined at polar and azimuth angles corresponding to the position of the scatterer.

3.14. DETERMINATION OF SCATTERING PARAMETERS

We assumed in this chapter that a target Sinclair matrix is known. **S** can be measured as noted in Section 3.3. It can also be found exactly or approximately from the Maxwell equations and the target's geometry and physical properties. Methods for calculating **S** or the scattering cross-section are discussed briefly in this section, and references are given to extensive treatments.

Models

To determine a target's scattering properties, a theoretical model of the target is constructed. If the scattering is from a target surface which can be described mathematically, the model is the mathematical description together with the electromagnetic properties of the scattering body. A few simple target shapes can be modeled in this manner, with their scattering parameters determined by methods noted briefly later in this section. Examples are a conducting sphere or a conducting plane of infinite extent. A complex body, such as an aircraft, can be modeled by an assembly of simple shapes, a tail group by an assembly of flat plates, the nose by a cone, etc. (Ruck et al., 1970). One coherently adds the scattered fields from the simple scatterers. The scattering from a complex body depends strongly on the radar-target aspect angle. A rough, irregular surface, such as the earth, can be modeled by facets, small, contiguous planar surfaces with random slopes

(Long, 1975, p. 79), whose scattered fields are combined coherently. Note that in modeling the rough surface with random slopes, no attempt is made to model a specific earth area, unlike the aircraft model whose component scattering shapes are chosen to correspond to a particular aircraft. It will be seen in later chapters that the measured Sinclair matrix elements are random variables for most targets of interest. It is therefore appropriate to model a rough surface with random slopes and to calculate the scattering from a complex target using different aspect angles.

Another model of a rough surface can be constructed by placing simple shapes, whose scattering can be determined, randomly on a plane or faceted surface. These models commonly use perfectly conducting surfaces, because surface roughness affects the scattered field more than its electrical properties do (Beckmann and Spizzichino, 1963, p. 5).

The target model may be mathematical, with no consideration given to its geometry or physical properties. An example would be an equation for the average cross-section of terrain as a function of aspect angle, with the equation chosen to approximate measured values for the terrain. Statistical models applicable to aircraft and similar targets have been developed by Marcum and Swerling (Barrett, 1987, p. 345).

If the rms value of surface height departures from an average value is very small compared to a wavelength, a surface is called smooth. If it is somewhat larger but small compared to a wavelength, the surface is slightly rough; if it is much larger than a wavelength, the surface is very rough. Some surfaces are compositely rough, with a slightly rough profile superimposed on a very rough surface. An example is a pasture with terrain height variations constituting the very rough component and grass the slight roughness.

Rice (1951) studied scattering by a slightly rough surface, using a perturbation approach with an expansion of the surface and the scattered fields in Fourier series, and his work was extended by Peake (Ruck, et al., 1970, p. 703). Very rough surfaces represented by the probability density of their surface slopes have been studied using geometric and physical optics (Ulaby and Elachi, 1990).

In remote sensing of the earth, waves may penetrate into vegetation and scatter from branches and leaves of the vegetation. Models have been developed to represent this volume scattering (Peake, 1959, 1965; Tsang et al., 1990; Tsang et al., 1985).

Reflection at an Interface

If an electromagnetic wave is incident on the plane interface between two materials, a reflected wave will exist in the first medium and a transmitted wave in the second. A plane of incidence is defined by the direction of the incident wave and a normal to the interface. Both the reflected and transmitted waves lie in this plane, and the angles of the waves, as shown in Fig. 3.4, are related by Snell's laws,

$$\theta_r = \theta_i \qquad \frac{\sin\theta_t}{\sin\theta_i} = \frac{k_1}{k_2}$$

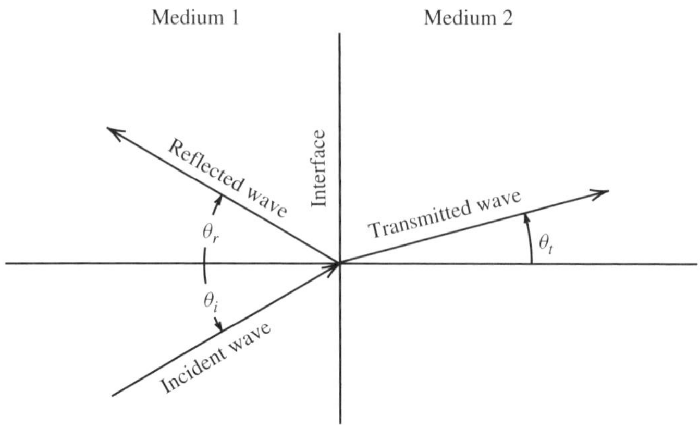

Fig. 3.4. Reflection at an interface.

A linearly polarized wave in region 1 with its electric field vector perpendicular to the plane of incidence is partially reflected with its electric field vector also perpendicular to the plane of incidence and with a Fresnel reflection coefficient,

$$\Gamma_{\perp} = \frac{E_r}{E_i} = \frac{Z_2 \cos\theta_i - Z_1 \cos\theta_t}{Z_2 \cos\theta_i + Z_1 \cos\theta_t}$$

If the incident field's electric vector lies in the plane of incidence, the reflected electric field vector also lies in the plane of incidence, and the reflection coefficient is

$$\Gamma_{\parallel} = \frac{E_r}{E_i} = \frac{Z_1 \cos\theta_i - Z_2 \cos\theta_t}{Z_1 \cos\theta_i + Z_2 \cos\theta_t}$$

Geometric Optics

If the wavelength is small relative to the radii of curvature of the body surface, scattering from a conductor or a nonconductor with a large Fresnel coefficient is principally from specular points, at which a normal vector at the surface bisects the angle between the incident wave direction and a ray from the point to the receiver and lies in the plane formed by those two directions. For vanishingly small wavelengths, electromagnetic energy is transported along a family of rays that can be traced through dielectric media and after reflections at conducting boundaries. Rays in an inhomogeneous medium are curved, but straight in a homogeneous medium.

For a general surface, let **n** be a normal to the surface at point P. A plane containing **n** intersects the surface to form a curve, which may be taken in a small region as a segment of a circle with radius ρ. Unit vector **u** is tangent to the surface and lies in the plane. If the plane is rotated around **n**, a rotation angle can be found which yields a maximum value of circle radius ρ. We designate ρ

and $\mathbf{u}$ for this plane ρ_1 and $\mathbf{u}_1$. If the plane is further rotated, a rotation angle is found for which the circle radius is minimum. It is designated as ρ_2 with a corresponding vector $\mathbf{u}_2$ lying in the rotated plane. For an arbitrary surface, $\mathbf{u}_1$ and $\mathbf{u}_2$ are perpendicular to each other and to $\mathbf{n}$. The two planes found in this manner are called the principal planes of the surface at point P; the circle radii, one maximum and one minimum, are the principal radii of curvature of the surface at point P. For perfectly conducting surfaces expressible as second-degree polynomials, the backscattering cross section is given by (Ruck et al., 1970, p. 541),

$$\sigma_s = \pi\rho_1\rho_2$$

The principles of geometric optics are not applicable to the determination of the Sinclair matrix elements.

A Differential Equation for the Fields

Operations on the time-invariant Maxwell equations yield

$$\begin{aligned}\nabla^2\mathbf{E} + k^2\mathbf{E} &= 0\\ \nabla^2\mathbf{H} + k^2\mathbf{H} &= 0\end{aligned}$$

where $k^2 = \omega^2\mu\epsilon$. These are the vector Helmholtz equations.

In rectangular coordinates, the vector Helmholtz equation can be separated into three scalar Helmholtz equations or scalar wave equations and the solution of a scalar equation used to construct a solution to the vector equation. The solution of the scalar equation is carried out after a coordinate system appropriate to the scattering problem of interest is selected. A well-known method of solving it is separation of variables. This is possible in eleven known coordinate systems (Ruck et al., 1970, p. 34). Solving the separated equation is not simple in most of the coordinate systems, and the determination of scattering in this manner requires numerical computations which can be extensive.

Integrals for the Fields

A useful identity for determining scattering parameters is Green's Theorem, which can be written as

$$\iiint_V (\Phi\nabla^2\Psi - \Psi\nabla^2\Phi)\,dv = -\iint_{S_1+S_2} (\Phi\mathbf{n}\cdot\nabla\Psi - \Psi\mathbf{n}\cdot\nabla\Phi)\,da$$

Φ and Ψ are scalar fields defined in V and on S_1 and S_2. The scattering body has surface S_1 and is illuminated by a wave whose sources we do not take into account. S_2 is a spherical surface with radius approaching infinity. Together the two surfaces form a closed surface that bounds volume V. $\mathbf{n}$ is a unit normal vector pointing into V.

Let Φ be

$$\Phi = \Phi(\mathbf{r}, \mathbf{r}') = G(\mathbf{r}, \mathbf{r}') = \frac{e^{-jkR}}{4\pi R} \qquad R \neq 0$$

where

$$R = |\mathbf{R}| = |\mathbf{r} - \mathbf{r}'|$$

$\mathbf{r}'$ is an observation point in V. G, the free-space Green's function, satisfies the scalar Helmholtz equation everywhere except at $\mathbf{r} = \mathbf{r}'$. To account for this, we complete the definition of G by requiring that it satisfy

$$(\nabla^2 + k^2)G = -\delta(\mathbf{r} - \mathbf{r}')$$

where δ is the spatial impulse function.

If Ψ and its partial derivative obey the strict radiation condition,

$$\lim_{r\to\infty} rF = 0$$

where F is Ψ or its partial derivative, Green's theorem yields

$$\Psi(\mathbf{r}') = \iint_{S_1} [\Psi(\mathbf{r})\mathbf{n} \cdot \nabla G(\mathbf{r}, \mathbf{r}') - G(\mathbf{r}, \mathbf{r}')\mathbf{n} \cdot \nabla \Psi(\mathbf{r})]\, da \qquad (3.41)$$

The integral in this equation is the Kirchhoff Integral. The sources that cause the scalar field in the surface integral are external to S_1 and do not appear in the equation for $\Psi(\mathbf{r}')$. In general, Ψ and its normal derivative on the surface of the scattering body are not known, and approximations to them are made.

The Stratton–Chu Integrals

Solutions for $\mathbf{E}$ and $\mathbf{H}$ can be constructed from (3.41) (Stratton, 1941, p. 392). They are

$$\mathbf{E}(\mathbf{r}') = \iint_{S_1} \left[\left(\mathbf{n} \cdot \nabla G(\mathbf{r}, \mathbf{r}')\right) \mathbf{E}(\mathbf{r}) - G(\mathbf{r}, \mathbf{r}')(\mathbf{n} \cdot \nabla)\mathbf{E}(\mathbf{r})\right] da$$

$$\mathbf{H}(\mathbf{r}') = \iint_{S_1} \left[\left(\mathbf{n} \cdot \nabla G(\mathbf{r}, \mathbf{r}')\right) \mathbf{H}(\mathbf{r}) - G(\mathbf{r}, \mathbf{r}')(\mathbf{n} \cdot \nabla)\mathbf{H}(\mathbf{r})\right] da$$

$\mathbf{E}$ and $\mathbf{H}$ in these equations are the total fields, but the integrals can be altered and separated to give the scattered field, related to the incident and total fields by $\mathbf{E}^t = \mathbf{E}^i + \mathbf{E}^s$. The resulting relationships are

$$\mathbf{E}^s(\mathbf{r}') = \iint_{S_1} [\nabla G \mathbf{E}^t(\mathbf{r}) \cdot \mathbf{n} + (\mathbf{n} \times \mathbf{E}^t(\mathbf{r})) \times \nabla G - j\omega\mu G(\mathbf{n} \times \mathbf{H}^t(\mathbf{r}))]\, da$$

and

$$\mathbf{H}^s(\mathbf{r}') = \iint_{S_1} [\nabla G \mathbf{H}^t(\mathbf{r}) \cdot \mathbf{n} + (\mathbf{n} \times \mathbf{H}^t(\mathbf{r})) \times \nabla G + j\omega\epsilon G(\mathbf{n} \times \mathbf{E}^t(\mathbf{r}))]\, da$$

These are the Stratton–Chu integrals (Stratton, 1941, p. 464; Ruck et al., 1970, p. 34). They simplify for good conductors to

$$\mathbf{E}^s(\mathbf{r}') = \iint_{S_1} [\nabla G \mathbf{E}^t(\mathbf{r}) \cdot \mathbf{n} - j\omega\mu G(\mathbf{n} \times \mathbf{H}^t(\mathbf{r}))]\, da$$

$$\mathbf{H}^s(\mathbf{r}') = \iint_{S_1} [(\mathbf{n} \times \mathbf{H}^t(\mathbf{r})) \times \nabla G]\, da$$

The Kirchhoff Approximation

The Stratton–Chu equations can be simplified by approximations. We divide a scatterer surface into two parts, a front surface that is directly illuminated, and a back surface that is shadowed by the scattering body. Integration over the back surface is neglected. If we assume that the scattered fields from each point on the scatterer surface are those that would be scattered from a plane surface at that point, we can use the Fresnel scattering coefficients to relate the scattered and incident waves. The approximation is better for short wavelengths, relative to the radii of surface curvature, than for long. These approximations lead to the Kirchhoff or physical optics solution to scattering.

A local coordinate system is formed at each point on the surface. The local components of $\mathbf{E}^t$ and $\mathbf{H}^t$ are then found from a knowledge of the local components of the incident wave. Finally, integrations are performed to find the scattered fields.

The Moment Method

If the equation of continuity in the form of a two-dimensional divergence,

$$\nabla_s \cdot \mathbf{J}_s = -j\omega\rho_s$$

and the relationship of the surface charge density to the outward-normal electricity field,

$$\mathbf{E}^t \cdot \mathbf{n} = \rho_s/\epsilon$$

are used with the two-dimensional divergence theorem and the boundary condition on the total magnetic field at the surface of a conductor, and if we note that near the conductor surface the total electric field vanishes and the scattered field is the negative of the incident field, the Stratton–Chu integral for the electric field becomes

$$\mathbf{E}^i(\mathbf{r}') = \frac{1}{j\omega\epsilon}\nabla' \iint_S \mathbf{J}_s \cdot (\nabla_s G)\, da + j\omega\mu \iint_S G\mathbf{J}_s\, da \tag{3.42}$$

The prime on the gradient operator means that the derivatives are with respect to $\mathbf{r}'$.

The incident electric field and G in (3.42) are known, but not the surface current density. It can be found by the method of moments adapted to the solution of electromagnetic scattering problems (Harrington, 1993). The surface current density is expanded in a set of basis functions, with unknown coefficients, chosen to represent $\mathbf{J}_s$ accurately and minimize computational effort. The summation and integration operations are interchanged and the integrals evaluated. The unknown coefficients of the summation can then be found from the known incident field by solving simultaneous equations.

REFERENCES

C. R. Barrett, Jr., "Target Models", in *Principles of Modern Radar*, J. L. Eaves and E. K. Reedy, eds., Van-Nostrand Reinhold, New York, 1987.

P. Beckmann and A. Spizzichino, *The Scattering of Electromagnetic Waves from Rough Surfaces*, Pergamon Press, Macmillan, New York, 1963.

W-M. Boerner et al., "Polarization Utilization in Radar Target Reconstruction", University of Illinois at Chicago, CL-EMID-NANRAR-81-01, January, 1981.

R. F. Harrington , *Field Computation by Moment Methods*, IEEE Press, New York, 1993.

W. A. Holm, "Polarimetric Fundamentals and Techniques", in *Principles of Modern Radar*, J. L. Eaves and E. K. Reedy, eds., Van-Nostrand Reinhold, New York, 1987.

R. A. Horn and C. R. Johnson, *Matrix Analysis*, Cambridge University Press, Cambridge, UK, 1990.

J. R. Huynen, *Phenomenological Theory of Radar Targets*, Drukkerij Bronder-Offset, Rotterdam, 1970.

M. W. Long, *Radar Reflectivity of Land and Sea*, Lexington Books, Lexington, MA, 1975.

E. Lüneburg, "Principles in Radar Polarimetry: The Consimilarity Transformations of Radar Polarimetry Versus the Similarity Transformations in Optical Polarimetry", *IECE* (Japan), *Trans. On Electronics*, Special Issue on Electromagnetic Theory, Vol. E-78C (10), Oct., pp. 1339–1345, 1995 (b).

W. H. Peake, "The Interaction of Electromagnetic Waves with Some Natural Surfaces", Antenna Laboratory, The Ohio State University, Report No. 898-2, 1959.

W. H. Peake, "Scattering from Rough Surfaces Such as Terrain", in *Antennas and Scattering Theory: Recent Advances*, Short Course Notes, The Ohio State University, August 1965.

M. C. Pease, III, *Methods of Matrix Algebra*, Academic, New York, 1965.

S. O. Rice, "Reflection of Electromagnetic Waves by Slightly Rough Surfaces", in *The Theory of Electromagnetic Waves*, M. Kline, ed., Wiley-Interscience, New York, 1951.

G. T. Ruck, D. E. Barrick, W. D. Stuart, and C. K. Krichbaum, *Radar Cross Section Handbook*, Plenum Press, New York, 1970.

J. A. Stratton, *Electromagnetic Theory*, McGraw-Hill, New York, 1941.

L. Tsang, K. H. Ding, and B. Wen, "Polarimetric Signatures of Random Discrete Scatterers Based on Vector Radiative Transfer Theory", in *Progress in Electromagnetics Research, PIER 3, Polarimetric Remote Sensing*, J. A. Kong, ed., Elsevier, New York, 1990.

L. Tsang, J. A. Kong, and R. T. Shin, *Theory of Microwave Remote Sensing*, Wiley-Interscience, New York, 1985.

F. T. Ulaby and C. Elachi, eds., *Radar Polarimetry for Geoscience Applications*, Artech House, Norwood, MA, 1990.

J. J. Van Zyl, H. A. Zebker, and C. Elachi, "Imaging Radar Polarization Signatures: Theory and Observation", *Radio Sci.*, 22(4), 529–543 (1987).

C. R. Wylie and L. C. Barrett, *Advanced Engineering Mathematics*, McGraw-Hill, New York, 1982.

PROBLEMS

3.1. The Sinclair matrix of a target for backscattering is

$$\begin{bmatrix} 3 & j \\ j & 1+j \end{bmatrix}$$

Find the polarization ratios of the radar antenna that give maximum and sub-maximum powers. Find the polarization ratios that give minimum power.

3.2. Find the scattering matrix in circular-component form that corresponds to the matrix of Problem 3.1.

3.3. A bistatic Sinclair matrix is given by

$$\begin{bmatrix} 3 & 1+j \\ j & 1+j \end{bmatrix}$$

Find the polarization ratio of the transmitting antenna that gives maximum scattered power density. Find the polarization ratio of the receiving antenna that gives the greatest received power.

3.4. Note that (3.11) and (3.15) for received power in circular and rectangular components, respectively, have the same form. Follow the procedure of Section 3.11 and find the circular polarization ratios Q_1 and Q_2 that give maximum and sub-maximum received powers for backscattering. Show all steps, including proof that maximum power can be received if the same antenna is used for transmitting and receiving.

3.5. Find the Sinclair matrices of these targets for backscattering to within a multiplying constant. The incident wave travels on the z-axis.
(a) A long straight wire perpendicular to the z-axis and tilted at 30° with respect to the x-axis. (b) A metal sphere. (c) A flat metal plate perpendicular to the z-axis.

3.6. What are the units of the Sinclair matrix?

3.7. Find the vector that is orthonormal to

$$\mathbf{h} = \frac{1}{\sqrt{3}} \begin{bmatrix} 1 \\ 1 + j \end{bmatrix}$$

CHAPTER 4

AN INTRODUCTION TO RADAR

Radar systems range from simple ones for motion detection to more complex systems designed to detect, track, and assist in identifying aircraft at great distances; to assist in the control of industrial processes; to provide meteorological information; to assist in the control of aircraft at airports; and to carry out surveillance and mapping of the earth for military and civilian purposes. Yet all operate on simple principles that are discussed here.

If a radar target is present when an electromagnetic wave is radiated, the wave is scattered (reflected) by the target. The scattered wave is detected by the radar receiver, and information about the target is inferred. The first radars did not take the polarization-altering properties of a target into account, but an important development of recent years has been the fully polarimetric radar. It sequentially transmits two orthogonally polarized waves and after each transmission simultaneously receives two orthogonally polarized waves. We are primarily concerned with polarimetric radars. Two other important developments following the initial development and use of radar are the *synthetic aperture radar*, which allows high-resolution imaging of the earth, and the *radar interferometer*, which allows an elevation profile of the earth to be constructed. The synthetic aperture radar and the radar interferometer may be fully polarimetric.

Radars may be classified as pulsed or continuous wave (CW). In its simplest form, transmitting a single-frequency wave, the CW radar can measure the radial velocity (velocity component along the line from radar to target) of a target by the Doppler effect, but not the target distance. If the frequency of the transmitted wave is varied, as in a frequency-modulated continuous-wave radar (FMCW), target distance can be measured.

Remote Sensing with Polarimetric Radar, by Harold Mott

In its simplest form, a pulse radar can determine the distance to a target by measuring the time required for a short transmitted pulse to travel to the target, be reflected, and return to the radar receiver. The transmitted pulse generally consists of many cycles of a sinusoidal wave of known frequency. If there is relative motion between radar and target, the frequency of the wave will be shifted by the Doppler effect; the radial velocity of the target can be found from the frequency shift. The radar can also determine the amplitude of the received signal. Some radars transmit a coded pulse or a wave whose frequency changes during the time interval of the pulse. Initially, however, we consider only a pulse made up of many cycles of a single-frequency sinusoidal wave.

Radars use antennas that transmit and receive more strongly in some directions than in others. They provide angular information about a target's location. The angular information can be made more precise if a narrow pencil-beam antenna is used or if four skewed antenna beams are utilized for receiving the signal scattered by the target. Radar frequencies range from a few megahertz (MHz), for ground-penetrating radars and those that use ionospheric refraction to extend the radar range, to tens of gigahertz (GHz). Laser radars, at infrared and visual frequencies, are also used. For remote sensing of terrain from airborne and spaceborne radars, the commonly-used frequencies are in the microwave region, from 1 to 12 GHz. The use of microwaves allows the radar to be smaller than one using a lower frequency, and the antenna is smaller for the same antenna beamwidth than it would be if a lower frequency were used. Microwaves can penetrate the earth's cloud cover without substantial loss. Scattering from terrain is frequency-dependent, and sometimes radars with different frequencies are paired to measure terrain scattering characteristics simultaneously. Essential elements in a radar are a transmitter, receiver, antennas, and a signal processor. The signal processor can be a human operator with a visual display unit at the output of the receiver; it can also be fully automated with an output that is stored or used to control some action. The separation between radar elements is not always readily made. Each element in an antenna array, for example, may be supplied by its own transmitter. Nevertheless, it is convenient to distinguish the elements of a radar system as we have, and the principles of operation deduced from this separation are applicable to any system configuration.

4.1. PULSE RADAR

In a simple form, a pulse radar transmits a time-sinusoidal electromagnetic wave for a short interval called the pulse duration time. The transmitter then is quiescent while the radar receiver is turned on to receive the wave reflected from a target. After a fixed time, the pulse repetition time, the cycle is repeated. Figure 4.1 shows on a time axis two pulses of the transmitter voltage and a voltage pulse produced in the radar receiver by a target occupying infinitesimal radial distance (a point target) if no noise or additional targets are present. Also shown is a received voltage pulse from a range-extended target. The transmitted sinusoidal

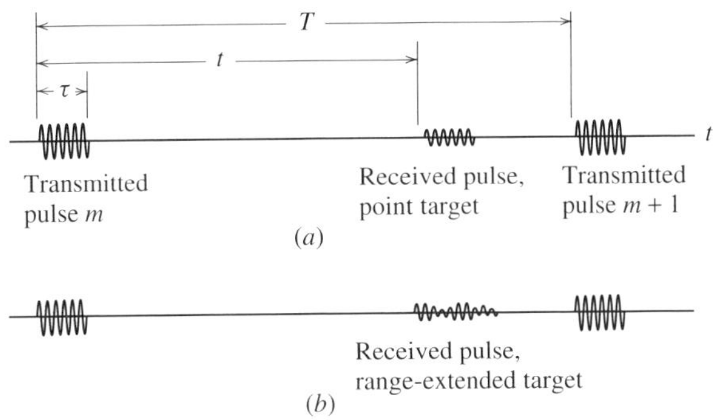

Fig. 4.1. Segment of radar pulse train.

wave has a frequency f_c . Only a few cycles of the wave are shown in Fig. 4.1, but each pulse may consist of hundreds or thousands of cycles. The pulse repetition frequency f_p is much lower than the radar frequency. It is the inverse of the pulse repetition time T. The pulse length of the wave reflected to the receiver is the same as that of the transmitted pulse, τ, for a point target.

The electromagnetic wave travels at the velocity of light, c, so if the time interval t between transmitted and received pulses is measured, the distance between radar and target can be determined from $R = ct/2$.

If the target distance (range) is so great that $t > T$ in Fig. 4.1, an ambiguity will occur. We cannot determine without more information whether the reflected wave was caused by a long-range target illuminated by pulse m or by a short-range target illuminated by pulse $m + 1$. Such ambiguities can be eliminated by choosing the pulse repetition time to be compatible with the target distances expected.

For a single-frequency pulse radar, the transmitter voltage and current are sinusoidal time functions during the transmission interval. We represent them by their corresponding exponential forms, $V = |V|e^{j\Theta_V}$ and $I = |I|e^{j\Theta_I}$ in the interval $-\tau/2 < t < \tau/2$. They are related by $I = V/Z$, with Z the complex impedance across which the voltage appears. The complex scalars V and I are to be multiplied by $\exp(j\omega t)$ and the real part of the product taken to find the corresponding sinusoidal time-varying voltage and current.

If antenna losses are neglected, the peak transmitted power is $W_p = \text{Re}\,(VI^*)/2$. The energy transmitted during one pulse is $E = W_p\tau$, and the average transmitted power is

$$W_{\text{av}} = \frac{E}{T} = \frac{\tau}{T} W_p \tag{4.1}$$

Several factors combine to cause the reflected pulse to differ from that shown in Fig. 4.1a. If the target is extended in range, reflections of different amplitude occur at different ranges; the voltage at the receiver will be a pulse that is longer than the transmitted pulse and has a nonconstant amplitude, as shown in

Fig. 4.1b. If there is a nonzero rate of change of distance between target and radar, the received sine wave will differ in frequency from that transmitted. For an air- or space-borne radar looking at terrain, the returned signal may be extended in time for a much greater interval than is shown in Fig. 4.1b.

Unwanted targets (clutter) at the same range can interfere with the signal from the desired target. Electrical noise, generated either inside the receiver or external to it, may cause receiver voltages greater than the voltage produced by a target pulse. This problem can be ameliorated by using a receiver that has a small bandwidth and by integrating the received pulses; that is, by summing over a number of pulses. The total energy of the summed pulses, as a function of the number summed, increases faster than the total integrated noise. In this way, targets can be detected which have single-pulse returns much smaller than the accompanying electrical noise.

If a human operator is part of the radar system, the received pulse, after rectification to eliminate the negative-going parts of the wave and after low-pass filtering, can be displayed visually on a cathode-ray tube or LCD screen. A visual display of this type also serves to integrate the received pulses because of the persistence of the image on the screen and in the eye.

A Coherent Radar System

In a *coherent radar*, integration is carried out by summing the received voltage pulses with the proper phase relationships. A simplified block diagram of a coherent radar is shown in Fig. 4.2. In a noncoherent radar, the phase is discarded before the voltages are summed.

The transmitter of a coherent radar employs a *stable oscillator* operating near the frequency of the transmitted wave and a *coherent oscillator* operating at the intermediate frequency of the superheterodyne radar receiver. Their outputs are up converted by Mixer 1. An electronic mixer is a signal multiplier followed by a filter. Its output before filtering is the product of two input cosine waves. This product can be considered the sum of a cosine wave whose frequency is the sum of the frequencies of the input waves, and a cosine wave whose frequency is the difference of the frequencies of the input waves. The sum frequency is retained by the up converter. The output is amplified by a power amplifier and pulse modulated to create the wave radiated by the transmitting antenna.

If the target has a small range extent, the returned signal is a pulse of approximately the same time duration as that transmitted. If there is relative motion between radar and target, the returned wave will be frequency shifted by the Doppler effect.

The returned wave is down converted by Mixer 2, whose inputs are the received wave and a signal from the stable oscillator, to give a cosine time function whose frequency, $f_0 + f_d$, is sufficiently close to f_0 that it can be amplified by the IF amplifier, which is a matched filter with a band-pass filter characteristic that rejects much wide-band noise and gives a maximum signal–noise ratio at the output.

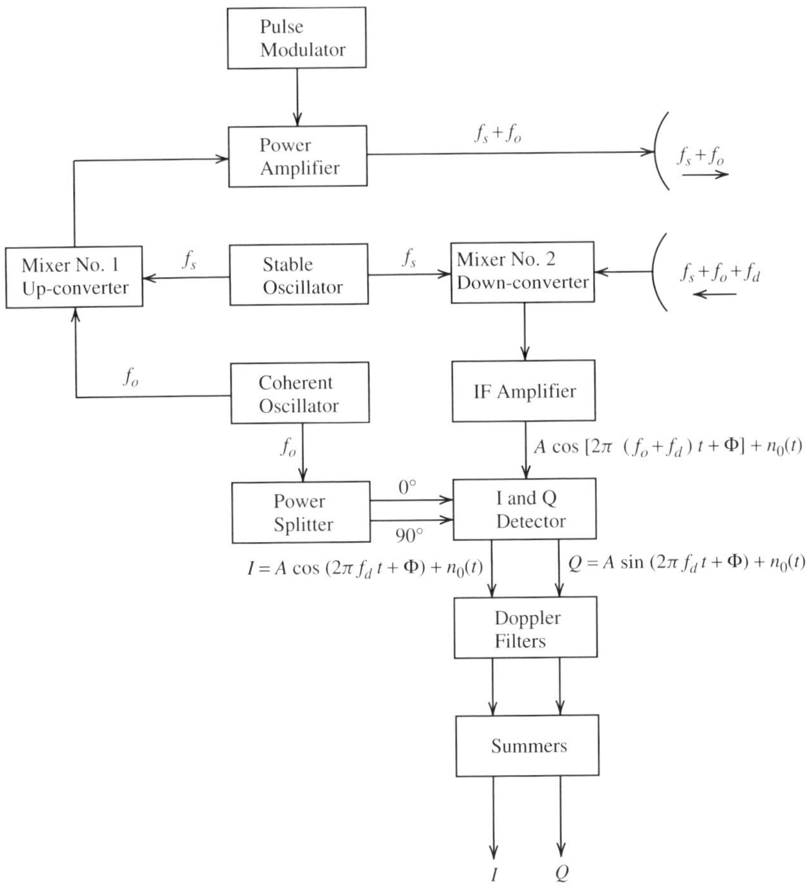

Fig. 4.2. Coherent radar system.

The IF amplifier output is a cosine wave of frequency $f_0 + f_d$ and unknown phase Φ. Phase Φ includes phase changes in the receiver, those caused by wave travel to and from the target, and those caused by reflection characteristics of the target. If Φ were known, a synchronous detector could recover the amplitude and Doppler frequency of the signal from the IF amplifier by mixing it with $\cos(2\pi f_0 t + \Phi)$. Φ is not known, however, and the I and Q detector (in-phase and quadrature) shown in Fig. 4.2 achieves the same result. The signal from the coherent oscillator is split and one of the resulting signals is shifted by $\pi/2$. The resulting signals, $\cos 2\pi f_0 t$ and $\sin 2\pi f_0 t$, are mixed with $A \cos[2\pi(f_o + f_d)t + \Phi]$ in the *I and Q detector*, which is effectively two mixers. The difference frequency is retained and the outputs of the detector, excluding noise, are

$$I = A \cos(2\pi f_d t + \Phi)$$
$$Q = A \sin(2\pi f_d t + \Phi)$$

Not shown in Fig. 4.2 are range gates. They can be placed in the IF section of the receiver, and their function is to break the received wave into time segments corresponding to range intervals. The output from a particular gate is sent to an I and Q detector corresponding to the chosen range interval. Range-gate width depends on the range accuracy desired, but the equivalent time-gate width is usually on the order of pulse duration (Skolnik, 1962, p 117).

Variations on the system shown in Fig. 4.2 are possible. It is common practice to use one antenna, rather than two, and switch it from transmitter to receiver. With the use of either one or two antennas, the receiver must be protected from the high-power pulse during the time of transmission. An RF amplifier can be used after the receiving antenna to improve the system noise performance. The I and Q detector of Fig. 4.2 can be replaced by an envelope detector, which is a rectifier and low-pass filter, without input from the coherent oscillator. Doppler information is lost. The oscillator, pulse modulator, and power amplifier can be replaced by a pulsed oscillator of high power. The requirements of remote sensing normally preclude these changes, and we will not consider them further.

Following the I and Q detector are *Doppler filters*. It is simple conceptually to think of these as banks of analog band-pass filters, one for the I signal and one for Q. The filters in a bank are supplied simultaneously by an input signal from the detector. If only one target is present in the range gate of interest, the filter tuned to the target's Doppler frequency will have an output and the remaining filters will not.

Following the Doppler filter is a *coherent summer* or *integrator*. It is the practice in radar to sum, or integrate, many received pulses to improve the signal-noise ratio. If they are summed with signal phases retained, the summation is coherent. If the signal phases are discarded, as they would be if the power of a pulse were found before summation, the integration is noncoherent.

The summer stores the received pulses and adds them when a specified number are stored. A random noise voltage $n_0(t)$ unavoidably appears in the receiver, and is shown in Fig. 4.2. Assume that it is the same in the two outputs of the Doppler filter corresponding to the I and Q inputs. Take the ideal case for which every received pulse is the same. Then the outputs of the Doppler filter for the nth pulse are

$$I_n = A_n \cos(2\pi f_d t + \Phi_n) + n_0(t)$$

$$Q_n = A_n \sin(2\pi f_d t + \Phi_n) + n_0(t)$$

The noise signals add in power, and the sum of N pulses from the summer is

$$I = N A_n \cos(2\pi f_d t + \Phi_n) + \sqrt{N n_0^2(t)}$$

$$Q = N A_n \sin(2\pi f_d t + \Phi_n) + \sqrt{N n_0^2(t)}$$

The I and Q signals can be combined as

$$I^2 + Q^2 = N^2 A_n^2 + 2Nn_0^2(t) + 2N^{3/2} A_n n_0(t)[\cos(2\pi f_d t + \Phi_n) + \sin(2\pi f_d t + \Phi_n)]$$

If the last term of this summation, which can be removed by rectifying and low-pass filtering, is neglected, the output signal-noise ratio is N times greater than that of one pulse. If an envelope detector is used instead of the I and Q detector, the signal–noise improvement from pulse integration is less than N.

The amplitude $I^2 + Q^2$ is the third characteristic to be measured by the radar. It gives the target's radar cross-section.

During the time in which pulse integration is being carried out, the target return may move from one range gate to another or from one Doppler filter to another. This must be accounted for in the pulse summation.

Doppler Frequency

The frequency of the received wave differs from that transmitted if the distance between radar and target is changing. Let the transmitted voltage during a pulse be

$$V_t = |V_t| e^{j2\pi f_c t}$$

At distance r from the transmitting antenna, an electric field intensity component is

$$\tilde{E} = E(r) e^{j2\pi f_c(t - r/c)}$$

If the target is at range R, the received voltage is

$$V = |V| e^{j\Phi_r} = |V| e^{j[2\pi f_c(t - 2R/c) + \Phi_t]}$$

where Φ_t is a constant phase added by the target.

The frequency of the received voltage is

$$f_r = \frac{1}{2\pi} \frac{d\Phi_r}{dt} = f_c - \frac{2}{\lambda} \frac{dR}{dt}$$

The frequency difference between transmitted and received waves is the Doppler frequency,

$$f_d = -\frac{2}{\lambda} \frac{dR}{dt}$$

We cannot determine the relative velocity of radar and target from the Doppler frequency. The quantity determined is the rate of change of range. This is the radial velocity of the target.

We noted range ambiguities for the pulse radar. If the radar is used to measure radial velocity, there are velocity ambiguities and blind speeds at which a target will not be detected. These ambiguities can be eliminated by proper radar design.

Targets

A coherently scattering target is one for which the received voltage, disregarding additive noise, for a pulse is independent of the pulse number in amplitude, frequency, and phase. This is the target we considered when we found a signal-noise improvement of N for N pulses integrated. More generally, amplitude A_n and phase Φ_n are random variables, and in an extreme case (see Appendix A) I_n^2 and Q_n^2 are summed. There is no signal–noise improvement with integration.

Virtually all targets of interest have properties that fall between these limiting cases.

4.2. CW RADAR

A radar that transmits a sinusoidal electromagnetic wave of constant amplitude and constant frequency is a continuous-wave, or CW, radar. It can measure the relative radial velocity of a target, using the Doppler effect, but not the target range. Such a radar is simple and inexpensive. Typical is a radar used to measure the speed of automotive traffic.

If the frequency of the transmitted wave is varied, the returned signal at a specified time will have one frequency if it is scattered from a target at range R_1 and a different value if from a target at R_2. The range can be obtained by mixing the received wave with the transmitted wave and measuring the frequency difference. The target's radial velocity can also be determined. Such a radar is called an FMCW (frequency-modulated continuous-wave) radar.

4.3. DIRECTIONAL PROPERTIES OF RADAR MEASUREMENTS

A radar uses an antenna, or antennas, with directional properties. The radiative properties of a radar antenna are best described graphically by its radiation pattern, which shows the radiated power density as a function of polar and azimuth angles. Figure 4.3 shows a section of a pattern of a typical radar antenna. The pattern is a function of two angles, and it might, for example, be the section of Fig. 4.3 rotated about its axis of symmetry. The included angle of a pattern section between the angles at which the power density is one-half the maximum is called the *half-power beamwidth*, or the *one-way half-power beamwidth*. In a monostatic radar that uses the same antenna for transmitting and receiving, the antenna pattern comes into play twice, and it is convenient to use the two-way pattern, which is the square of that shown. Figure 4.3 shows the main lobe of the antenna pattern section and small side lobes, which unavoidably appear in useful radar antenna patterns. If the antenna pattern is rotationally symmetric about a center line, the antenna is said to have a "pencil beam." If the pattern is broader in one angular direction than in the orthogonal direction, it is a "fan beam" antenna. Both are useful in radar. If the pattern is narrow, the antenna has a high gain and a large receiving area. If the antenna has a narrow pattern, we can determine

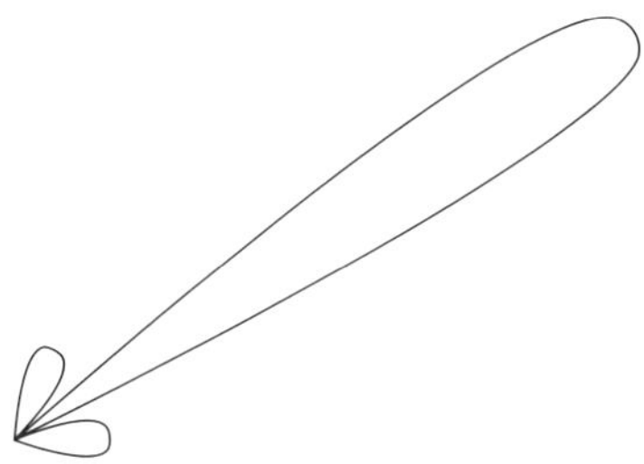

Fig. 4.3. Section of radiation pattern.

approximately the direction of a target from the known pointing direction of the antenna when it receives a relatively strong signal from the target.

The antenna pattern of Fig. 4.3 has a half-power beamwidth of many degrees, but if the antenna is made sufficiently large, the beamwidth can be reduced to tenths of a degree. An approximate relationship between beamwidth in radians and antenna width is

$$\theta_{\text{half}} = \lambda/D \tag{4.2}$$

where λ is the wavelength and D the antenna dimension measured in the plane of the antenna pattern section. In this equation, θ_{half} is the difference between the angles at which radiated power density is half its maximum value, or, for a symmetric beam, twice the angle between the direction of maximum power and the direction at which power density falls to half its maximum. A large antenna, or array, compared to a wavelength, is necessary to produce a narrow beam. We will see later that a large antenna can be synthesized by moving a smaller antenna along a known trajectory. A radar that uses this technique is called a synthetic aperture radar (SAR).

4.4. RESOLUTION

Resolution describes the radar's ability to cause two closely spaced targets to appear as two targets rather than one. If the resolution distance is small, we speak of a high-resolution radar. *Range resolution* refers to the ability to separate two targets in range and *angle resolution* to the ability to separate targets in angle.

Single-Frequency Pulse

Consider two point targets at ranges R_1 and R_2, as shown in Fig. 4.4a. One target does not obscure the other by shadowing it. Single-frequency pulses are radiated and received by the radar, and the received pulse envelopes appear as in Fig. 4.4b, where $t_1 = 2R_1/c$ and $t_2 = 2R_2/c$.

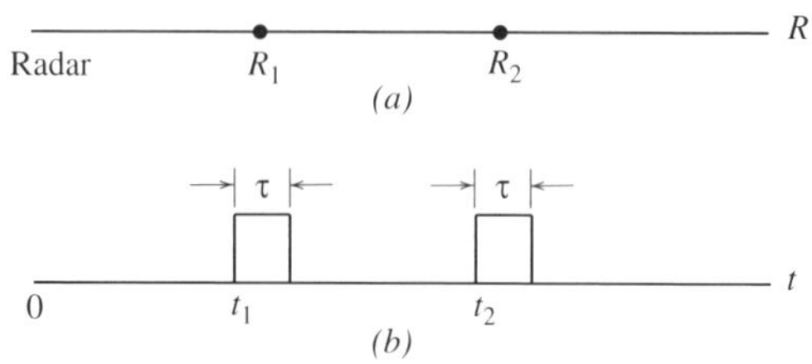

Fig. 4.4. Point targets and received pulse envelopes.

A simple criterion for establishing that the radar sees two targets rather than one is that $t_2 \geq t_1 + \tau$, or $R_2 - R_1 \geq c\tau/2$. The difference in range for equality of the two sides of this equation is the range resolution distance. If we take the radar system bandwidth B as the inverse of the pulse length, the range resolution distance is

$$\delta R = \frac{c\tau}{2} = \frac{c}{2B} \tag{4.3}$$

A specific radar may not be able to resolve targets separated by this resolution distance, and if the targets are other than point targets of equal cross section, the resolution of the radar is affected, but this simple criterion is nonetheless useful.

It is clear from (4.3) that range resolution distance can be made smaller by decreasing pulse-width τ. If this is done, the average power, given by (4.1), may drop below an acceptable value unless the peak power becomes unacceptably large. These two powers limit the range resolution for a single-frequency pulse.

Spread-Spectrum Radar

The peak power–average power constraint on range resolution can be overcome if a pulse is transmitted that does not obey the inverse relationship between pulse length and bandwidth, $\tau = 1/B$. The last term of (4.3) implies that resolution distance is determined by the bandwidth of the pulse, and it is desirable to consider this possibility.

For a pulse of a sinusoidal signal, the bandwidth can be increased by changing the frequency or phase of the sinusoid while the pulse is generated. For example, a pulse of length τ can be divided into N segments of length τ/N. If the frequency is increased by a finite step from one segment to the next, the signal bandwidth can be increased beyond $1/\tau$. The oscillator used to down-convert the received wave is similarly stepped in frequency.

In another spectrum-spreading technique, the phase is changed discretely within the transmitted pulse. The pulse is divided into N segments, and the phase is kept constant for S_1 segments, switched to another phase, normally a reversed phase, for S_2 segments, back to the first phase for S_3 segments, and so on. The numbers S_1, S_2, and so on are selected by a code, commonly a pseudo-random or pseudo-noise code (Eaves and Reedy, 1987, p. 483). In the receiver, the scattered wave is correlated with a delayed replica of the transmitted wave.

The correlated signal output in both cases has a significant peak that is narrower than the transmitted pulse. For that reason, radars using spread-spectrum techniques are sometimes called *pulse-compression radars*. A widely used pulse-compression radar uses a linear frequency-modulated pulse, or *chirp* pulse. We will develop the output for a radar using this pulse and show that range resolution is not determined by pulse length. Range resolution is therefore unconstrained by the ratio of peak power to average power of the transmitted pulse.

The Linear FM Pulse

A linear FM pulse has a frequency that is a linear function of time,

$$f = f_c + Kt \qquad -\tau/2 < t < \tau/2$$

and the form

$$g(t) = \cos 2\pi(f_c t + Kt^2/2) \qquad -\tau/2 < t < \tau/2 \tag{4.4}$$

Constant K is chosen so that the difference between the highest and lowest frequencies of the wave is much greater than $1/\tau$. The signal bandwidth is then given by

$$B = f_{\max} - f_{\min} = K\tau \tag{4.5}$$

The signal returned from a point target at range R is given by

$$h(t) = \cos 2\pi \left[f_c(t - 2R/c) + \frac{K}{2}(t - 2R/c)^2 \right]$$
$$-\tau/2 + 2R/c < t < \tau/2 + 2R/c \tag{4.6}$$

No phase change is considered except that due to distance R, which is not known in advance of receiving the signal. Doppler frequency f_d is not shown explicitly in $h(t)$ but is inherent in the variation of range R.

Assume that in the range interval being considered there is only one point target. The returned signal of (4.6) is processed by correlation, using a delayed replica of the transmitted pulse. With a delay time of t', the correlation integral is

$$V_{gh}(t') = \int uv\, dt \tag{4.7}$$

where

$$u = \cos 2\pi \left[f_c(t - t') + \frac{K}{2}(t - t')^2 \right]$$

$$v = \cos 2\pi \left[f_c(t - 2R/c) + \frac{K}{2}(t - 2R/c)^2 \right]$$

In this correlation integral, we use the transmitted center frequency f_c, but the received and correlating signals can be down converted to any frequency before the correlation is performed.

The product of cosines in (4.7) can be written as the cosine of the sum of the arguments plus the cosine of the difference of the arguments. The sum of the arguments contains a doubled frequency term that can be removed by a low-pass filter before the integration. Then the correlation integral reduces to

$$V_{gh}(t') = \frac{1}{2}\mathrm{Re}\left[e^{j\Phi(t')} \int e^{j2\pi K(t'-2R/c)t}\, dt \right] \tag{4.8}$$

where

$$\Phi(t') = 2\pi \left[f_c(t' - 2R/c) - \frac{K}{2}\left(t'^2 - (2R/c)^2\right) \right] \tag{4.9}$$

We make the simplifying assumption that R is constant during the pulse duration.

The envelopes of the pulses to be correlated are illustrated in Fig. 4.5. There are four cases to consider. Two cases, those for which $t' < -\tau + 2R/c$ and $t' > \tau + 2R/c$, give zero correlations. The two cases that lead to nonzero correlations are

Case 1: $-\tau + 2R/c < t' < 2R/c$

$$V_{gh}(t') = \frac{1}{2}\mathrm{Re}\left[e^{j\Phi(t')} \int_{-\tau/2+2R/c}^{\tau/2+t'} e^{j2\pi K(t'-2R/c)t}\, dt \right]$$

$$= \frac{1}{2}\tau \cos 2\pi [f_c(t' - 2R/c)] \frac{\sin \pi K[\tau(t' - 2R/c) + (t' - 2R/c)^2]}{\pi K \tau (t' - 2R/c)}$$

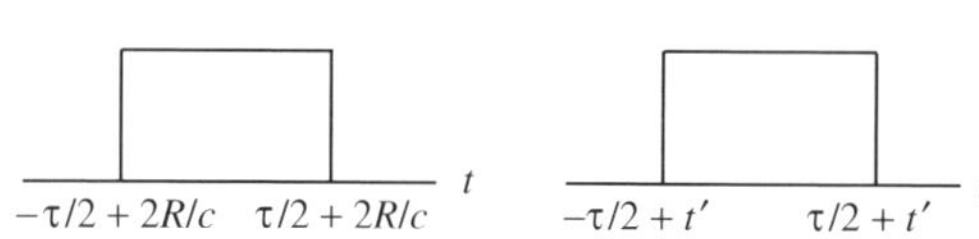

Fig. 4.5. Pulses to be correlated.

Case 2: $2R/c < t' < \tau + 2R/c$

$$V_{gh}(t') = \frac{1}{2}\text{Re}\left[e^{j\Phi(t')}\int_{-\tau/2+t'}^{\tau/2+2R/c} e^{j2\pi K(t'-2R/c)t}\,dt\right]$$

which leads to the same result as Case 1. The time t' of interest to us in the modulating term is that where $t' \approx 2R/c$. Then we assume $(t' - 2R/c) \ll \tau$, and the correlation becomes

$$V_{gh}(t') = \frac{1}{2}\tau\frac{\sin\pi K\tau(t'-2R/c)}{\pi K\tau(t'-2R/c)}\cos 2\pi f_c(t'-2R/c)$$
$$-\tau + 2R/c < t' < \tau + 2R/c \qquad (4.10)$$

This function has maximum amplitude at $t' = 2R/c$. Measurement of time at which the processed signal is maximum gives range R to the point target. This voltage is similar to that for the received wave using a single-frequency pulse, except for the pulse envelope shape. The pulse has a multilobed structure, but it is the main lobe centered at $2R/c$ that is important. Received power from this main lobe will decrease to half its greatest value when $(\sin u_h/u_h)^2 = 0.5$, which occurs at $u_h \approx 0.44\pi$. The pulse width of the processed signal can then be taken as

$$|t' - 2R/c| = 2\frac{u_h}{\pi K\tau} \approx \frac{1}{K\tau}$$

The signals received from two point targets at ranges R_1 and R_2 will reach their maximum values at $t_1' = 2R_1/c$ and $t_2' = 2R_2/c$, respectively. If the difference in these two times is equal to or greater than the correlated pulse width, it is probable that the two targets will be recognized as separate targets. Then the time resolution is $t_2' - t_1' = 1/K\tau$, and the range resolution distance is

$$R_2 - R_1 = \delta R = \frac{c}{2}(t_2' - t_1') = \frac{c}{2K\tau}$$

If the bandwidth of the transmitted signal from (4.5) is used, the resolution distance becomes $\delta R = c/2B$. The resolution expressed in terms of bandwidth is the same as for the constant-frequency pulse. Resolution of a constant-frequency pulse can be increased by making the pulse narrower, which decreases the average power of the radar unless the peak power of the transmitted pulse is increased, a solution that may be unsatisfactory. The range resolution of the linear FM pulse depends only on the bandwidth, so the transmitted peak power can be independently chosen.

Correlation of a Single-Frequency Pulse

We considered previously the reception of a single-frequency pulse with rectangular envelope by down conversion in frequency and the use of an I and Q detector. The pulse may also be correlated with a delayed replica of the transmitted pulse to give an output with a different envelope. The single-frequency pulse is given by (4.4) with $K = 0$, so the correlation can be determined by setting $K = 0$ in (4.8) and (4.9). As with the linear FM pulse, two cases give non-zero correlations. The output for these cases is

$$V_{gh}(t') = \begin{cases} \frac{1}{2}(\tau + t' - 2R/c)\cos 2\pi f_c(t' - 2R/c) & -\tau + 2R/c < t' < 2R/c \\ \frac{1}{2}(\tau - t' - 2R/c)\cos 2\pi f_c(t' - 2R/c) & 2R/c < t' < \tau + 2R/c \end{cases} \tag{4.11}$$

The envelope of the correlated output is triangular with width 2τ and peak at $t' = 2R/c$.

Range-Extended Target

When the radar looks at a range-extended target, the returned signal is a time function extending from the time corresponding to the return from the nearest part of the target to that from the farthest part. In the correlation process, this continuous return can be divided by range gating into intervals if information about the signal intensity as a function of range is to be obtained. The correlation process is carried out for each time segment, which corresponds to each range interval.

Angle Resolution

Consider two identical point targets, with the first at the center of the radar antenna beam and the second at the half-power (3 dB) angle of the one-way antenna pattern. The power received from the second target will be 6 dB less than from the first because the antenna pattern is a factor in both transmitting and receiving. This angular separation might be sufficient to allow us to distinguish the targets as two rather than one, but a conservative definition of angle resolution is that it is the same as the included angle between half-power directions of the one-way antenna pattern.

4.5. IMAGING RADAR

If a radar target occupies a polar and azimuth angular space greater than that between the resolution angles of the radar, it is possible to construct an image of the target, limited to the front surface of the target as it would be seen from one point in space, or from a narrow range of angles. If received power is mapped, with brightness corresponding to power, the map is analogous to the visual image formed by an eye.

A radar map of an extended target is superior, in some respects, to the image that would be acquired visually. The human eye is nonpolarimetric and most visual light sources are nonpolarimetric. A polarimetric radar, on the other hand, can obtain target images at many polarizations of the radar, and the images thus obtained may be quite different. The differences may be significant in identifying the target. A radar interferometer can determine ranges to various areas of an extended target. This range determination is not possible visually, even with the binocular capability of human eyes. Finally, microwaves can penetrate cloud cover in remote sensing of the earth, and this is not possible visually. Visual wavelengths are much shorter than those of radar, and this may allow better visual resolution than can be obtained by radar.

4.6. THE TRADITIONAL RADAR EQUATION

Figure 3.1 illustrates a bistatic radar and target, with transmitting and receiving antennas separated. If the two antennas are colocated, the radar is a monostatic or backscattering radar. In many radars, the same antenna is used for transmitting and receiving, and this is a special case of a monostatic radar. In the configuration shown, the power density incident on the target is

$$\mathcal{P}_i = \frac{W_t G_t(\theta_t, \phi_t)}{4\pi r_1^2}$$

where W_t is the power accepted by, and G_t the gain of, the transmitting antenna. The transmitted signal may be pulsed or continuous, but its characteristics do not affect this development. The wave striking the target is reradiated in a directional manner, and a portion of the reradiated, or scattered, power is intercepted by the receiving antenna. The power received depends on the transmitted power, the antenna gains, and the *radar cross-section* of the target (IEEE, 1983). The radar cross-section may be bistatic or monostatic.

It is desirable to define another target cross section, the *scattering cross-section* (IEEE, 1983) before defining the radar cross-section. The scattering cross-section of a target is the projected area of an equivalent target that intercepts a power equal to the equivalent-target area multiplied by the power density of an incident plane wave and reradiates it isotropically to produce at the receiving antenna a power density equal to that produced by the real target. The following reasoning makes clear the reasons for the definition in terms of a fictitious equivalent target: An observer at the receiver can determine the power density of the scattered wave at the receiver and from it the radiation intensity of the scattered wave in the direction of the receiver. The observer does not know how the target scatters the incident wave without more information than can be obtained by one measurement; yet to describe the target an agreed-on assumption is necessary.

This assumption is that of isotropic scattering. With it, an observer calculates that the total scattered power is 4π times the radiation intensity in the direction of the receiver. It is reasonable to say that this total power is scattered as a result of a target with area σ_s intercepting an incident plane wave with power density established at the target by the radar transmitter. The transmitting antenna characteristics enter into the definition of scattering cross-section since the power scattered in a particular direction depends on the incident wave. The definition is in terms of a power density or radiation intensity, not received power, so it does not depend on a receiving antenna.

The radar cross-section σ_r of a target is defined somewhat like the scattering cross-section except that both transmitting and receiving antennas are specified. Only that part of the scattered wave is considered that can be received by the specified antenna. The definitions of both cross-sections will be clearer as the radar equations are developed.

The power intercepted by the equivalent target with scattering cross-section σ_s is

$$W_{\text{int}} = \sigma_s \mathcal{P}_i = \frac{W_t G_t(\theta_t, \phi_t)\sigma_s}{4\pi r_1^2}$$

where W_t is the transmitter power and G_t the gain of the transmitting antenna. If this power is scattered isotropically, the power density at the receiving antenna is

$$\mathcal{P}_r = \frac{W_{\text{int}}}{4\pi r_2^2} = \frac{W_t G_t(\theta_t, \phi_t)\sigma_s}{(4\pi r_1 r_2)^2} \tag{4.12}$$

If we consider only that part of the received power density that is effective in producing power in the receiver load, it is given by the same equation, using the radar cross-section σ_r rather than the scattering cross-section,

$$\mathcal{P}_r' = \frac{W_t G_t(\theta_t, \phi_t)\sigma_r}{(4\pi r_1 r_2)^2}$$

The power to an impedance-matched load at the receiver is

$$W_r = \mathcal{P}_r' A_{er}(\theta_r, \phi_r) = \frac{W_t G_t(\theta_t, \phi_t) A_{er}(\theta_r, \phi_r)\sigma_r}{(4\pi r_1 r_2)^2}$$

This is a common form of the radar equation. In defining the radar cross-section, the polarization characteristics of both transmitting and receiving antennas must be specified and it may be used only for cases involving those states. Typically, radar cross-sections are specified as HH (horizontal receiving, horizontal transmitting antennas), HV (horizontal, vertical), LR (left, right circular), and so on. The equation is not sufficient to describe completely the scattering behavior of a target because it neglects the target's polarimetric behavior.

If the transmitter and receiver for a radar are at the same site, and if the same antenna is used for transmitting and receiving, the relation between gain and effective receiving area for the antenna causes the radar equation to be

$$W_r = \frac{W_t G^2(\theta, \phi)\lambda^2\sigma_r}{(4\pi)^3 r^4}$$

4.7. THE POLARIMETRIC RADAR EQUATION

Neither the scattering nor radar cross-section can characterize the polarimetric behavior of a target. Since both are widely used, however, they will be related to each other and to the target Sinclair matrix.

If (3.4) for the scattered electric field is used and if the incident field at the target is given in terms of the effective length of the transmitting antenna, using (2.37), the scattered power density is

$$\mathcal{P}_r = \frac{Z_0 I^2}{32\pi\lambda^2 r_1^2 r_2^2}|\mathbf{Sh}_t|^2$$

If this is equated to (4.12), the result is

$$W_t G_t \sigma_s = \frac{\pi Z_0 I^2}{2\lambda^2}|\mathbf{Sh}_t|^2$$

The transmitter power W_t is equal to $R_t I^2/2$, where R_t is the transmitting antenna resistance. If this is used, and if it is noted from Section 2.12 that $G_t R_t = \pi Z_0 |\mathbf{h}_t|^2/\lambda^2$, the scattering cross-section becomes

$$\sigma_s = \frac{|\mathbf{Sh}_t|^2}{|\mathbf{h}_t|^2}$$

Note that σ_s depends on the transmitting antenna, but not on the receiving antenna and not on the radar–target distance. The cross-section found by this equation is valid for both backscattering and bistatic scattering.

Equation 3.7 gives the received power in terms of the target's Sinclair matrix, and equating it to the received power in terms of the radar cross-section leads to

$$W_t G_t A_{er}\sigma_r = \frac{\pi Z_0^2 I^2}{8\lambda^2 R_a}|\mathbf{h}_r^T \mathbf{Sh}_t|^2$$

With substitutions for W_t and G_t, this reduces to

$$A_{er}\sigma_r|\mathbf{h}_t|^2 = \frac{Z_0}{4R_a}|\mathbf{h}_r^T \mathbf{Sh}_t|^2$$

It can also be seen from Section 2.12 that $A_{er}R_a = Z_0|\mathbf{h}_r|^2/4$. If this is used, the radar cross-section becomes

$$\sigma_r = \frac{|\mathbf{h}_r^T \mathbf{S}\mathbf{h}_t|^2}{|\mathbf{h}_t|^2|\mathbf{h}_r|^2} \tag{4.13}$$

This equation is valid for backscattering and bistatic scattering. The ratio of the two cross-sections is the polarization efficiency of the receiving antenna,

$$\frac{\sigma_r}{\sigma_s} = \frac{|\mathbf{h}_r^T \mathbf{S}\mathbf{h}_t|^2}{|\mathbf{S}\mathbf{h}_t|^2|\mathbf{h}_r|^2} = \rho$$

One other form of the radar equation for backscattering should be noted. If (4.13), specialized to backscattering with the same antenna used for transmitting and receiving, is substituted into (3.7), similarly specialized, and the resulting equation compared to the equation for maximum copolarized power developed in Section 3.11, it will be seen that the maximum value of the radar cross-section is $|\gamma_1|^2$, where $|\gamma_1|^2$ is the largest eigenvalue of $\mathbf{S}^*\mathbf{S}$. Since this largest eigenvalue corresponds to optimum polarization for the antenna, the polarization efficiency is one, and the maximum value of the radar cross-section is equal to the maximum value of the scattering cross-section. A backscatter polarization efficiency with the same antenna transmitting and receiving was defined in section 3.11 as

$$\rho_s = \frac{|\hat{\mathbf{h}}^T \mathbf{S}\hat{\mathbf{h}}|^2}{|\gamma_1|^2}$$

where the circumflex indicates the normalized value. It can be seen now that the numerator is the radar cross-section for backscattering and the denominator its greatest value.

4.8. A POLARIMETRIC RADAR

Measurement of the Sinclair matrix requires a radar that sequentially transmits two orthogonally polarized waves and simultaneously receives two orthogonal field components. In effect, two receivers, each with an antenna, are needed. One transmitter can be used if it is switched from an antenna to an orthogonally-polarized one.

Antenna Matrices

Suppose we wish to transmit a linear x-polarized wave. We employ an x-polarized antenna fed by current I_x acting through a radar channel that we designate the x channel. To transmit a y-polarized wave, we use a y-polarized antenna, current I_y, and the radar's y channel. This may be done sequentially to measure $\mathbf{S}$ or concurrently if we wish to transmit a wave of other than x or y polarization.

The wave transmitted by the antenna of the x channel may not have a perfect x polarization because the antenna may be imperfect and transmit also a y-polarized component or because part of the x-channel signal may feed through to the y channel and be radiated from the y-channel antenna. We can determine the incident field at a target from this two-channel radar and account for system imperfections by combining two equations of the form (3.3) with currents I_x and I_y in the x and y channels. The field is

$$\mathbf{E}^i = \frac{jZ_0}{2\lambda r_1} e^{-jkr_1} \mathbf{T}\mathbf{I}_c$$

where the *transmitting current vector* is

$$\mathbf{I}_c = \begin{bmatrix} I_x \\ I_y \end{bmatrix}$$

and $\mathbf{T}$ is the *transmitting-antenna system matrix*. It has also been called the transmitter distortion matrix (Whitt et al., 1990). Matrix $\mathbf{T}$ combines the effective lengths of the two system antennas and accounts for antenna imperfections and leakage from one channel to the other. If the x- and y-channel antennas radiate only their designed polarizations and there is no signal leakage, $\mathbf{T}$ is diagonal with h_x and h_y as elements.

The wave scattered to the radar produces voltages in the two receiver channels, and they may be combined in a voltage vector,

$$\mathbf{V} = \begin{bmatrix} V_x \\ V_y \end{bmatrix}$$

If $\mathbf{E}^s$ is the wave scattered by the target to the radar, the voltages in the receiver channels can be determined by combining two equations of the form (2.40) as

$$\mathbf{V} = \mathbf{R}\mathbf{E}^s$$

where $\mathbf{R}$ is the *receiving antenna system matrix*, called also the distortion matrix of the receiving antennas (Whitt et al., 1990). Like the transmitting antenna system matrix, $\mathbf{R}$ accounts for antenna imperfections and signal leakage. If two equations of the form (3.6) for the scattered wave from a target with Sinclair matrix $\mathbf{S}$ are combined, the vector received voltage is

$$\mathbf{V} = \frac{jZ_0}{2\sqrt{4\pi}\,\lambda r_1 r_2} e^{-jk(r_1+r_2)} \mathbf{R}\mathbf{S}\mathbf{T}\mathbf{I}_c$$

This equation is valid for backscattering or bistatic scattering.

Calibration

The transmitting and receiving antenna system matrices can be determined by measuring three targets with known Sinclair matrices. If both **T** and **R** are diagonal, measurement of two targets is sufficient (Whitt et al., 1990).

4.9. NOISE

Electromagnetic noise generated in a receiver or entering the receiver from external sources limits radar range. We distinguish noise from clutter, which is an unwanted return to a radar caused by the radar transmission. Noise is unrelated to the radar transmission. It may be polarized or unpolarized and may have a narrow or wide bandwidth (See Mott, 1992, p. 98, for a more detailed discussion).

Brightness

Power received from a radiating body can be described by a term from radio astronomy, the *brightness* of the body. The power flux density per steradian Hertz, with units $\mathrm{W/m^2 Hz\ rad^2}$, reaching a receiver from a radiating source is the brightness b of the part of the radiator that is within an antenna beam. The power flux density, with units of $\mathrm{W/m^2}$, at the receiver is

$$\mathcal{P} = \iint_{\Omega} \int_f b(\theta, \phi, f)\, df\, d\Omega = \int_\theta \int_\phi \int_f b \sin\theta\, d\theta\, d\phi\, df$$

where Ω is the lesser of the antenna beam or the solid angle at the receiver subtended by the radiating body, and f defines the bandwidth of the noise. An impedance-matched, polarization-matched receiving antenna with effective area $A_e(\theta, \phi, f)$ will receive from the radiating body a total power,

$$W = \iint_{\Omega} \int_f b(\theta, \phi, f) A_e(\theta, \phi, f)\, df\, d\Omega \tag{4.14}$$

If the incident wave is unpolarized, as it is for much noise, an antenna of any polarization receives half the power available in the wave, and received power is one-half that given by (4.14). If b and A_e are independent of frequency in the bandwidth B of interest, the received power simplifies to

$$W = \frac{B}{2} \iint_{\Omega} b(\theta, \phi, f_0) A_e(\theta, \phi, f_0)\, d\Omega \tag{4.15}$$

where f_0 is the center frequency of the receiver band. In this equation and the following developments the noise is assumed to be unpolarized.

Two cases are of interest: If the solid angle subtended by the radiating region is greater than that of the antenna beam, integration is carried out only over the beam solid angle. This might be noise from the sky with a high-gain antenna. In another case, if the radiating body is so small that the effective area of the antenna is constant over solid angle Ω_s subtended at the receiver by the target, the received power is

$$W = \frac{1}{2}\int_f b_0(\theta_0, \phi_0, f) A_e(\theta_0, \phi_0, f) \Omega_s \, df$$

where an average brightness is used. It is defined as

$$b_0(\theta_0, \phi_0, f) = \frac{1}{\Omega_s}\iint \Omega_s b(\theta, \phi, f_0) \, d\Omega$$

Finally, if A_e and b_0 are independent of frequency in bandwidth B centered at f_0,

$$W = \frac{B}{2} b_0(\theta_0, \phi_0, f_0) A_e(\theta_0, \phi_0, f_0) \Omega_s$$

This situation is typified by radiation from a star or the sun received by an antenna whose half-power beamwidth is greater than the solid angle subtended by the radiating body.

The product

$$S = b_0 \Omega_s = \iint b \, d\Omega$$

has the units of W/m^2 Hz and is called here the *spectral flux density*. Measurements of power reaching an antenna from extended regions of the sky are normally expressed in terms of brightness, or brightness temperature, and from discrete objects in terms of spectral flux density (Skolnik, 1962, p. 461).

Thermal Power Received and Power Produced by a Resistor

Bodies at temperatures above absolute zero radiate electromagnetic energy over a wide frequency range, including the microwave range. They also absorb or reflect incident electromagnetic waves. It has been shown that good absorbers are also good radiators. An object that absorbs all incident energy and reflects none is called a *blackbody*, even at nonvisible-light frequencies. The brightness of a blackbody is given by Planck's radiation law,

$$b = \frac{2hf^3}{c^2}\frac{1}{e^{hf/kT} - 1}$$

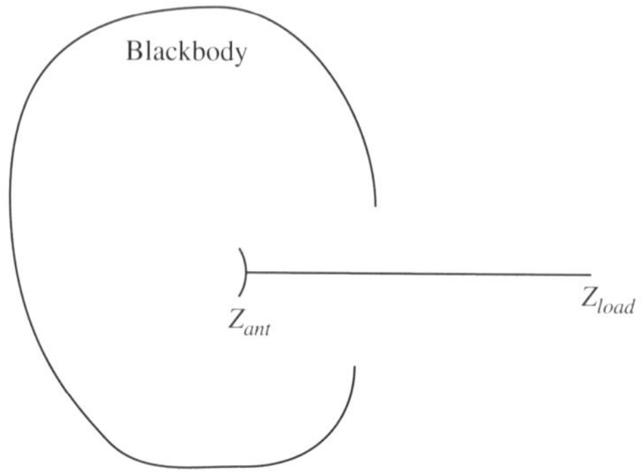

Fig. 4.6. Antenna and blackbody radiator.

where h is Planck's constant ($= 6.63 \times 10^{-34}$ Js), k is Boltzmann's constant (1.38×10^{-23} J/K), f is the frequency, and T is the temperature. At radio frequencies, $hf << kT$, and the exponential can be approximated by two terms of a Taylor's series. The brightness then is given by the Rayleigh–Jeans radiation law,

$$b = \frac{2kTf^2}{c^2} = \frac{2kT}{\lambda^2} \tag{4.16}$$

Consider a lossless antenna with effective area A_e and impedance $R + jX$ surrounded by a blackbody at temperature T, as in Fig. 4.6. The antenna is connected by a lossless line to an impedance-matched load at temperature T. The radiation is unpolarized, so half the available power is received. The power delivered to the matched load is, from (4.14) and (4.16),

$$W = \frac{1}{2} \iint_{\Omega} \int_f \frac{2kT}{\lambda^2} A_e(\theta, \phi, f)\, df\, d\Omega$$

If A_e is independent of f over the frequency band B, the load power is

$$W = \frac{kT}{c^2} \iint_{\Omega} A_e(\theta, \phi, f_0)\, d\Omega \int_{f_0 - B/2}^{f_0 + B/2} f^2\, df = \frac{kTBf_0^2}{c^2} \iint_{\Omega} A_e(\theta, \phi, f_0)\, d\Omega$$

The effective area of the antenna can be replaced by its gain G, using the relationship developed in Section 2.7, and the received power is

$$W = kTB \iint_{\Omega} \frac{G(\theta, \phi, f_0)}{4\pi}\, d\Omega$$

For a lossless antenna, G is related to the radiation intensity in a specific direction by

$$G(\theta, \phi, f_0) = \frac{U(\theta, \phi, f_0)}{U_{\mathrm{av}}}$$

Then

$$\iint_{\Omega} \frac{G(\theta, \phi, f_0)}{4\pi} \, d\Omega = \iint_{\Omega} \frac{U(\theta, \phi, f_0)}{4\pi U_{\mathrm{av}}} \, d\Omega = 1$$

It follows that

$$W = kTB \tag{4.17}$$

This equation shows that the received power depends only on the bandwidth and blackbody temperature, not on the location of the frequency band. This power is referred to as white noise, "noise" because of the random phase relations of its components, and "white" because of its independence of frequency, which makes it similar to "white" light, having (in the visible region) spectral components of approximately equal strength. This frequency independence is characteristic of the received power, not of the brightness or spectral flux density of the radiation.

Antenna Noise Temperature

The electromagnetic noise reaching an antenna may not be closely related to the actual temperature of the source. An example is manmade noise from nonthermal sources. It is useful to assign an equivalent temperature to such sources. This can be done in the RF region by using the Rayleigh–Jeans equation. Then the *brightness temperature* of an object is defined by

$$T_b = \frac{b\lambda^2}{2k}$$

The brightness temperature may be greater than the actual temperature of the radiating object, for a nonthermal source, or less, for an object that radiates thermally with lower efficiency than a blackbody. The brightness temperature of the sky may range from a few degrees for an antenna pointed parallel to the galactic axis to a much larger value for an antenna pointed in the direction of the galactic center. The sun is an extremely bright radiator, not only in the visible range of frequencies but in the microwave region.

An equivalent temperature can be assigned to an antenna. Let an antenna be exposed to an assemblage of radiating objects with overall brightness $b(\theta, \phi, f)$. The antenna and load are impedance matched, but the load resistance may be at an arbitrary temperature, since thermal equilibrium is not required. The power in small bandwidth B is given by (4.15). It is also given by (4.17) with T replaced by an *equivalent antenna temperature*, T_a. Equating the two powers results in

$$T_a = \frac{W}{kB} = \frac{1}{2k} \iint_{\Omega} b(\theta, \phi, f_0) A_e(\theta, \phi, f_0)\, d\Omega$$

The antenna temperature in terms of the brightness temperature is

$$\begin{aligned} T_a &= \iint_{\Omega} \frac{T_b(\theta, \phi, f_0)}{\lambda^2} A_e(\theta, \phi, f_0)\, d\Omega \\ &= \frac{1}{4\pi} \iint_{\Omega} T_b(\theta, \phi, f_0) G(\theta, \phi, f_0)\, d\Omega \end{aligned} \tag{4.18}$$

where G is the antenna gain. Distance to a radiating body does not appear in the equations involving brightness or brightness temperature since brightness is spectral power density per steradian and does not involve distance. If, however, there is attenuation in the region between source and antenna, distance is a factor.

The greatest natural contributors to antenna noise temperature are the sun, the moon, cosmic noise from radio stars and ionized interstellar gas, the warm earth, and a lossy atmosphere. The sun is by far the largest contributor for a high-gain antenna pointed toward it.

Noise Figure

The noise figure of a linear two-port system, such as an amplifier, a lossy line, or a wave propagation path that attenuates the wave, is defined as

$$F = \frac{S_{\text{in}}/N_{\text{in}}}{S_{\text{out}}/N_{\text{out}}}$$

where S_{in} and N_{in} are the available signal and noise powers at the input, and S_{out} and N_{out} are available signal and noise powers at the output. The signal–noise power ratio is S/N. The two-port network is assumed to be matched to the source impedance at the input and to the load impedance at the output.

The input noise power is $N_{\text{in}} = kTB$, where B is the bandwidth. If the network gain is G (which may be greater than or less than one), input signal power and input noise power are multiplied by G at the output. In addition, the network adds noise N. With this input noise power and the additional noise power generated in the network, the noise figure becomes

$$F = \frac{S_{\text{in}}}{S_{\text{out}}} \frac{N_{\text{out}}}{N_{\text{in}}} = \frac{1}{G} \frac{GN_{\text{in}} + \Delta N}{N_{\text{in}}} = 1 + \frac{\Delta N}{kTBG}$$

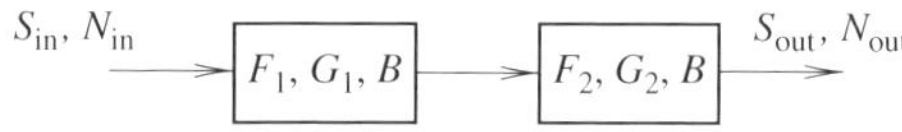

Fig. 4.7. Cascaded networks.

In order to standardize the definition of noise figure, T is specified as $T = T_0 = 290$ K in measurements of network noise figure, so that the standard definition of noise figure is

$$F = 1 + \frac{\Delta N}{kT_0 BG} \tag{4.19}$$

Consider two impedance-matched networks in cascade, as in Fig. 4.7. They have the same bandwidth B, but different gains and noise figures. The output signal is $S_{\text{out}} = G_1 G_2 S_{\text{in}}$. The noise at the input of the second network is $F_1 G_1 k T_0 B$. This noise is multiplied by G_2 and to it is added, from (4.19),

$$\Delta N_2 = (F_2 - 1)kT_0 BG_2$$

Then the overall noise figure of the cascaded networks is

$$F = \frac{1}{G_1 G_2} \frac{F_1 G_1 G_2 k T_0 N + (F_2 - 1)kT_0 BG_2}{kT_0 B} = F_1 + \frac{F_2 - 1}{G_1}$$

This procedure can be extended to N cascaded, impedance-matched networks to give the overall noise figure,

$$F = F_1 + \frac{F_2 - 1}{G_1} + \cdots + \frac{F_N - 1}{G_1 G_2 \ldots G_{N-1}}$$

Effective Noise Temperature of a Two-Port Network

The input noise to a two-port network is $kT_0 B$, and it is multiplied by the gain G to give a component of the output noise power. The added noise power, N, can be regarded as being produced by an input resistance at temperature T_e connected to a noise-free equivalent network with the same gain. At the output, this added noise is $\Delta N = kT_e BG$. From (4.19), the noise figure is

$$F = 1 + \frac{kT_e BG}{kT_0 BG} = 1 + \frac{T_e}{T_0} \tag{4.20}$$

Temperature T_e is the effective noise temperature of the network. For cascaded networks with effective temperatures $T_1, T_2, \ldots, T_N$, the overall effective temperature can be shown, in the same manner as the overall noise figure, to be

$$T_e = T_1 + \frac{T_2}{G_1} + \ldots + \frac{T_N}{G_1 G_2 \ldots G_{N-1}}$$

Noise Figure and Temperature of a Lossy Medium

Consider once more an antenna surrounded by a blackbody. The region between the blackbody and antenna is filled with an absorbing material that attenuates a wave traveling through it. The blackbody, antenna, and absorbing material are in thermal equilibrium at temperature T. The noise power available from the blackbody is kTB; after this power passes through the lossy region, the noise power at the antenna is $kTBG$, where G is the gain, smaller than one, resulting from travel through the lossy region. The power absorbed by the region is $kTB - kTBG$. This power is then radiated by the lossy region, and ΔN in the equation for noise figure is

$$\Delta N = kTB(1 - G) \qquad G \leq 1$$

The noise figure of the lossy region is, therefore,

$$F = 1 + \frac{\Delta N}{kT_0BG} = 1 + \frac{T}{T_0}\frac{1 - G}{G}$$

where T is the actual temperature of the system. If this expression is equated to (4.20), it will be seen that

$$T_e = T\left(\frac{1}{G} - 1\right) \qquad G \leq 1$$

It can be seen from this equation that the effective temperature of a lossy region may be greater than its real temperature.

The region under consideration may be the atmosphere, whose loss mechanism is the absorption of energy by water vapor or oxygen, a radome, a device such as a mixer or lossy transmission line, or the lossy antenna itself.

Example

The earth is a good absorber of microwave radiation and a good radiator at microwave frequencies. Let us take it as a perfectly efficient radiator at 290 K. In remote sensing of terrain by an air- or space-borne radar, the earth is seen by the main antenna beam and all sidelobes. This earth noise is attenuated by the atmosphere and by a radome, but these attenuating paths will in turn generate noise. As a good approximation, we neglect the attenuation and the subsequent generated noise. We take the brightness temperature of the earth, T_b, as 290 K. The earth is seen by the entire antenna beam, and if this is used, the effective antenna temperature T_a given by (4.19) is the earth temperature, 290 K.

As a contrast, the equivalent antenna temperature for a 40-dB gain antenna pointed at the sky but not directed toward the sun, moon, radio star, or galactic center has been estimated at approximately 80 K (Mott, 1992, p. 111). This includes atmospheric, radome, and antenna losses and sidelobe reception from the warm earth. The sun is an extremely bright radiator, with X-band temperatures of thousands of degrees. If an earth-sensing radar sees a specular reflection of the sun, as it may from a lake, the equivalent antenna temperature is much greater than 290 K.

Noise power does not decrease with radar height above the earth, so for a high-altitude radar, noise power from the warm earth is a significant hindrance to earth sensing.

Most internal noise in a receiver is generated by the first active element after the antenna; if internal noise is to be kept below external noise, the noise figure of the first element should be small. Lane has noted (Eaves and Reedy, 1987, p. 201) that mixers with noise figures of 1 dB at 1-GHz and 5 dB at 95 GHz are available. The 1-GHz noise figure corresponds to an effective temperature of 75 K. A maser amplifier following the antenna has a lower effective temperature.

REFERENCES

J. L. Eaves and E. K. Reedy, eds., *Principles of Modern Radar*, Van Nostrand Reinhold, New York, 1987.

H. Mott, *Antennas for Radar and Communications: A Polarimetric Approach*, Wiley-Interscience, New York, 1992.

IEEE Trans. AP **31**, (6) November, 1983.

M. I. Skolnik, *Introduction to Radar Systems*, 2nd ed., McGraw-Hill, NY, 1962.

M. W. Whitt, F. T. Ulaby, and K. Sarabandi, "Polarimetric Scatterometer Systems and Measurements", Chapt. 5 in *Radar Polarimetry for Geoscience Applications*, F. T. Ulaby and C. Elachi, eds., Artech House, Norwood, MA, 1990.

PROBLEMS

4.1. The Sinclair matrix, in x and y components, of a target in a backscattering configuration, is

$$\begin{bmatrix} 3 & j \\ j & 1+j \end{bmatrix}$$

One antenna is used for transmitting and receiving. Find the target's radar cross-section if: (a) the antenna is linearly polarized in the y direction and (b) the antenna polarization ratio is $P = j$.

4.2. Find the scattering cross-section of the target of Problem 4.1. Find the receiving antenna's polarization efficiency.

4.3. A radar operating at 10 GHz transmits a constant-frequency pulse of length 100 ns. Find the range resolution of the radar. If the transmitted pulse is linear FM with transmitted frequency varied from 10 to 10.03 GHz, find the range resolution. The radar antenna is a planar array with 40 x 40 elements spaced two wavelengths apart in rows and columns. Find the resolution transverse to a ray from the antenna, in a plane that contains a row of the array, if the target is 10 km from the radar.

4.4. A monostatic radar with an antenna having an effective length of $0.5\mathbf{u}_x$ sees a target at a distance of 20 km on the z-axis. The antenna has an internal impedance of $50 + j30\ \Omega$ and a radiation efficiency of 0.96. The receiver load impedance is the conjugate of the antenna impedance. The radar transmits an average power of 20 kW at a frequency of 10 GHz. The target Sinclair matrix is

$$\begin{bmatrix} 3 & j \\ j & 1+j \end{bmatrix}$$

Find the average received power. If the receiver load impedance is $40 - j30\ \Omega$, find the average received power.

4.5. A radar located at the earth's surface at 30° N latitude transmits a 10-GHz wave in a narrow beam. A rectangular coordinate system located at the radar has x-axis pointing east, y axis pointing north, and z-axis perpendicular to the earth's surface. The radar beam points at a polar angle, with respect to its coordinate system, of 60° and an azimuth angle of 45°. An aircraft traveling at 400 mph passes from south to north through the radar beam, at a distance of 5 km. Find the frequency of the wave scattered back to the radar.

4.6. The effective noise temperature of an antenna at 10 GHz is 80 K and the noise bandwidth is 1 MHz. It receives a signal from a transmitter at a distance of 20 km. The antennas have gains of 20 dB, and atmospheric loss between transmitter and receiver is 0.1 dB/km. What power must the transmitter radiate in order to give a signal-noise ratio of 5 dB at the receiver?

4.7. A radar operating at 20 GHz observes a target at 4 km. Atmospheric attenuation is 0.1 dB/km and the combined loss of antenna and radome is 0.6 dB. Find the effective temperature of the antenna. If the target has a radar cross-section of 2 m^2 and the radar has a gain of 20 dB and a bandwidth of 1 MHz, find the transmitter power necessary to give a signal–noise ratio of 0 dB.

CHAPTER 5

SYNTHETIC APERTURE RADAR

A synthetic aperture radar (SAR) is a radar that moves on a known trajectory, at intervals along the path transmitting a pulse and receiving the scattered wave from a target. The received signals from the target for a large number of transmitted pulses are combined in such a manner that a synthetic antenna array is formed. The combined signal is that which would be received by a linear array as long as the radar path.

The measured properties of the scattered wave are intensity, time of return, rate of change of the return time, and, for a polarimetric SAR, polarization. The intensity of the wave provides a high-resolution map of an extended target. Time of signal return and Doppler frequency provide positional information that allows the map to be constructed. The synthetic aperture radar normally operates in the microwave region, and its operation is unaffected by time of day or cloud cover. It can map regions of the earth otherwise virtually inaccessible.

Only brief mention is made here of signal processing in the SAR, since that would lead to mathematical complexities inappropriate to this work.

5.1. CREATING A TERRAIN MAP

The process of creating a terrain map discussed here is based on an adaptation of a coherent radar described in Section 4.1. There it was convenient to think of the radar as an analog device with Doppler filters. In the following developments, the Doppler filters are seen to be unnecessary, but the concept of directing the output of a range gate–Doppler filter pair to the integrator for a chosen terrain cell leads to a clear understanding of how a terrain map is formed. While the analog implementation leads to a good understanding, signal processing in the SAR, in general, is carried out digitally. Trebits (1987, p. 502) discusses digital signal processing and provides a block diagram of a digital SAR processor. Trebits also gives an introduction to optical processing of synthetic aperture radar signals.

The beamwidth of a linear antenna array is inversely proportional to the array length. The resolution distance of two azimuth-separated targets, the minimum separation at which they can be distinguished one from the other, is dependent on the beamwidth, and array length must be increased to decrease the minimum discernible separation distance. A fixed antenna array cannot be made long enough to give the desired resolution for the surveillance of terrain by an air- or space-borne radar, However, a long array can be synthesized by processing the signals received at known positions of the radar as it moves along its path. The target return at one antenna position does not arrive at the same time as that at another position, nor do the two returns result from the same transmitted radar pulse, but appropriate storage and processing of the signals can account for these differences.

An assumption in this discussion is that terrain surveillance is the purpose of the synthetic aperture radar use. For some illustrations, the earth is flat and the radar platform moves in a straight line at constant velocity and height above it, but the discussion is general, and independent of the flat-earth examples. It is necessary to compensate for any variation in the radar platform motion from the mean trajectory, and we will discuss that briefly.

The terrain to be surveyed is a swath to the right and/or left of a line on the surface that parallels the radar platform track. The swath is bounded by the near and far limits of the antenna beam, as shown in Fig. 5.1. It is common to use a real antenna array, rather than a single antenna, as the synthetic aperture element. The real array has length L in the direction of radar platform motion and width D perpendicular to the motion, with, in general, $L > D$. This gives a fan beam, narrow in a plane containing the radar track and wider in a plane perpendicular to the track. The ground footprint of such an antenna is shown in Fig. 5.1. It is not essential that the antenna beam be perpendicular to the radar trajectory, but that is a common mode of operation and the one discussed here.

The end product of the SAR operation is a map of the terrain. The terrain is divided into cells, here called terrain or resolution cells, that are smaller than the antenna footprint of Fig. 5.1, and the radar cross-sections of the cells are mapped into an image of the ground area. Polarimetric radars yield maps of the scattering matrix elements of each cell.

The radar periodically transmits a broadband pulse and receives the scattered return from the ground area within the antenna beam. This return can be separated to obtain the signal from each terrain cell by recognizing that the signal from a cell can be distinguished from signals from other cells by use of its distance from the radar and the Doppler frequencies of the signals. Figure 5.2 shows three point targets. A synthetic aperture radar moves by those targets along the straight-line

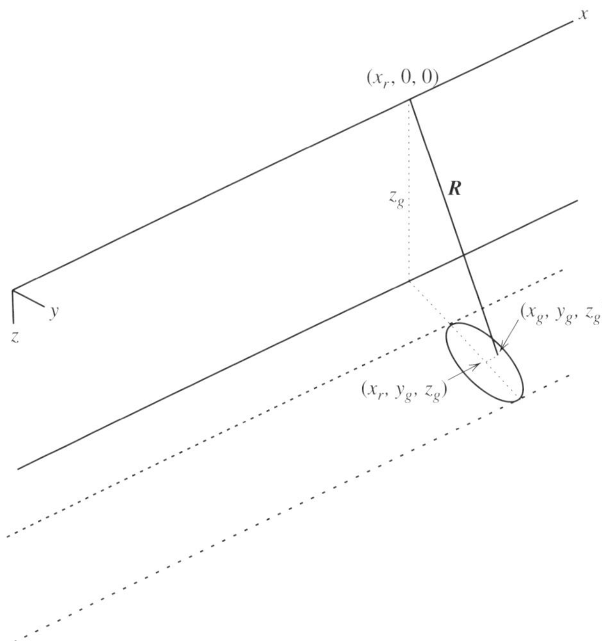

Fig. 5.1. Footprint of SAR antenna.

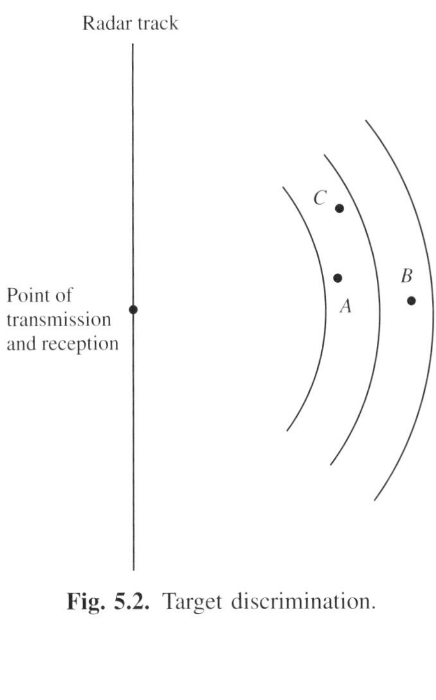

Fig. 5.2. Target discrimination.

track shown. The radar transmits a short pulse and receives returns from targets A, B, and C. Target A is distinguished from target B by the time of arrival of the scattered signals. Target C is approximately the same distance from the radar as is target A, so the signals from them cannot be separated by time of arrival. However, the relative radial velocity between radar and target at a particular radar position is different for targets A and C, and the received signals have different Doppler frequencies that can be used to distinguish between the two targets.

The Terrain Map

The process of forming a terrain map can be illustrated by an adaptation of the coherent radar described in Section 4.1, although there are many differences in signal processing between it and a modern SAR. There is also a significant difference in the targets for a terrain-sensing radar and the radar discussed in Section 4.1. In the previous discussion, we were primarily interested in a target giving, for each pulse, an output from one range gate and one Doppler filter. In terrain sensing, each transmitted pulse will result in an output from many range gates. Moreover, the output of one range gate will be a signal with a spectrum of Doppler frequencies, and many Doppler filters associated with a specific range gate will have nonzero outputs. The part of the signal that corresponds to a chosen terrain cell must be selected and directed to the proper pulse integrator in order to create the terrain map.

The terrain to be mapped is divided into subareas whose cross-track and along-track dimensions depend on the range- and azimuth-resolution distances of the radar. The coordinates of each terrain cell are known and so are the radar trajectory and velocity. Let the center of a terrain cell be at (x_g, y_g, z_g) and the radar at $(x_n, 0, 0)$ when it transmits pulse n. The range and Doppler frequency of a signal from the terrain cell can be calculated from knowledge of the radar path and velocity. If range gate i and Doppler filter j correspond to the calculated range and frequency values, the signal from the gate–Doppler pair ij is directed to an integrator for the specified terrain cell.

When the radar transmits pulse $n + 1$, the calculated range and Doppler frequency for the desired terrain cell may no longer correspond to the gate-Doppler pair ij. The output of the ij pair is therefore directed to the integrator for a different terrain cell, and the input for the cell at (x_g, y_g, z_g) comes from a different gate–Doppler pair. The signals, when properly directed, are summed, and a map of intensity versus cell position is constructed from the sum.

More efficient signal processing techniques are used in synthetic aperture radars than indicated by this example, but this system is discussed here because it is easily understood.

Time Notation

It is convenient to use two variables, t and s, to represent time. If the radar is at $x_r = 0$ when $s = 0$ and if the radar moves at constant velocity v, then $x_r(s) = vs$.

We refer to time s as the "along-track" time. Periodically, with period T, the radar transmits a pulse of length τ centered at $s_n = nT$, where n is the pulse number. For each transmitted pulse, it is convenient to measure time from the reference time at which the pulse is transmitted. We therefore define a "cross-track" or "range" time t as $t = s - s_n = s - nT$.

Doppler Frequency

Let the position of the radar be $(x_r, 0, 0)$ and the coordinates of a point on the ground be (x_g, y_g, z_g). The slant range vector from radar to ground point is

$$\mathbf{R} = \mathbf{u}_x(x_g - x_r) + \mathbf{u}_y y_g + \mathbf{u}_z z_g$$

The transmitted frequency is f_c, with wavelength λ. The radial velocity of a target with respect to the radar is

$$v_r = -\frac{\mathbf{v}\cdot\mathbf{R}}{R} = -\frac{v\mathbf{u}_x\cdot\mathbf{R}}{R} = -v\frac{x_g - x_r}{R}$$

where R is the slant range from radar to target. The Doppler frequency observed by the radar is

$$f_d = -\frac{2v_r}{\lambda} = 2v\frac{x_g - x_r}{R\lambda} \tag{5.1}$$

Target Location

Two curves will help to explain how the SAR discriminates between targets. To draw them, it is convenient to let the radar be at $(0, 0, 0)$ and the target at (x_g, y_g, z_g), for which $R = R(0)$. The Doppler frequency is

$$f_{d0} = \frac{2vx_g}{\lambda R(0)} = \frac{2vx_g}{\lambda\sqrt{x_g^2 + y_g^2 + z_g^2}} \tag{5.2}$$

If this equation is rewritten as

$$\left[\left(\frac{2v}{\lambda f_{d0}}\right)^2 - 1\right]x_g^2 - y_g^2 - z_g^2 = 0$$

it has the form $Ax_g^2 + By_g^2 + k = 0$, and if $A > 0,\ B < 0$, and $k \neq 0$, it represents a hyperbola on the z_g plane. With the correspondences,

$$A = \left(\frac{2v}{\lambda f_{d0}}\right)^2 - 1 \qquad B = -1 \qquad k = -z_g^2$$

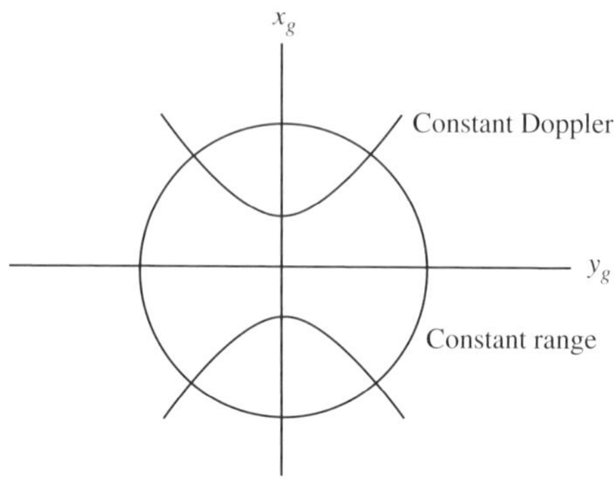

Fig. 5.3. Constant range and constant Doppler curves.

we need only show that $A > 0$ to show that the Doppler frequency equation is a hyperbola. By using (5.2) in the equation for A, we get

$$A = \left(\frac{R(0)}{x_g}\right)^2 - 1 > 0$$

A target scattering a wave with Doppler frequency f_{d0} lies on a hyperbola symmetric about the line $(x_g, 0, z_g)$. The two branches of the hyperbola are also symmetric about the line $(0, y_g, z_g)$. One branch is for positive Doppler and the other for negative. One chooses the right or left side of the hyperbola [about the $(x_g, 0, z_g)$ line] from knowledge of the direction of the antenna beam.

The other measurement required is that of slant range $R(0)$. A curve of constant $R(0)$ is a circle on the z_g plane and the intersection of circle and hyperbola uniquely determines the target location. The results found here for $x_r = 0$ at $s = 0$ are readily generalized to any position of the radar.

Figure 5.3 shows these curves on the z_g plane.

5.2. RANGE RESOLUTION

In Section 4.4, the range-resolution distance for a single-frequency pulse of time duration τ was found to be $\delta R = c\tau/2 = c/2B$, where B is the bandwidth required to pass the rectangular pulse without significant alteration. This is the minimum range difference for which two point targets are recognized as two, rather than being grouped together as one target. In ground mapping by an SAR, this is the slant-range resolution.

We considered in the same section the transmission of a linear FM pulse and found the range-resolution distance to be given by the same equation, $\delta R = c/2B$. For this pulse, however, B is the difference between maximum and minimum transmitted frequencies.

Ground-Range Resolution

From Fig. 5.1, it may be seen that resolution distance along the ground perpendicular to the radar trajectory is the slant-range resolution divided by the sine of the angle between the vertical and a ray to a point on the surface. At a point on a line perpendicular to the SAR trajectory,

$$\delta y_g = \frac{\delta R}{\sin \gamma} = \frac{c}{2B \sin \gamma}$$

Alternative Interpretation of the Correlated Output for a Linear FM Pulse

Delay time t' of (4.7) is the same as time t referenced to the transmission of the radar pulse if the correlation is carried out in real time. Then an alternative form of (4.10) is

$$V_{gh}(t) = \frac{1}{2} \frac{\sin \pi K \tau (t - 2R/c)}{\pi K \tau (t - 2R/c)} e^{j2\pi f_c (t-2R/c)} \quad -\tau + 2R/c < t < \tau + 2R/c$$

We can now interpret V_{gh} as a time function. Here, we use the exponential form rather than the cosine of (4.10).

In (4.10) we compared time delay variable t' to $2R/c$, the time required for a transmitted signal to reach a target at distance R and return to the radar. Now let us interpret t' as the time required for a wave to reach a hypothetical target at reference range R' and return, $t' = 2R'/c$. Then another useful form for the output voltage is

$$V_{gh}(t) = \frac{1}{2} \frac{\sin \pi K \tau (2R/c - 2R'/c)}{\pi K \tau (2R/c - 2R'/c)} e^{j2\pi f_c (t-2R/c)} \quad -\tau + 2R/c < t < \tau + 2R/c$$

5.3. AZIMUTH RESOLUTION

The output of a receiver correlator for the nth transmitted pulse was found in Section 4.4 to be

$$V_n = A_n(t, R_n) \cos 2\pi f_c (t - 2R_n/c) \quad -\tau + 2R_n/c < t < \tau + 2R_n/c \tag{5.3}$$

where $A_n(t, R_n)$ is a slowly varying time function given for a single-frequency rectangular pulse by

$$A_n(t, R_n) = \begin{cases} \frac{1}{2}(\tau + t - 2R_n/c) & -\tau + 2R_n/c < t < 2R_n/c \\ \frac{1}{2}(\tau - t + 2R_n/c) & 2R_n/c < t < \tau + 2R_n/c \end{cases} \tag{5.4}$$

and for a linear FM pulse by

$$A_n(t, R_n) = \frac{1}{2}\tau \frac{\sin \pi K\tau(t - 2R_n/c)}{\pi K\tau(t - 2R_n/c)} \qquad -\tau + 2R_n/c < t < \tau + 2R_n/c \quad (5.5)$$

Equations 5.3–5.5 are for point targets. A synthetic aperture radar examining terrain will receive the waves from many scatterers within the range interval used in the correlation process, and the waves will be combined to give a voltage that is different from that predicted by the equations. In this section, we are concerned with determining the idealized azimuth resolution, however, so we consider only a point target.

The received voltage V_n for the nth transmitted pulse is

$$V_n = A_n e^{j2\pi f_c(s - s_n - 2R_n/c)} \quad (5.6)$$

where s_n is the time at which pulse n is transmitted. If the transmitted pulses are synchronized, s_n can be removed from the voltage equation.

In a discussion of a coherent radar in Section 4.1, signal processing was discussed in terms of range gates and Doppler filters. The concept of range gates can still be used because V_n is nonzero for a point target only in a finite range interval about the target. Doppler filters are unnecessary, however, if the phases of all V_n are shifted to a common phase and the resulting signals added coherently.

The location of the point target is unknown. Instead, a terrain cell is selected by choosing a cross-track and along-track location of a reference point within the cell. The minimum slant range, which is the range when the ray from radar to reference point is perpendicular to the radar track, is denoted by R'_0. The slant range to the reference point when the nth pulse is transmitted is R'_n, and the range to the target is R_n.

We shift the phase of each received pulse by multiplying (5.6), with s_n removed, by $\exp\{-j2\pi f_c(2R'_0/c - 2R'_n/c)\}$. The N phase-shifted pulses are added coherently to give

$$V = Ae^{j2\pi f_c(s - 2R'_0/c)} \sum_{-(N-1)/2}^{(N-1)/2} e^{j2\pi f_c(2R'_n/c - 2R_n/c)}$$

where we note that slowly varying time function A_n for a point target is independent of n. N was arbitrarily chosen as odd.

Let the reference point be at $(x_{\text{ref}}, y_{\text{ref}}, z_{\text{ref}})$ and the target be at

$$(x_{\text{tar}}, y_{\text{tar}}, z_{\text{tar}}) = (x_{\text{ref}} + \delta x, y_{\text{ref}} + \delta y, z_{\text{ref}} + \delta z)$$

The radar is at $(x_{\text{ref}} + nvT, 0, 0)$, where v is the radar velocity and T the time between transmitted pulses.

The ranges R_n and R'_n to target and reference point, respectively, are related by

$$R_n \approx R'_n - \frac{nvT\delta x}{R'_0} + \frac{y_{\text{ref}}\delta y + z_{\text{ref}}\delta z}{R'_0}$$

where the minimum range to the reference point is used in the denominators. V then becomes

$$\begin{aligned} V &= Ae^{j2\pi f_c(s-2R'_0/c-2y_{\text{ref}}\delta y/R'_0c-2z_{\text{ref}}\delta z/R'_0c)} \sum_{-(N-1)/2}^{(N-1)/2} e^{j4\pi f_c nvT\delta x/R'_0c)} \\ &= NAe^{j2\pi f_c(s-2R_0/c)} \frac{\sin(2\pi f_c NvT\delta x/R'_0c)}{N\sin(2\pi f_c vT\delta x/R'_0c)} \end{aligned} \tag{5.7}$$

If S is the time during which the radar sees the target and X is the distance traveled by the radar during that time (see Fig. 5.4), then $X = vS \approx NvT$. The output is then

$$V = NA\frac{\sin\left[\dfrac{2\pi}{\lambda}\dfrac{X}{R'_0}(x_{\text{tar}} - x_{\text{ref}})\right]}{N\sin\left[\dfrac{2\pi}{\lambda}\dfrac{X}{NR'_0}(x_{\text{tar}} - x_{\text{ref}})\right]} e^{j2\pi f_c(s-2R_0/c)}$$

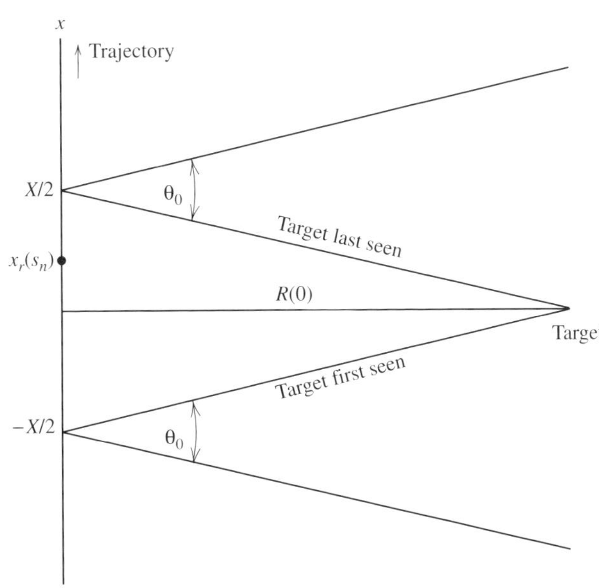

Fig. 5.4. Target viewing distance.

The amplitude of V is multilobed with a peak at $x_{tar} = x_{ref}$. Other amplitude maxima occur for nonzero values of $\delta x = x_{targ} - x_{ref}$, but the first maximum of this kind has a power approximately 13 dB below that at $\delta x = 0$. Succeeding maxima are even smaller. Another lobe having the same amplitude as that at $\delta x = 0$ occurs when the argument of the sine term in the denominator of (5.7) is equal to π. This occurs near $\delta x = NL/2$, where L is the antenna dimension in the x direction. This will generally lie outside the azimuth processing window, and we will not consider it further.

Let point targets 1 and 2 be located at $x_{tar\,1}$ and $x_{tar\,2}$. If the reference point for azimuth processing is chosen to coincide with target 1, V will have maximum amplitude for the return from target 1 and a smaller value for the return from target 2. To ensure that they are seen as separate targets, let the output from target 2 be zero. Then,

$$\frac{2\pi}{\lambda}\frac{X}{R_0'}(x_{tar\,2} - x_{tar\,1}) = \pi$$

or

$$x_{tar\,2} - x_{tar\,1} = \frac{\lambda}{2}\frac{R_0'}{X} \tag{5.8}$$

The one-way antenna beamwidth is X/R_0'. It is also λ/L. Then

$$x_{tar\,2} - x_{tar\,1} = \frac{\lambda}{2}\frac{1}{\theta_0} = \frac{L}{2} \tag{5.9}$$

This is the azimuth resolution distance of the SAR. It is optimistic, since the point target amplitude will be constant over a smaller angle than the one-way half-power angle, and that will require a larger value of δx to make the numerator sine function of (5.7) zero. In addition, two azimuth-extended targets will be more difficult to distinguish than two point targets. Nonetheless, the resolution distance found in this manner is useful.

The SAR output voltage was developed by shifting the phase of each pulse before the coherent addition of the returns was carried out. An SAR that processes received pulses is this way is described as *focused*. Signal processing is simpler if the phase-shift operation is omitted. The needed phase correction is greater near the ends of the synthetic aperture, and its omission becomes significant for a long aperture. A shorter aperture must be used if phase correction is omitted, and azimuth resolution distance is increased; it is $\sqrt{2\lambda R_0}$ (Elachi, 1987, p. 206).

Azimuth Resolution Limit

From (5.9), it appears that the azimuth resolution distance can be made as small as desired by decreasing antenna length L. However, if L is made smaller, the antenna beamwidth in the along-track direction increases. In turn, the Doppler

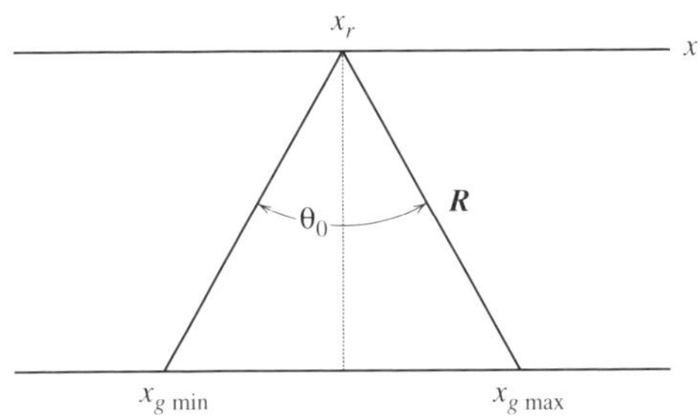

Fig. 5.5. Positions for minimum and maximum Doppler.

bandwidth of the received signals increases, and if the Doppler frequency is to be measured accurately, the radar prf must be increased. The maximum and minimum Doppler frequencies, from (5.1) and Fig. 5.5, are

$$f_{d\,\max} = \frac{2v\sin(\theta_0/2)}{\lambda} \approx \frac{v\sin\theta_0}{\lambda} \approx \frac{v}{\lambda}\frac{\lambda}{L} = \frac{v}{L}$$

$$f_{d\,\min} = -\frac{v}{L}$$

If the transmitted frequency is f_c, maximum and minimum received frequencies are $f_c + f_{d\,\max}$ and $f_c + f_{d\,\min}$, and the bandwidth is

$$B = f_{d\,\max} - f_{d\,\min} = 2v/L$$

The sampling theorem for a bandpass signal of bandwidth B and greatest frequency $f_{\max}$ requires that it be sampled at a rate $f_s = 2f_{\max}/m$ if the signal is to be perfectly recovered. In this equation, m is the largest integer not exceeding $f_{\max}/B$ (Ziemer and Tranter, 1995, p. 93). In using the sampling theorem, we are assuming that the use of a pulse radar is equivalent to sampling. Since $f_{\max}/B$ is a large number for a typical radar, m will be approximately equal to $m' = f_{\max}/B$, and we therefore use m' to find the sample rate,

$$f_p = f_s \approx \frac{2f_{\max}}{m'} = \frac{2f_{\max}}{f_{\max}/B} = 2B$$

which leads to a radar prf,

$$f_p = 4v/L \tag{5.10}$$

Let the time required for the radar to move through a distance equal to the antenna length L be Δs, so that $L = v\Delta s$. From this equation and (5.10) the pulse repetition time is $T = 1/f_p = \Delta s/4$. This time must be equal to or less than the time for the radar to move through distance $L/4$. The antenna length must therefore satisfy

$$L \geq 4v/f_p$$

and the azimuth resolution distance has a lower bound

$$x_{\text{tar}\,2} - x_{\text{tar}\,1} = L/2 = 2v/f_p$$

From this equation, it is seen that the radar prf limits the azimuth resolution obtainable.

Pulse Repetition Frequency

It was shown in the previous subsection that the pulse repetition frequency imposes a lower bound on the azimuth resolution obtainable. The converse is that azimuth resolution, or equivalently antenna length L, imposes a lower bound on the prf,

$$f_p \geq \frac{4v}{L}$$

An upper bound can be found from the requirement that the returned signal from the far swath boundary be received before the next pulse is transmitted. If the prf is increased in order to improve azimuth resolution, the swath width will decrease. The far swath boundary coordinate is $y_{g\,\text{max}}$ and the maximum slant range is

$$R_{\text{max}} = \left(z_g^2 + y_{g\,\text{max}}^2\right)^{1/2}$$

The constraint on the time of signal reception then gives

$$f_p \leq \frac{c}{2R_{\text{max}}}$$

Antenna Width

From Fig. 5.6, if the swath width on the ground, W_g, is small compared to z_g,

$$C \approx (y_{g\,\text{max}} - y_{g\,\text{min}}) \cos \phi_m \approx W_g \cos \phi_m$$

Take

$$R_m = \frac{R_{\text{max}} + R_{\text{min}}}{2}$$

as the approximate distance from radar to the center of the swath. Then,

$$C \approx R_m(\phi_{\text{max}} - \phi_{\text{min}}) = R_m \phi_0$$

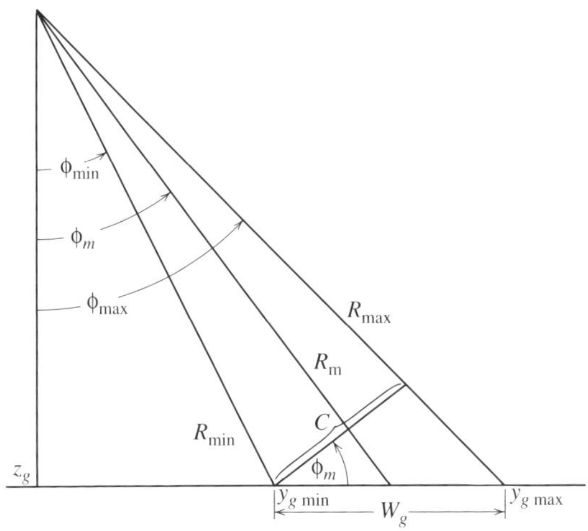

Fig. 5.6. Ground swath width.

If we wish the half-power angle of the antenna beam to coincide with the swath boundaries, we set ϕ_0 equal to the half-power angle. or $\phi_0 \approx \lambda/D$. If this substitution is made in one of the expressions for C and if the two expressions are equated,

$$D = \frac{R_m \lambda}{W_g \cos \phi_m}$$

This is the optimum antenna dimension. A greater value will give an antenna beam that does not cover the desired swath, and a smaller value will result in a too-broad beam that wastes power.

Fading and Speckle

In developing (5.7) for a point target, the amplitude of V_n in (5.3) was taken to be independent of n, and its phase was shifted appropriately before the coherent summation. A terrain cell may contain more than one scattering center, and V_n, the received voltage from one terrain cell for one pulse, must be considered a summation of voltages with differing amplitudes and phases. The relative amplitudes and phases of the waves from the scattering centers will differ for different viewing angles, and V_n is a complex random variable. The variation in amplitude from one pulse to the next is called *fading*. The coherent summation of the V_n in azimuth processing to give the voltage for each terrain cell will also yield a random variable, and a map of received power will have pixels that are not equally bright, even if the terrain mapped has the same scattering properties everywhere. The brightness variation is called *speckle*. Noise will also add random components to the received voltage. Appendix A treats fading and speckle in more detail.

5.4. GEOMETRIC FACTORS

The calculation of the radar to reference-point distance $R'_n(s)$ from the known SAR trajectory is straightforward and relatively easy for a stationary target on a flat earth with a constant-velocity SAR on a linear trajectory. The SAR platform may, however, be a satellite in an elliptical orbit, and the target a ground patch on a rotating earth that is approximately an oblate spheroid. The determination of $R'_n(s)$ for this system is relatively complex. A discussion is given by Curlander and McDonough (1991, p. 572).

The Effect of Terrain Variations

The map produced by an SAR is based on a smooth earth, although not necessarily a flat one. For a flat earth, slant range and ground distance y_g, when a ray to the target is perpendicular to the radar track, are related by

$$y_g = \frac{R}{\sin \gamma} \tag{5.11}$$

where γ is the angle between vertical and a ray from radar to target. Slant range and ground resolutions obey the same relationship,

$$\delta y_g = \frac{\delta R}{\sin \gamma} \tag{5.12}$$

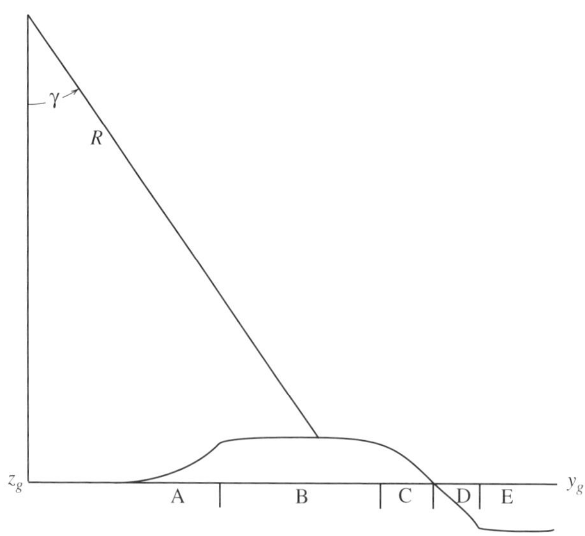

Fig. 5.7. Terrain profile.

A terrain map with slant range and azimuth coordinates can be converted by (5.11) to a map with ground range and azimuth coordinates.

If the terrain has height variations as illustrated by Fig. 5.7, the terrain map will be distorted. If slant range to the flat earth is used as a map coordinate, the radar returns for regions A, B, and C of the figure come from terrain points at smaller values than predicted by (5.11). The reverse is true for regions D and E. It may also be shown that resolution distance on the earth's surface is greater than δy_g of (5.12) in region A and smaller in regions C and D. If the terrain slope in regions C and D is great, targets will be shadowed and cannot be seen. It has been noted that for mountainous ranges, ground-range terrain maps appear to be more distorted than slant range maps (Ulaby et al., 1986).

5.5. POLARIMETRIC SAR

A polarimetric capability can be added to an SAR to give the scattering matrix elements of each resolution cell of an extended target. It was noted in Section 4.8 that a dual-channel radar is necessary for measuring the Sinclair matrix of a target. The matrix is measured by sequentially transmitting two orthogonally polarized waves and, for each transmission, simultaneously receiving two orthogonally polarized waves. The received waves can be processed by range correlation, as discussed in Section 4.4, and in azimuth, as discussed in Section 5.3. The relative phases of the received voltages must be measured, and this requires phase coherence of all transmitted signals and coherent addition of received signals.

The measured voltages from which the Sinclair matrix is constructed are random variables, subject to fading and speckle. Two limiting cases can be considered: For a point target with a Sinclair matrix, but no spatial extent, all elements of the Sinclair matrix are independent of the transmitted pulse. The summation of N pulses for each element in azimuth processing gives the Sinclair matrix of the target. In a second limiting case, the element values are uncorrelated, and a Sinclair matrix cannot be found. The waves from real targets will generally have characteristics between these limiting cases. If the target is more like the first case, the Sinclair matrix can be found, while if the target is more like the second case, it cannot be.

5.6. SAR ERRORS

A brief outline of errors inherent in SAR systems and comments on their alleviation is given here.

System Frequency Stability

Effect on Range Measurement. If carrier frequency f_c changes between the time of transmission and the time at which the returned pulse is received and correlated with a delayed version of the transmitted pulse, the range correlation is performed between signals of different frequencies. The returned signal,

$$\cos 2\pi \left[f_c(t - 2R/c) + \frac{K}{2}(t - 2R/c)^2 \right]$$

which represents a linear FM pulse with K nonzero or a constant-frequency pulse with K zero, is correlated by the procedures of Section 4.4 with a pulse having frequency $f_c + \delta f$. If the correlation is carried out, it will be seen that the cosine functions of (4.10) and (4.11) are altered by δf and the envelopes are altered in shape, but not position. For typical SAR parameters, range measurement is unaffected by oscillator frequency stability.

Effect on Azimuth Measurement. A drift in oscillator frequency that takes place between transmission and reception is indistinguishable from the Doppler frequency of a target displaced in azimuth. Consider a point target located at $x_{\text{tar}} = x_r$ where x_r is the radar position. Its Doppler frequency, from (5.1), is zero. If the local oscillator drifts by δf between transmission and reception, the target Doppler frequency will appear to have shifted by this amount, and from (5.1) the target position appears to be $x_r + \lambda R\delta f/2v$. If the maximum allowable error in the target position is taken as the azimuth resolution $L/2$, the allowable oscillator frequency shift is $vL/\lambda R$. If one examines the properties of a typical space-borne radar (Kramer, 1996), it will be seen that the required oscillator frequency stability is achievable. The required oscillator stability is not as great for an airborne radar as for a space-borne radar because of the shorter range.

Motion Errors: Translation

The inability to compensate fully for motion errors of the SAR antenna is the primary limitation on SAR resolution (Ulaby et al., 1986). Air turbulence causes an aircraft to deviate from the ideal trajectory of straight, level flight; even without turbulence, perturbations to the ideal flight path occur. The motion of a spacecraft diverges less from the ideal than does an aircraft, but nonetheless divergence occurs. Consideration of SAR platform motions in terms of their effect on Doppler frequencies provides an estimate of allowable platform motions.

For the SAR antenna at (x_r, y_r, z_r) and the ground target at (x_g, y_g, z_g), with separation R, the phase of the returned signal with respect to that transmitted is

$$\Phi = -\frac{4\pi}{\lambda} R = -\frac{4\pi}{\lambda} \left[(x_g - x_r)^2 + (y_g - y_r)^2 + (z_g - z_r)^2 \right]^{1/2}$$

and the Doppler frequency is

$$f_d = \frac{1}{2\pi} \frac{d\Phi}{dt} = -\frac{2}{\lambda} \left[\frac{(x_g - x_r)}{R} \frac{dx_r}{dt} + \frac{(y_g - y_r)}{R} \frac{dy_r}{dt} + \frac{(z_g - z_r)}{R} \frac{dz_r}{dt} \right]$$

Let the platform motion be a linear function of time, and let the radar position be (0, 0, 0). Then

$$f_d = -\frac{2}{\lambda}\left[\frac{x_g}{R}v_x + \frac{y_g}{R}v_y + \frac{z_g}{R}v_z\right]$$

where v_x, v_y, and v_z are the velocity components of the platform. The x component of the velocity gives the desired Doppler frequency, but the y and z components are error terms. It can be seen that lateral velocities must be small relative to the x-directed velocity if Doppler frequency is to be accurate. If v_x is not constant, the measured Doppler frequency will also be affected.

Motion Errors: Rotation

The SAR platform is subject to roll, a rotation around the x-axis of the vehicle, pitch, a rotation around the y-axis, and yaw, a rotation around the z-axis. Since the SAR antenna in general is not on the rotational axes, rotation of the flight vehicle causes a translation of the antenna. We have discussed translation and consider only rotational effects here.

The typical SAR antenna has a broad beam in a plane transverse to the flight path and a narrow beam in a plane that includes the line of motion of the SAR platform. See Fig. 5.1 for the footprint of the antenna, but note that the ellipse shown in the figure has a smaller axial ratio than the normal SAR antenna.

Roll of the SAR-carrying vehicle moves the antenna footprint transversely to the flight path. Roll does not affect the SAR processing, but the gain of the antenna for a particular resolution cell will be affected. Since the antenna beam is broad in the direction of footprint movement, only a large amount of roll is apt to be significant.

The effect of yaw on SAR performance is more significant than the effect of roll because of the shape of the antenna footprint. The SAR azimuth beamwidth may be only a few degrees, and yaw of the SAR platform may cause the selected resolution cell to be illuminated weakly. An appropriate remedy is to stabilize the antenna so that it does not yaw when the vehicle yaws. Note that translational motion of the antenna occurs when the platform rotates, even if the antenna itself is stabilized against rotation.

When the radar platform pitches, the desired resolution cell may not be illuminated for some transmitted pulses, and the received signal is degraded. As with yaw, it is desirable to reduce the effects of pitch by stabilizing the antenna in addition to stabilizing the vehicle itself. As with yaw, pitch is accompanied by a translation of the antenna.

Error Compensation

In the discussion of the frequency stability required of the SAR system, we recognized that the frequency of the oscillator used to provide the signal with which to correlate the signal returned from the target has the same effect on the output as the target's Doppler frequency. Then the frequency of the local oscillator can be varied to compensate for the effects of undesired SAR motions.

The first step to assure proper SAR performance is to stabilize the SAR platform by means of an inertial navigation system, in order to prevent, to the greatest extent possible, lateral motions, changes in longitudinal velocity, and roll, pitch, and yaw. In addition to platform stabilization, the SAR antenna can be mechanically stabilized against roll, pitch, and yaw. Integrating accelerometers mounted on the antenna will provide information about velocity and position for the stabilization. The antenna-mounted integrating accelerometers can also provide signals carrying information about antenna translational motions that are not mechanically compensated. The residual errors are correctable by electronic means.

Another sensing and compensating system has been used for aircraft. We discussed the pulse nature of the radar as though the pulses are transmitted with a constant interpulse period. If the platform velocity varies, some radars transmit pulses at equal space intervals, not equal time intervals. A Doppler navigation system has been utilized to provide velocity information and establish the times of pulse transmission to ensure equal space intervals (Cutrona et al., 1961).

Finally, it has been pointed out that while motion parameters obtained from inertial navigation units are adequate for low-resolution mapping, the INU output errors cause significant image degradation for high-resolution mapping. Practical synthetic aperture radar systems therefore use motion compensation based on the mapping data. The signal processing methods, called autofocus algorithms, for this type of compensation are outside the scope of our discussion, but an overview with appropriate references has been given by Hawkins (1996, p. 165).

5.7. HEIGHT MEASUREMENT

A *synthetic aperture radar interferometer* can be used to create a terrain height map. The interferometer uses one antenna to transmit a signal toward a target, and two antennas, with a known geometric relationship, to receive the scattered wave. The phase difference between the received signals can be used with the geometry of the interferometer to determine the height of a terrain resolution cell above a reference plane. A radar interferometer is shown in Fig. 5.8. In the figure, antenna 1 is at $(0, 0, 0)$, antenna 2 at $(0, B\cos\alpha, -B\sin\alpha)$, and a point target at (x, y, z). The geometry is such that rays from the antennas to the target are effectively parallel. Distances R_1 and R_2 are sufficiently close to the same value that, except in their difference, $R_1 \approx R_2 \approx R$. Antenna baseline B is many wavelengths, and if received voltages in receivers at the two antennas are added or subtracted, many narrow beams will be formed in the yz plane.

Let antenna 1 radiate and both antennas receive. This gives a broader transmitted beam in the yz plane than would exist if both antennas transmitted simultaneously. Relative to the transmitted wave, the phases of the received

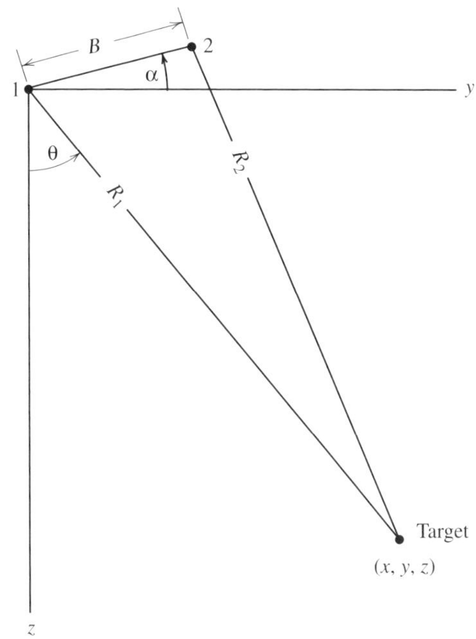

Fig. 5.8. Radar interferometer.

waves at antennas 1 and 2 are

$$\Phi_1 = -\frac{2\pi}{\lambda}(R_1 + R_1) = -\frac{4\pi R_1}{\lambda}$$

$$\Phi_2 = -\frac{2\pi}{\lambda}(R_1 + R_2)$$

Baseline B is much less than R_1, so that

$$R_2 \approx R_1 - \frac{B}{R_1}(y\cos\alpha - z\sin\alpha)$$

From this equation and Fig. 5.8,

$$R_1 - R_2 = B\sin(\theta - \alpha) \tag{5.13}$$

The phase difference between the two received signals is

$$\Phi = \Phi_2 - \Phi_1 = \frac{2\pi}{\lambda}(R_1 - R_2) \tag{5.14}$$

If the equations for $R_1 - R_2$ and Φ are combined, the target angle and vertical distance from the radar can be written as

$$\theta = \alpha + \sin^{-1}\left(\frac{\lambda\Phi}{2\pi B}\right) \tag{5.15}$$

$$z = R\cos\theta \tag{5.16}$$

These equations cannot be used immediately to find z for the target. Range R can be measured by the time of return of the scattered pulse, but the phase is measured modulo 2π. Let Ψ represent the measured phase. In order to find Φ and, from it, z, it is necessary to add the proper multiple of 2π to Ψ.

Here, we consider in a preliminary manner the process of determining Φ. Let a point target at (x, y, z) in Fig. 5.8 be target a. Assume a nearby point target b whose range R_b and phase Φ_b are known. Let the difference in measured phases be $\delta\Psi = \Psi_a - \Psi_b$. If $\delta\Psi$ is small, the difference in measured phases is the same as the difference $\Phi_a - \Phi_b$. Therefore the phase of target a is $\Phi_b + \delta\Psi$. Phase Ψ is called the *wrapped phase* and Φ the *unwrapped phase*, and the process of going from the wrapped to the unwrapped phase is called phase unwrapping. It proceeds from a reference target to a nearby unknown target, and thence to the next target. In order for phase unwrapping to succeed, the phase difference between one target and the next must be less than π (Ghiglia and Pritt, 1998, p. 19).

The determination of range and vertical distance z is more difficult if the target is an area on the ground. It will be taken for granted that the radar interferometer uses synthetic aperture radar principles and that signal processing eliminates scattering from those targets outside the terrain cell of interest.

The target is on the earth's surface and lies in a selected cell identified by its across-track and along-track coordinates as shown in Fig. 5.9. The center point of the selected cell is designated as $(x_{\text{ref}}, y_{\text{ref}}, z_{\text{ref}})$. The vertical distance from the radar to an average terrain level is z_{ref}. For the nth transmitted pulse, slant ranges $\overline{R}_{1n}$ and $\overline{R}_{2n}$ can be calculated. The minimum slant ranges $\overline{R}_{10}$ and $\overline{R}_{20}$ from the radar to the reference point, occurring when the x coordinates of radar and reference point are equal, can also be calculated.

When antenna 1 radiates pulse n, the received voltages in the antennas are

$$V_{1n} = A_n e^{j2\pi f_c(s - 2R_{1n}/c)} \tag{5.17}$$

$$V_{2n} = A_n e^{j2\pi f_c(s - R_{1n}/c - R_{2n}/c)} \tag{5.18}$$

where A_n is a slowly varying time function for the envelope of the range-correlated received voltages. It is taken as independent of n, and A is used in its place.

In accord with the azimuth integration procedure of Section 5.3, V_{1n} and V_{2n} are shifted in phase by multiplying V_{1n} by $\exp\{-j2\pi f_c(2\overline{R}_{10}/c - 2\overline{R}_{1n}/c)\}$ and V_{2n} by $\exp\{-j2\pi f_c(\overline{R}_{10}/c - \overline{R}_{1n}/c + \overline{R}_{20}/c - \overline{R}_{2n}/c)\}$. The resulting phase-shifted voltages are coherently summed.

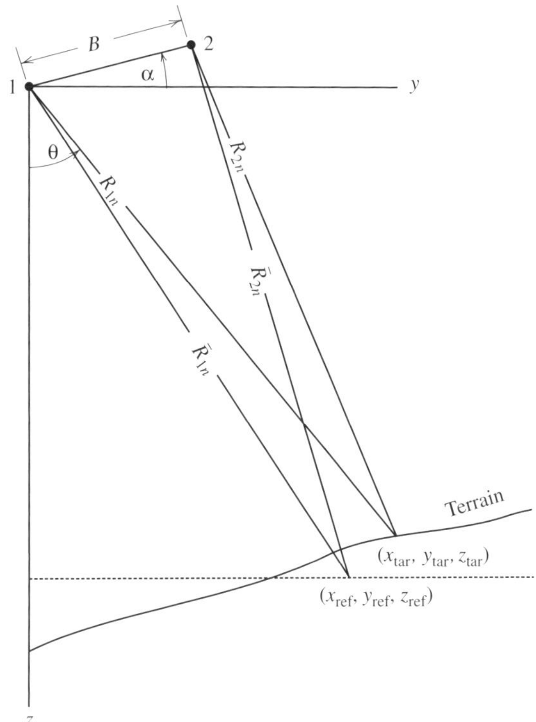

Fig. 5.9. SAR interferometer.

Voltage V_{1n} in (5.17) is the same as V_n in (5.6), with s_n removed, and V_{2n} differs only slightly in form. Then, if the integration process of Section 5.3 is followed, the coherent sum of N pulses of V_{1n} is given by (5.7), and the sum of the V_{2n} is given by a slight variation of the same equation,

$$V_1 = NA \frac{\sin(2\pi f_c N v T \delta x / \overline{R}_0 c)}{N \sin(2\pi f_c v T \delta x / \overline{R}_0 c)} e^{j2\pi f_c (s - 2R_{10}/c)} \tag{5.19}$$

$$V_2 = NA \frac{\sin(2\pi f_c N v T \delta x / \overline{R}_0 c)}{N \sin(2\pi f_c v T \delta x / \overline{R}_0 c)} e^{j2\pi f_c (s - R_{10}/c - R_{20}/c)} \tag{5.20}$$

In deriving these equations the approximation was made that, in the envelope, $\overline{R}_{10} = \overline{R}_{20} = \overline{R}_0$. The phase difference between the two coherently summed voltages can be found from the product $V_1^* V_2$ and is

$$\Phi = \Phi_2 - \Phi_1 = 2\pi f_c (R_{10}/c - R_{20}/c) = \frac{2\pi}{\lambda}(R_{10} - R_{20}) \tag{5.21}$$

Slant ranges R_{10} and R_{20} are the minimum ranges to the target, occurring when a ray from radar to target is perpendicular to the radar track. The radar uses the same number of pulses, $(N - 1)/2$, on either side of this position in the summation. Examination of Fig. 5.9 shows that R_{10} and R_{20} also obey (5.13), or

$$R_{10} - R_{20} = B \sin(\theta - \alpha)$$

It follows that (5.15) and (5.16) can be used to obtain θ and z_{tar},

$$\theta = \alpha + \sin^{-1}\left(\frac{\lambda\Phi}{2\pi B}\right) \tag{5.22}$$

$$z_{\text{tar}} = R_0 \cos\theta \tag{5.23}$$

where $R_0 \approx R_{10} \approx R_{20}$.

The voltages V_{1n} and V_{2n} in (5.17) and (5.18) are those received from a point target. Amplitude A_n is independent of n and is the same for both receiving antennas. The phase difference (5.21) is also independent of n. A terrain cell may have many scattering centers, and the received voltages from one cell differ from one SAR position to the next. V_{1n} and V_{2n} are then complex random variables. Further, since the interferometer antennas see the target from a different angle, the voltages are decorrelated. A measure of the quality of the phase measurements is the magnitude of the complex cross-correlation,

$$|R_{12}| = \frac{|\langle V_{1n}^* V_{2n}\rangle|}{\langle V_{1n}^* V_{1n}\rangle\langle V_{2n}^* V_{2n}\rangle} \tag{5.24}$$

The angle brackets indicate a summation over N pulses. It is readily seen that for the point target this coefficient is 1.

The sum voltages V_1 and V_2 are random, and it follows that vertical distance z_{tar} is random. The variance of the measured vertical distance is small relative to true terrain elevation differences if (5.24) is near 1.

An approximation is used for R_0 in (5.23), and the value of z_{tar} is also approximate. It is the difference in terrain elevation from cell to cell that is of interest, however, and the approximation for R_0 has no effect on the difference.

Two-Pass Interferometry

Interferometric measurements can be made by one antenna that is transported over the terrain on a known path and then is carried over the same terrain on a second, parallel, path. The path separation is the interferometer baseline. The same geometry that was used for the two-antenna, single-pass interferometer can be used for the one-antenna, two-pass interferometer. The antenna on its first pass is treated as antenna 1 and on the second as antenna 2. With the two-antenna system, only one antenna transmits and both receive. With the single-antenna system the antenna must be used as a transmitter on both passes over the terrain to be mapped. The received signals after coherent summation are given by (5.19) for antenna 1 and by the same equation, with R_{20} replacing R_{10}, for antenna 2. The phase difference is

$$\Phi = \frac{4\pi}{\lambda}(R_{10} - R_{20})$$

which is twice that of the interferometer using only one antenna for transmission.

Differential Interferometry

Changes in the earth's surface over time can be measured by differential interferometry. Two phase-difference maps are created, with a time lapse between their creation. For the first map, phase difference Φ for one terrain cell corresponds to the range difference of (5.14). A second phase-difference map (interferogram) is obtained with the interferometer at a later time. If no change in the cell has occurred, the phase difference for the cell will be the same for the two measurements. If an elevation change has occurred that affects the entire cell, the range difference of (5.14) will be altered by the change, and phase difference Φ' of the second interferogram will differ from Φ. The difference between Φ and Φ' is proportional to the elevation change. Phase maps of an area made before and after some geological occurrence are subtracted to determine elevation changes that have occurred. A single antenna can be used for the second measurement if a second map is created by differencing the phase measured by the single antenna and that measured by one of the antennas used to generate the first interferogram. The difference between $(R_1 - R_2)$ for the first measurement and the corresponding value $(R_1' - R_2')$ for the second cannot be uniquely determined if it is greater than a wavelength, and this imposes an upper limit on the elevation changes that can be measured. The lower limit is in the neighborhood of a few centimeters. If the scattering properties of the terrain cell are altered between the first and second images because of relative position changes of the scatterers in the cell or changes in their scattering characteristics, temporal decorrelation makes phase comparison between the two images more difficult.

5.8. POLARIMETRIC INTERFEROMETRY

Polarization of the interferometer antennas was not explicitly considered in the previous discussion, but it may be significant. For example, an area covered by grain may appear to have different elevations for linear-vertical and linear-horizontal polarizations. If the Sinclair matrix is measured by the interferometer, any antenna polarization can be synthesized. Phase-difference maps can be constructed for all possible states of the two interferometer antennas by using a voltage corresponding to an element of the scattering matrix, transformed if desired to another polarization basis (Cloude and Papathanassiou, 1998; Papathanassiou, 1999). Some received voltages will be less affected by clutter and noise than others and lead to more accurate elevation maps. The Sinclair matrix, apart from its utility in obtaining better elevation information, is useful in target identification and so are elevation differences found by using different antenna polarizations.

5.9. PHASE UNWRAPPING

The vertical distance between a terrain cell and the interferometric radar, from (5.15) and (5.16) with antenna baseline tilt angle α taken to be zero, is

$$z_{\text{tar}} = R \cos\left[\sin^{-1}\left(\frac{\lambda\Phi}{2\pi B}\right)\right]$$

Φ is the phase difference of the received signals at the two antennas of the interferometer, R the slant range from interferometer to ground cell, and B the spacing between antennas of the interferometer.

A plot of measured, or wrapped, phases is shown in Fig. 5.10. The range of phase measurements is $0 - 2\pi$, and all phase values on the plot are divided by 2π. A pixel on this plot corresponds to a terrain cell. In order to find the elevation for all terrain cells, the phases must be unwrapped. If two assumptions are made, that in a plot of the unwrapped phase there is no pixel-to-pixel phase increment greater than π and that the unwrapped phase of one pixel is known, the unwrapped phases of all pixels can be found.

Figure 5.11 is a plot of the unwrapped phases corresponding to the wrapped phases of Fig. 5.10. It was constructed by integrating the directional derivative of the wrapped phase from the (1, 1) pixel, for which the unwrapped phase was assumed to be known, and recognizing that the gradient of the unwrapped phase is the same as that of the wrapped phase. Integration of the directional derivative is effectively the addition of pixel-to-pixel phase increments.

It has been shown that one-dimensional phase unwrapping can be carried out by summing the wrapped phase differences from one pixel to the next, provided that no increment of the unwrapped phase between adjacent pixels exceeds π radians (Itoh, 1982), and we assume the process valid for the two-dimensional

0.5	0.8	0.1	0.4	0.7	0.0
0.7	0.0	0.3	0.6	0.9	0.2
0.9	0.2	0.5	0.8	0.1	0.4
0.1	0.4	0.7	0.0	0.3	0.6
0.3	0.6	0.9	0.2	0.5	0.8
0.5	0.8	0.1	0.4	0.7	0.0

Fig. 5.10. Wrapped phases.

0.5	0.8	1.1	1.4	1.7	2.0
0.7	1.0	1.3	1.6	1.9	2.2
0.9	1.2	1.5	1.8	2.1	2.4
1.1	1.4	1.7	2.0	2.3	2.6
1.3	1.6	1.9	2.2	2.5	2.8
1.5	1.8	2.1	2.4	2.7	3.0

Fig. 5.11. Unwrapped phases.

case. In the integration, the apparent phase increments of the plot of wrapped phase may not be the true phase increments. In moving from pixel (1, 2) to pixel (1, 3) in Fig. 5.10, for example, the apparent increment is -0.7. This is not allowed by the first assumption of phase unwrapping, so 2π was added to the increment, making it $+0.3$ rather than -0.7. Any path can be followed in the integration; the phases of Fig. 5.11 are independent of the path taken.

The requirement that phase increments between adjacent pixels not be greater than π can be met for the most part by selecting terrain cell dimensions appropriate to the terrain slope, but atypical terrain can prevent the requirement from being met in some regions. We noted that V_1 and V_2 may not be completely correlated because of different look angles from antennas to target. This will lead to an error in measured phase and may keep the requirement on maximum phase-increment magnitude from being met (Cloude and Papathanassiou, 1998). Failure to meet that requirement is a major difficulty of phase unwrapping. Figure 5.12 illustrates this problem. On the plot, the integrals of the directional derivatives of the phase are not independent of the path of integration. The sums of the phase increments in traversing the plot from reference pixel (8, 2) to pixel (3, 6) are:

Path 1: $-0.2 - 0.4 + 0.0 + 0.4 + 0.1 + 0.0 + 0.1 + 0.0 + 0.2 = +0.2$
Path 2: $-0.2 - 0.4 + 0.0 + 0.1 + 0.1 - 0.2 - 0.2 - 0.1 + 0.1 = -0.8$
Path 3: $0.1 + 0.1 + 0.1 + 0.0 + 0.1 + 0.0 - 0.2 - 0.1 + 0.1 = +0.2$

Two possibilities exist to explain the different values obtained in the integration: Integration of the differential derivative does not give the correct unwrapped phase, or certain integration paths are not allowed. The first possibility is not in accord with the example of Figs. 5.10 and 5.11. To examine the second, closed paths are formed from the three open paths 1, 2, and 3, and integrations in a counterclockwise direction carried out. The results are

Closed Path A: Path 2 and Path 1: $-0.8 - (+0.2) = -1.0$
Closed Path B: Path 3 and Path 2: $0.2 - (-0.8) = +1.0$
Closed Path C: Path 3 and Path 1: $0.2 - (+0.2) = 0.0$

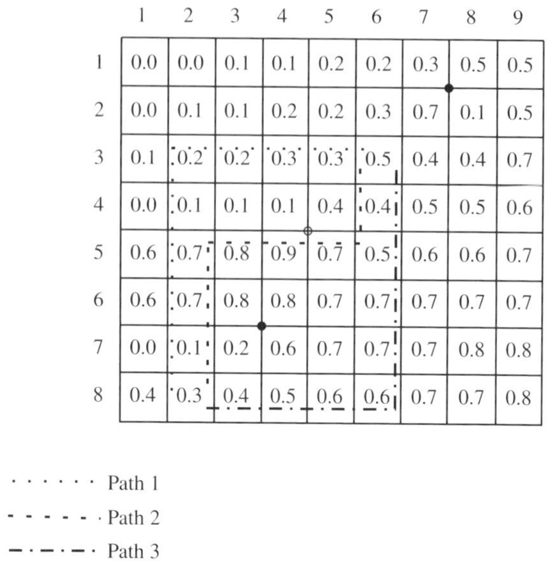

Fig. 5.12. Line integrations.

Singular Points and Residues

In complex-variable theory, a nonzero value for the integral around a closed path indicates a singularity inside the path, with a residue, and this analogy will be carried forward here. To isolate the singularities that are apparent from the results of the three closed-path integrations, line integrations were carried out around all 2×2 pixel groups on the plot. The integral values found are

Group 1: Pixels 6,3; 7,3; 7,4; 6,4: Integral $= +1.0$
Group 2: Pixels 4,4; 5,4; 5,5; 4,5: Integral $= -1.0$
Group 3: Pixels 1,7; 2,7; 2,8; 1,8: Integral $= +1.0$
All other groups of four: Integral $= 0$

Study of these four-pixel groups and other groups that can be constructed indicates the following: If a group of four pixels has an odd number of phase increments between adjacent pixels with magnitudes greater than 0.5 when they are traversed rotationally, a nonzero residue exists. If the number of phase increments with magnitudes greater than 0.5 is even, the residue is zero. Three singularities exist on the phase plot of Fig. 5.12 and each is shown as a dot or circle at the common junction of the four pixels of a group. The location of a singularity cannot be made more precise than the junction point and need not be. The singularities coincide with the center points of groups 1, 2, and 3, and the residues, or charges, associated with each singularity are the integral values given above. If one integrates around any closed path on the phase plot, it will be seen that the integral is the sum of the residues associated with the singularities within the closed curve.

Examination of other integration paths from pixel (8, 2) to pixel (3, 6) shows the following: For paths that do not cross (an odd number of times) a line between the center points of groups 1 and 2 and do not cross a line between the center of group 3 and any boundary of the plot, the values of all integrals are the same. For paths that cross a line between the centers of groups 1 and 2 and do not cross a line between the center of group 3 and a boundary, an integral is path dependent. For paths that cross a line between the center of group 3 and a boundary and do not cross a line between the centers of groups 1 and 2, an integral is path dependent.

Branch Cuts

It can be concluded from these observations that if branch cuts are placed between certain singularities and between a singularity and the plot boundary and if the path of integration does not cross a branch cut an odd number of times, the unwrapped phases can be found by integrating the directional derivative of the wrapped phase from a reference pixel to the pixel of interest. The prohibition against crossing a line between singular points, when applied to a phase plot with finite pixel sizes, means that whole pixels lying between the singularities must not be crossed, and in this sense it is a connected group of pixels that forms the branch cut. Pixels forming a branch cut must be contiguous, and a corner of each end-pixel of the branch cut must coincide with the center of the four-pixel groups being connected. Thought will show that branch cuts should be placed between singularities having residues of opposite sign. It will also show that a singular point should be connected to one and only one other singularity. A singular point lying near a boundary and not connected to another singularity can be connected to the boundary. Further study of Fig. 5.12 shows that the pixels forming the branch cuts have an indeterminate wrapped phase, with one value if the integration path approaches the pixel from one direction and another value if the path approaches it from another direction. It is therefore desirable, other things being equal, to choose a branch cut with the smallest number of pixels.

Not all phase-unwrapping procedures utilize branch cuts and forbidden paths of integration, but it has been noted that branch-cut methods are the most successful in determining the unwrapped phases (Cusack et al., 1995).

A phase map may have many thousands of pixels, with hundreds or thousands of nonzero residues, and it is necessary to use computer algorithms to locate the singularities with nonzero residues and connect them with branch cuts. A reasonable approach to choosing residue pairs for connection by a branch cut is to connect a singularity having a nonzero residue to its nearest neighbor of opposite sign. This procedure is well suited for computer implementation. The positions of singularities with nonzero residues are examined in an orderly manner. When the nearest neighbor with residue of opposite sign to the singularity being examined is found, the two singularities are paired and removed from further consideration. If the residue under examination is closer to the region boundary than to another

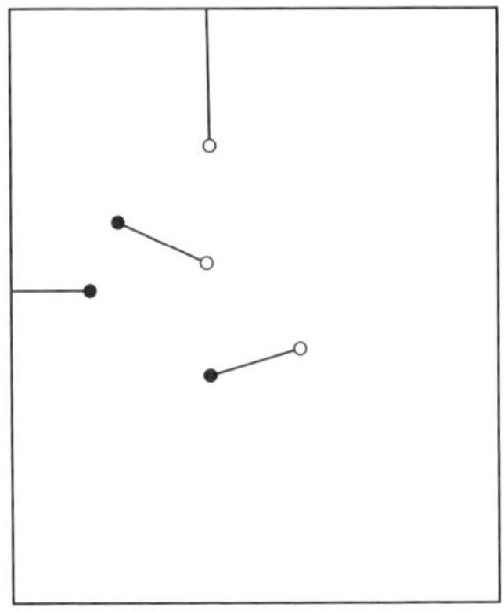

Fig. 5.13. Nearest-neighbor algorithm by column.

residue of opposite sign, it is connected to the boundary and removed from further consideration. Figure 5.13 shows the result of applying this algorithm to a simple set of singularities. Positive residues are shown as filled-in circles and negative as open circles. In applying the algorithm, the residues were considered starting with the leftmost column and taking the residues from top to bottom. Note that the topmost singularity is connected to the boundary rather than to a singularity that is closer to it than is the boundary. If the problem had been treated beginning with the top row and considering the residues in each row left to right, this would not have been the case. The nearest-neighbor algorithm does not yield unique branch cuts. The lack of uniqueness of the nearest-neighbor algorithm can be overcome by modifying the process. Cusack et al., (1995) outline steps for doing so.

A procedure by Goldstein et al. (1988) for finding branch cuts has been recommended as the first to use in many problems (Ghiglia and Pritt, 1998, p. 175). In applying the method, a convention is adopted that a singularity is identified by the pixel at the upper left of the four-pixel group with the singularity at the center. A "box" of 3×3 pixels is placed around the representative pixel and searched for another non-zero residue. If none is found, the box is enlarged to 5×5 and searched, and so on. If a residue is found, the action taken depends on the sign of the singularity residue and whether it has previously been connected to another singularity. The procedure is amenable to an efficient computer algorithm.

Some pixels are of higher quality than others; their phase values are less corrupted by noise and correspond more closely to the values that would exist on an ideal, continuous phase plot. Bone (1991) considered pixel quality and excluded low-quality pixels from the integration path. He found the second differences of the phase at each pixel and excluded those pixels whose second differences exceeded a threshold. In the algorithm, residues need not be found, but examples show that the excluded pixels are grouped around singularities and the paths between them.

Ghiglia and Pritt (1998) discuss other phase unwrapping procedures in detail.

REFERENCES

D. J. Bone, "Fourier Fringe Analysis: The Two-Dimensional Phase Unwrapping Problem", *Appl. Op.*, **30**(25), 3627–3632 (1991).

S. R. Cloude and K. R. Papathanassiou, "Polarimetric SAR Interferometry", *IEEE Trans. GRS*, **36**(5), 1551–1565 (September 1998).

J. C. Curlander and R. N. McDonough, *Synthetic Aperture Radar: Systems and Signal Processing*, Wiley, New York, 1991.

R. Cusack, J. M. Huntley, and H. T. Goldrein, "Improved Noise-Immune Phase-Unwrapping Algorithm", *Appl. Op.*, **28**(5), 781–789 (1995).

L. J. Cutrona, W. E. Vivian, E. N. Leith, and G. O. Hall, "A High-Resolution Radar Combat-Surveillance System", *IRE Trans. Mil Elect.*, **5**, 127–131 (1961).

C. Elachi, *Introduction to the Physics and Techniques of Remote Sensing*, John Wiley & Sons, Inc., New York, 1987.

D. C. Ghiglia and M. D. Pritt, *Two-Dimensional Phase Unwrapping, Theory, Algorithms, and Software*, Wiley-Interscience, New York, 1998.

R. M. Goldstein, H. A. Zebker, and C. L. Werner, "Satellite Radar Interferometry: Two-Dimensional Phase Unwrapping", *Radio Sci.*, **23**(4), 713–720 (July–August, 1998).

D. W. Hawkins, *Synthetic Aperture Imaging Algorithms: with application to wide bandwidth sonar*, Ph. D. thesis, University of Canterbury, Christchurch, NZ, October, 1966.

K. Itoh, "Analysis of the Phase Unwrapping Problem", *Appl. Op.*, **21**(14), 2470 (July, 1982).

H. J. Kramer, *Observation of the Earth and its Environment, Survey of Missions and Sensors*, 3rd enlarged ed., Springer-Verlag, Berlin, 1996.

K. R. Papathanassiou, *Polarimetric SAR Interferometry*, Ph. D. thesis, Technical University Graz, January 1999.

R. N. Trebits, "Synthetic Aperture Radar", in *Principles of Modern Radar*, J. L. Eaves and E. K. Reedy, eds., Van Nostrand Reinhold, New York, 1987.

F. T. Ulaby, R. K. Moore, and A. K. Fung, *Microwave Remote Sensing, Active and Passive*, Artech House, Norwood, MA, 1986.

R. E. Ziemer and W. H. Tranter, *Principles of Communications*, 4th ed., Wiley, New York, 1995.

PROBLEMS

5.1. A synthetic aperture radar operating at 10 GHz, height of 10 km, and velocity of 180 m/s has an antenna with length 2 m along the radar trajectory, and width D to be determined. It is desired to examine a ground swath from 1 to 3 km on one side of a line below the radar path. Determine the optimal antenna dimension D. Choose a pulse repetition frequency f_p and give reasons for selecting it.

5.2. A synthetic aperture radar operates at a frequency of 10 GHz, height of 10 km, and velocity of 180 m/s. If the radar position is taken as $0, 0, 0$, find the frequency of the wave scattered from a terrain patch at $x_g,\ y_g = 200\,\text{m},\ 2\,\text{km}$.

5.3. A focused SAR is designed so that signals are processed in azimuth for a shorter time than that for which the radar sees the target. If the azimuth-processing time is half the time for which the target lies within the half-power beamwidth of the radar antenna, find the azimuth resolution in terms of antenna dimension L.

5.4. If the radar of Problem 5.2 has a lateral motion of 2 m/s when a pulse is transmitted and received, find the frequency error of the received pulse due to the lateral motion. What error in along-track position results from this motion?

5.5. An interferometer operating at 10 GHz and height of 10 km, moving in the x direction, has a baseline of 3 m. The terrain cell of interest is at $x_g, \ y_g = 0, \ 2\,\text{km}$. Find the phase difference between the signals received at the interferometer antennas.

5.6. The vehicle carrying a 10 GHz interferometer in the x-direction rolls at a rate of 0.2 rad/s. The antenna baseline is 2 m. A target is 10 km distant in the yz plane. Find the error in the difference phase at the interferometer antennas.

5.7. It was shown in Section 5.6 that the phase and Doppler frequency of a the signal received by a synthetic aperture radar are affected by translational motions of the radar antenna and that y- and z-directed motions lead to errors in Doppler frequency. Analyze the signals to an interferometer to determine if translational motions in the x, y, and z directions of the two receiving antennas, with both antennas moving in the same manner, affect the phase difference between the received signals.

CHAPTER 6

PARTIALLY POLARIZED WAVES

Single-frequency waves are *completely polarized*. Orthogonal components of the field vector are completely correlated, and the end point of the field vector traces an ellipse. Many waves encountered are not monochromatic. Light reaching the earth from the sun covers a wide frequency spectrum and its orthogonal components are random and incompletely correlated. Other sources than the sun also produce polychromatic waves whose orthogonal field components are not completely correlated. Such waves are *partially coherent* (Beran and Parrent, 1964). Unlike that of a monochromatic wave, the tip of the electric field vector of a partially coherent wave does not trace an elliptical path as time increases, and the wave is *partially polarized*.

If its bandwidth is small relative to the mean frequency, the wave is *quasi-monochromatic*. The waves to be considered in this work are either monochromatic and completely coherent or quasimonochromatic. In radar, transmitted waves are monochromatic and completely polarized, or nearly so. Partially polarized waves arise primarily from scattering. The scattered wave from foliage, for example, is partially polarized. If the foliage is stirred by wind, the orthogonal components of the scattered wave vary with time, and not in the same manner. The position variation with time gives rise to a varying Doppler frequency that broadens the bandwidth of the scattered wave. The differently varying horizontal and vertical returns cause the scattered wave to be partially polarized. Such a target incoherently scatters and *depolarizes* the incident wave and is a *depolarizing target*. Another target that can be considered to be depolarizing is a distributed target, one extended in space and considered to be a collection

Remote Sensing with Polarimetric Radar, by Harold Mott

of separated scattering centers. If, as is common in radar, the target is examined by multiple pulse transmissions with different radar-target orientations from pulse to pulse, the addition of the contributions from each subscatterer gives a different received voltage for each radar-target orientation, with horizontal and vertical voltage components varying differently. From one pulse to the next, the polarization state of the received wave varies, just as it would if the radar-target orientation were fixed and the target were varying. The scattered wave can then be considered partially polarized. The phase difference between two separated subscatterers is linear with frequency, and the effect of changes in the radar-target orientation is more significant at higher frequencies. A target may not depolarize an incident wave at one frequency, but cause a significant depolarization at a higher frequency.

Time-average power is normally measured instead of a field intensity. The time averaging period may determine whether a wave is considered to be partially polarized or not. With a fast processor, the scattered wave can be considered completely polarized with slowly changing polarization characteristics. To a slow processor the same wave appears partially polarized.

6.1. REPRESENTATION OF THE FIELDS

A real component of a monochromatic plane wave traveling in the z direction,

$$\tilde{E}^r(z,t) = a(z)\cos[\omega t + \Phi(z)] \tag{6.1}$$

is customarily represented by a complex time-invariant form,

$$E(z) = a(z)e^{j\Phi(z)} \tag{6.2}$$

obtained by adding

$$j\tilde{E}^i(z,t) = ja(z)\sin[\omega t + \Phi(z)]$$

to (6.1) and suppressing the $\exp(j\omega t)$ multiplier. In this section, we will see that a similar representation can be used for polychromatic waves.

Analytic Signals

Let a real transverse component of a multifrequency wave traveling in the z direction be

$$\tilde{E}^r(z,t) = a(t)\cos[\overline{\omega}t + \Phi(z,t)] \tag{6.3}$$

$\tilde{E}^r$ can be regarded as a sample function of a random process with mean frequency $\overline{\omega}$. Amplitude a is positive and, in a lossless region, independent of z.

Equation 6.3 is particularly useful if the wave is quasi-monochromatic. For such signals, $a(t)$ varies slowly when compared to $\cos \overline{\omega} t$, and $\Phi(z, t)$ changes slowly compared to $\overline{\omega} t$. Then, $a(t)$ is the envelope of a wave that approximates a cosine time function, and $\Phi(z, t)$ is the associated phase. The methods discussed in this chapter require the quasimonochromatic constraint.

A requirement of narrow bandwidth is that the bandwidth must be much smaller than the center frequency, or $\Delta\omega << \overline{\omega}$, where $\Delta\omega/2$ is the maximum radian frequency deviation from the mean. This allows frequency-dependent parameters to be evaluated at the mean frequency of the wave, which causes the methods of handling partially polarized waves to be approximate to a small degree. To develop another constraint, consider a target of length d illuminated by a wave with extreme frequencies ω_1 and ω_2. If these two frequency components of the incident wave are in phase at the leading edge of the target, the phase difference of the reflected components is zero for the wave reflected from the leading edge and

$$\Delta\Phi = 4\pi d \left(\frac{1}{\lambda_1} - \frac{1}{\lambda_2} \right) = 4\pi d \left(\frac{f_1}{c} - \frac{f_2}{c} \right)$$

for the wave reflected from the trailing edge. To keep $\Delta\Phi$ small compared to 2π requires that $\Delta f \ll c/2d$. In this work, waves incident on a target are monochromatic and this constraint is not relevant.

Amplitude $a(t)$ and phase $\Phi(z, t)$ in (6.3) are arbitrary. $\Phi(z, t)$ may be chosen at will, subject only to the constraint that $a(t)$ be positive, and $a(t)$ assumes an appropriate form to yield the correct value for $\tilde{E}^r(z, t)$. This arbitrariness can be eliminated and a unique representation for a polychromatic wave developed.

A sufficient condition for the existence of Fourier and Hilbert integrals is that the function to be transformed be square integrable. If it is not, we nevertheless assume the existence of the transform (Beran and Parrent, 1964, p. 13). To obtain it, a truncated function equal to the field component in the interval $-T \leq t \leq T$, and zero elsewhere, is defined. The truncated function is transformed, and the limit as $T \to \infty$ is taken to be the transform of the untruncated field component. A real function that has a Fourier transform can be written as a Fourier cosine integral; thus for one component of the electric field intensity,

$$\tilde{E}^r(z, t) = \int_0^\infty g(\omega) \cos[\omega t + \Theta(z, \omega)]\, d\omega \tag{6.4}$$

To obtain the time-invariant form (6.2), we added a sine function shifted in phase by $\pi/2$ to a cosine function. By analogy, we use the Fourier sine integral,

$$\tilde{E}^i(z, t) = \int_0^\infty g(\omega) \sin[\omega t + \Theta(z, \omega)]\, d\omega$$

and add it to (6.4) after a phase shift of $\pi/2$, obtaining

$$\tilde{E}(z, t) = \tilde{E}^r(z, t) + j\tilde{E}^i(z, t) = \int_0^\infty g(\omega) e^{j[\omega t + \Theta(z, \omega)]}\, d\omega$$

Functions $\tilde{E}^r(z,t)$ and $\tilde{E}^i(z,t)$ are not independent, and it can be shown that they are related by the Hilbert transform,

$$\tilde{E}^i(z,t) = \frac{1}{\pi}\text{Prin}\int_{-\infty}^{\infty} \frac{\tilde{E}^r(z,t')}{t'-t}\,dt'$$

The notation signifies the Cauchy principal value of the integral; that is,

$$\text{Prin}\int_{-\infty}^{\infty} \frac{f(t')}{t'-t}\,dt' = \lim_{\epsilon\to 0}\int_{-\infty}^{t-\epsilon} \frac{f(t')}{t'-t}\,dt' + \lim_{\epsilon\to 0}\int_{t+\epsilon}^{\infty} \frac{f(t')}{t'-t}\,dt'$$

It has been noted that this integration is more readily performed in the complex z plane with the substitution $z = x + jy$ for t (Beran and Parrent, 1964, p. 16).

The function $\tilde{E}(z,t)$ is the *analytic signal* associated with the real signal $\tilde{E}^r(z,t)$ (Gabor, 1946). The analytic signal representation of a real time function is formed by adding its Hilbert transform after multiplication by j. The Hilbert transform of the wave component of (6.3) is

$$\tilde{E}_u^i(z,t) = a_u(t)\sin[\overline{\omega}t + \Phi_u(z,t)] \qquad u = x,\ y$$

Coefficients a_x and a_y are treated as independent of distance since we are primarily concerned with plane waves.

The analytic signals for the components of the quasimonochromatic plane wave are

$$\tilde{E}_u(z,t) = a_u(t)e^{j[\overline{\omega}t+\Phi_u(z,t)]} \qquad u = x,\ y \tag{6.5}$$

or, if the phase variation with z is made implicit,

$$\tilde{E}_u(t) = a_u(t)e^{j[\overline{\omega}t+\Phi_u(t)]} \qquad u = x,\ y \tag{6.6}$$

From (6.5),

$$a_x(t) = \left\{\left[\tilde{E}_x^r(z,t)\right]^2 + \left[\tilde{E}_x^i(z,t)\right]^2\right\}^{1/2} = \left[\tilde{E}_x(z,t)\tilde{E}_x^*(z,t)\right]^{1/2} = \left|\tilde{E}_x(z,t)\right|$$

$$a_y(t) = \left|\tilde{E}_y(z,t)\right|$$

Also,

$$\Phi_x(z,t) = \tan^{-1}\left(\frac{\tilde{E}_x^i(z,t)}{\tilde{E}_x^r(z,t)}\right) - \overline{\omega}t = -\tan^{-1}\left(j\,\frac{\tilde{E}_x - \tilde{E}_x^*}{\tilde{E}_x + \tilde{E}_x^*}\right) - \overline{\omega}t \tag{6.7}$$

$$\Phi_y(z,t) = -\tan^{-1}\left(j\,\frac{\tilde{E}_y - \tilde{E}_y^*}{\tilde{E}_y + \tilde{E}_y^*}\right) - \overline{\omega}t \tag{6.8}$$

These equations show that the envelope amplitudes $a_x(t)$ and $a_y(t)$ are independent of the choice of phases. Further, since $\overline{\omega}$ is the mean frequency of the random process of which $\tilde{E}_x^r$ and $\tilde{E}_y^r$ are sample functions, the phases Φ_x and Φ_y are completely specified by (6.7) and (6.8). We have obtained by use of the analytic signal a unique representation of a real quasimonochromatic wave.

The wave components of (6.6) can be written, with time variation $\overline{\omega}t$ implicit, as

$$E_u(t) = a_u e^{j\Phi_u(t)} \qquad u = x,\ y \tag{6.9}$$

The notation used here is a tilde for a field that includes an $\exp(j\omega t)$ term and the same letter without the tilde for a field that omits it. Equations 6.5 and 6.6 are the analytic signal representations of the wave, but often we will use the form (6.9) and refer to it as the analytic signal.

Time Averages of the Fields

A quantity proportional to the time average of the product of real field components is often measured by the radar. Consider the time average,

$$\overline{\tilde{E}_x^r(z,t)\tilde{E}_y^r(z,t)} = \lim_{T\to\infty} \frac{1}{2T}\int_{-T}^{T} \tilde{E}_x^r(z,t)\tilde{E}_y^r(z,t)\,dt$$

Substituting (6.3) in this expression causes it to become

$$\overline{\tilde{E}_x^r(z,t)\tilde{E}_y^r(z,t)} = \frac{1}{2}\overline{a_x(t)a_y(t)\cos[\Phi_y(z,t) - \Phi_x(z,t)]}$$

If the analytic signal representation of the wave is used, we find

$$\operatorname{Re}\overline{\tilde{E}_x^*(z,t)\tilde{E}_y(z,t)} = \overline{a_x(t)a_y(t)\cos[\Phi_y(z,t) - \Phi_x(z,t)]}$$

and see that

$$\overline{\tilde{E}_x^r(z,t)\tilde{E}_y^r(z,t)} = \frac{1}{2}\operatorname{Re}\overline{\tilde{E}_x^*(z,t)\tilde{E}_y(z,t)} = \frac{1}{2}\operatorname{Re}\overline{\tilde{E}_x^*(t)\tilde{E}_y(t)}$$

In the last term of this equation, distance z is implied and the fields of (6.9) are used. Other product terms give similar results.

6.2. REPRESENTATION OF PARTIALLY POLARIZED WAVES

Consider the measurement of the intensity of a partially polarized wave. The wave, in right-handed coordinates $\xi\eta\zeta$, is incident on a receiving antenna described in xyz coordinates, as shown in Fig. 2.12. The components of the wave are $E_\xi(t)$ and $E_\eta(t)$, and the voltage received by an antenna of effective length $\mathbf{h}$ is

$$V(t) = \mathbf{h}^T(x, y)\mathbf{diag}(-1, 1)\mathbf{E}(\xi, \eta, t)$$

The space coordinates in parentheses refer to the coordinate system in which the field and effective length are measured. The time-averaged power to a load is

$$\begin{aligned} W &= \frac{1}{8R_a}\overline{V(t)V^*(t)} \\ &= \frac{1}{8R_a}\left[|h_x|^2\overline{|E_\xi(t)|^2} - h_x h_y^* \overline{E_\xi(t)E_\eta^*(t)} - h_x^* h_y \overline{E_\xi^*(t)E_\eta(t)} + |h_y|^2\overline{|E_\eta(t)|^2}\right] \end{aligned} \tag{6.10}$$

Coherency Vector of a Quasimonochromatic Plane Wave

The time averages of (6.10) can be ordered, and we do so by forming a *coherency vector*,

$$\mathbf{J}(\xi, \eta) = \begin{bmatrix} J_{\xi\xi} \\ J_{\xi\eta} \\ J_{\eta\xi} \\ J_{\eta\eta} \end{bmatrix} = \begin{bmatrix} \overline{|E_\xi(t)|^2} \\ \overline{E_\xi(t)E_\eta^*(t)} \\ \overline{E_\xi^*(t)E_\eta(t)} \\ \overline{|E_\eta(t)|^2} \end{bmatrix} = \overline{\mathbf{E}(t) \otimes \mathbf{E}^*(t)} \tag{6.11}$$

For convenience, distance z is omitted from the coherency vector unless it is essential to a development. It is clear from the definition of the coherency vector elements as time averages that two quasimonochromatic waves with different center frequencies or different bandwidths can have equal coherency vectors, but this does not detract from the usefulness of the vector.

Born and Wolf (1965, p. 545) order the elements of the coherency vector in a *coherency matrix* of the wave. We write it here in xyz coordinates for a wave traveling in the z direction,

$$\mathbf{J}_m = \begin{bmatrix} J_{xx} & J_{xy} \\ J_{yx} & J_{yy} \end{bmatrix} = \begin{bmatrix} \overline{|E_x(t)|^2} & \overline{E_x(t)E_y^*(t)} \\ \overline{E_x^*(t)E_y(t)} & \overline{|E_y(t)|^2} \end{bmatrix} = \overline{\mathbf{E}(t)\mathbf{E}^\dagger(t)}$$

For $\mathbf{X}$ any vector,

$$\mathbf{X}^\dagger \mathbf{J}_m \mathbf{X} = (\mathbf{X}^\dagger \mathbf{E})(\mathbf{X}^T \mathbf{E}^*)^T = (\mathbf{X}^\dagger \mathbf{E})(\mathbf{X}^T \mathbf{E}^*) = (\mathbf{X}^\dagger \mathbf{E})(\mathbf{X}^\dagger \mathbf{E})^* \geq 0$$

This is the defining relationship for a positive semidefinite matrix and we see that $\mathbf{J}_m$ is Hermitian positive semidefinite.

The determinant of a positive semidefinite matrix is non-negative. Therefore,

$$\det(\mathbf{J}_m) = J_{xx} J_{yy} \left(1 - |\mu_{xy}|^2\right) \geq 0 \tag{6.12}$$

where

$$\mu_{xy} = \frac{J_{xy}}{\sqrt{J_{xx}}\sqrt{J_{yy}}} \tag{6.13}$$

The cross-covariance of complex random processes $X(t)$ and $Y(t)$ is (Peebles, 1980, p. 138)

$$C_{X,Y}(t, t+\tau) = E\left[(X(t) - E[X(t)])\,(Y(t+\tau) - E[Y(t+\tau)])^*\right] \tag{6.14}$$

The complex correlation coefficient at $\tau = 0$ is (Urkowitz, 1983, p. 291)

$$\rho = \frac{C_{X,Y}(t)}{\sqrt{\text{var}[X(t)]}\sqrt{\text{var}[Y(t)]}}$$

If we assume ergodicity and zero mean for random processes E_x and E_y, comparison of these equations shows that μ_{xy} is the correlation coefficient for E_x and E_y at $\tau = 0$. It follows from (6.12) that $|\mu_{xy}| \leq 1$.

For a wave traveling in the z direction, the coherency vector, if we use the signal form, (6.9), is

$$\mathbf{J} = \begin{bmatrix} J_{xx} \\ J_{xy} \\ J_{yx} \\ J_{yy} \end{bmatrix} = \begin{bmatrix} \overline{|E_x(t)|^2} \\ \overline{E_x(t)E_y^*(t)} \\ \overline{E_x^*(t)E_y(t)} \\ \overline{|E_y(t)|^2} \end{bmatrix} = \begin{bmatrix} \overline{a_x^2(t)} \\ \overline{a_x(t)a_y(t)e^{-j(\Phi_y(t)-\Phi_x(t))}} \\ \overline{a_x(t)a_y(t)e^{j(\Phi_y(t)-\Phi_x(t))}} \\ \overline{a_y^2(t)} \end{bmatrix} \tag{6.15}$$

Note: Since the coherency vector elements are sometimes written in a square matrix form, it is desirable to use double subscripts in the vector form.

The coherency vector of a wave radiated by an antenna of effective length $\mathbf{h}$ is

$$\mathbf{E} \otimes \mathbf{E}^* = \frac{jZ_0 I}{2\lambda r}\mathbf{h} \otimes \frac{-jZ_0 I}{2\lambda r}\mathbf{h}^* = \frac{Z_0^2 I^2}{4\lambda^2 r^2}\mathbf{h} \otimes \mathbf{h}^*$$

It is desirable to define a coherency vector for the antenna without the scalar multiplier of this equation and without time averaging and we do so,

$$\mathbf{J}_A = \mathbf{h} \otimes \mathbf{h}^* \tag{6.16}$$

If a wave with coherency vector $\mathbf{J}$ is incident on an antenna with antenna coherency vector $\mathbf{J}_A$, the power received in an impedance-matched load will be seen in Section 6.3 to be

$$W = \frac{1}{8R_a}\mathbf{J}_A^T(x, y)\mathbf{diag}(1, -1, -1, 1)\mathbf{J}(\xi, \eta) \tag{6.17}$$

where the coordinates refer to Fig. 2.12.

Unpolarized Waves

The received power obtained by expanding (6.17) is

$$W = \frac{1}{8R_a}\left[|h_x|^2 J_{\xi\xi} - 2\mathrm{Re}(h_x h_y^* J_{\xi\eta}) + |h_y|^2 J_{\eta\eta}\right]$$

If the wave is unpolarized, the horizontal and vertical components of the wave have equal power densities and are uncorrelated. It follows from (6.11) that $J_{\xi\eta} = J_{\eta\xi}^* = 0$ and $J_{\xi\xi} = J_{\eta\eta}$. Then the received power is independent of the effective length of the receiving antenna if the magnitude of the effective length is constant. The coherency vector in xyz coordinates is

$$\mathbf{J} = \begin{bmatrix} J_{xx} \\ 0 \\ 0 \\ J_{yy} \end{bmatrix} = Z_0\mathcal{P}\begin{bmatrix} 1 \\ 0 \\ 0 \\ 1 \end{bmatrix} \tag{6.18}$$

where $\mathcal{P}$ is the Poynting vector density of the wave.

Complete Polarization

The amplitude a and phases Φ_x and Φ_y of (6.15) are time independent for monochromatic waves, and the coherency vector is

$$\mathbf{J} = \begin{bmatrix} a_x^2 \\ a_x a_y e^{-j(\Phi_y - \Phi_x)} \\ a_x a_y e^{j(\Phi_y - \Phi_x)} \\ a_y^2 \end{bmatrix}$$

Substitution of these elements in (6.13) gives $\mu_{xy} = \exp(-j\Phi)$. Since the magnitude of μ_{xy} is 1, the monochromatic wave is completely polarized and the x and y field components are completely correlated.

We can also have complete polarization for nonmonochromatic waves. Suppose that a_x, a_y, Φ_x, and Φ_y depend on time in such a way that the ratio of amplitudes and the difference in phase are independent of time; that is

$$\frac{a_y(t)}{a_x(t)} = C_1 \tag{6.19}$$

$$\Phi_y - \Phi_x = C_2 \tag{6.20}$$

with C_1 and c_2 constant. The tip of the electric vector traces an ellipse as time varies. The ellipse size will vary slowly as the amplitudes change slowly, but no significant change occurs during one trace of the ellipse. If the polarization ellipse is based on the polarization state of the wave and not on the electric field, it is constant. The coherency vector is the same as for a monochromatic wave with components

$$E_x = \sqrt{\overline{a_x^2(t)}}e^{j\Phi_x} \tag{6.21}$$

$$E_y = C_1\sqrt{\overline{a_x^2(t)}}e^{j(\Phi_x + C_2)} \tag{6.22}$$

These wave components are completely correlated and the wave is completely polarized. It is clear from (6.12) that for a completely polarized wave, and only for such a wave, $\det(\mathbf{J}_m) = 0$

Linear Polarization

For linear polarization, the wave must satisfy the requirements for complete polarization, and in addition, $\Phi_y - \Phi_x = 0, \pm\pi, \pm 2\pi, \ldots$. Then the coherency vector for a monochromatic linearly polarized wave is

$$\mathbf{J} = \begin{bmatrix} a_x^2 \\ (-1)^m a_x a_y \\ (-1)^m a_x a_y \\ a_y^2 \end{bmatrix} \qquad m = 0, 1, 2, \ldots$$

with a similar form for a quasimonochromatic, completely polarized wave that can be obtained from (6.21) and (6.22).

Circular Polarization

For circular polarization, the rectangular component amplitudes of the electric field are equal, and $\Phi_y - \Phi_x = \pm\pi/2$ for left- (upper sign) and right-circular waves. The coherency vectors for left- (upper sign) and right-circular waves are

$$\mathbf{J} = Z_0\mathcal{P}\begin{bmatrix} 1 \\ \mp j \\ \pm j \\ 1 \end{bmatrix}$$

Degree of Polarization

Suppose that N independent electromagnetic waves propagate in the same direction. The electric field intensity components at any point in space are additive, with the sum

$$\tilde{E}_{x,y}(t) = E_{x,y}(t)e^{j\bar{\omega}t} = \sum_{1}^{N} \tilde{E}_{x,y}^{(n)}(t) = \sum_{1}^{N} E_{x,y}^{(n)}(t)e^{j\bar{\omega}t} = e^{j\bar{\omega}t}\sum_{1}^{N} E_{x,y}^{(n)}(t)$$

Then,

$$E_{x,y}(t) = \sum_{1}^{N} E_{x,y}^{(n)}(t)$$

The coherency vector elements of the wave are

$$\begin{aligned} J_{kl} &= \overline{E_k(t)E_l^*(t)} = \sum_{n}\sum_{m} \overline{E_k^{(n)}(t)E_l^{(m)*}(t)} \\ &= \sum_{n} \overline{E_k^{(n)}(t)E_l^{(n)*}(t)} + \sum_{n\neq m} \overline{E_k^{(n)}(t)E_l^{(m)*}(t)} \end{aligned}$$

where k and l independently represent x and y. Since the waves are independent, each term in the last summation is zero, and

$$J_{kl} = \sum_{n=1}^{N} J_{kl}^{(n)}$$

where the $J_{kl}^{(n)}$ are the coherency vector elements of the nth wave. Therefore, for independent waves propagating in the same direction, the coherency vector elements of the individual waves can be added to obtain the overall coherency vector. It is clear that we may also treat a wave as the sum of independent waves.

A quasimonochromatic plane wave can be considered the sum of an unpolarized wave and a completely polarized wave. To see this, note that the Hermitian matrix $\mathbf{J}_m$ can be unitarily diagonalized and has positive real eigenvalues λ_1 and λ_2, with $\lambda_1 \geq \lambda_2$. Then,

$$\mathbf{U}^{-1}\mathbf{J}_m\mathbf{U} = \begin{bmatrix} \lambda_1 & 0 \\ 0 & \lambda_2 \end{bmatrix} = \begin{bmatrix} \lambda_2 & 0 \\ 0 & \lambda_2 \end{bmatrix} + \begin{bmatrix} \lambda_1 - \lambda_2 & 0 \\ 0 & 0 \end{bmatrix}$$

This implies

$$\begin{aligned} \mathbf{J}_m &= \mathbf{U}\lambda_2\mathbf{I}\mathbf{U}^{-1} + \mathbf{U}\begin{bmatrix} \lambda_1 - \lambda_2 & 0 \\ 0 & 0 \end{bmatrix}\mathbf{U}^{-1} \\ &= \lambda_2\mathbf{I} + (\lambda_1 - \lambda_2)\begin{bmatrix} |U_{11}|^2 & U_{11}U_{21}^* \\ U_{11}^*U_{21} & |U_{21}|^2 \end{bmatrix} = \mathbf{J}_m^{(1)} + \mathbf{J}_m^{(2)} \end{aligned} \qquad (6.23)$$

$\mathbf{J}_m^{(1)}$ corresponds to the coherency vector, (6.18), of an unpolarized wave, and it is readily seen from (6.23) that $\det(\mathbf{J}_m^{(2)}) = 0$; therefore, $\mathbf{J}_m^{(2)}$ represents a completely-polarized wave. The first coherency matrix of (6.23) is $\mathbf{J}_m^{(1)} = A\mathbf{I}$ and the second is

$$\mathbf{J}_m^{(2)} = \begin{bmatrix} B & D \\ D^* & C \end{bmatrix} \tag{6.24}$$

Since the eigenvalues of $\mathbf{J}_m$ are nonnegative and the determinant of $\mathbf{J}_m^{(2)}$ zero, it is required that $A \geq 0$ and $BC - DD^* = 0$. A is the smaller eigenvalue of $\mathbf{J}_m$ and is given by

$$A = \frac{1}{2}(J_{xx} + J_{yy}) - \frac{1}{2}\left[(J_{xx} + J_{yy})^2 - 4(J_{xx}J_{yy} - J_{xy}J_{yx})\right]^{1/2} \tag{6.25}$$

To find the remaining elements of the coherency matrices in (6.24), set

$$\begin{aligned} J_{xx} &= A + B \\ J_{xy} &= J_{yx}^* = D \\ J_{yy} &= A + C \end{aligned} \tag{6.26}$$

D is given by (6.26), and B and C are readily found from (6.25) and (6.26) to be

$$\begin{aligned} B &= \frac{1}{2}(J_{xx} - J_{yy}) + \frac{1}{2}\left[(J_{xx} + J_{yy})^2 - 4(J_{xx}J_{yy} - J_{xy}J_{yx})\right]^{1/2} \\ C &= \frac{1}{2}(J_{yy} - J_{xx}) + \frac{1}{2}\left[(J_{xx} + J_{yy})^2 - 4(J_{xx}J_{yy} - J_{xy}J_{yx})\right]^{1/2} \end{aligned}$$

The Poynting vector magnitude of the wave is

$$\mathcal{P}_t = \frac{1}{2Z_0}(J_{xx} + J_{yy})$$

and that of the polarized part of the wave is

$$\mathcal{P}_p = \frac{1}{2Z_0}(B + C) = \frac{1}{2Z_0}\left[(J_{xx} + J_{yy})^2 - 4(J_{xx}J_{yy} - J_{xy}J_{yx})\right]^{1/2}$$

The ratio of the power densities of the polarized part and the total wave is the *degree of polarization* of the wave. It is given by

$$R = \frac{\mathcal{P}_p}{\mathcal{P}_t} = \left[1 - \frac{4(J_{xx}J_{yy} - J_{xy}J_{yx})}{(J_{xx} + J_{yy})^2}\right]^{1/2} \tag{6.27}$$

Now,

$$J_{xx}J_{yy} - J_{xy}J_{yx} \leq J_{xx}J_{yy} \leq \frac{1}{4}(J_{xx} + J_{yy})^2$$

Therefore, $0 \le R \le 1$. Consider the two extreme values of R. For $R = 1$, (6.27) requires that $J_{xx}J_{yy} - J_{xy}J_{yx} = 0$, which is the condition for complete polarization. Then $|\mu_{xy}| = 1$, and the x and y wave components are completely correlated. It follows from (6.12) that the determinant of $\mathbf{J}_m$ is zero. The equation formed by setting $R = 0$ in (6.27) can be satisfied only by $J_{xx} = J_{yy}$ and $J_{xy} = J_{yx} = 0$. It follows that $|\mu_{xy}| = 0$, and E_x and E_y are uncorrelated. The converse is not true. If the field components are uncorrelated, $J_{xy} = J_{yx} = 0$ and $|\mu_{xy}| = 0$. Then,

$$R = \frac{|J_{xx} - J_{yy}|}{J_{xx} + J_{yy}}$$

If the wave is unpolarized, we must also have $J_{xx} = J_{yy}$. It is easy to see why this is so. A wave with a large x component, for example, cannot be substantially depolarized by the addition of a small uncorrelated y component.

The coherency vector of a wave can be transformed to a new polarization basis by the unitary matrix used previously to transform the fields. The electric field $\mathbf{E}$ of a plane wave can be expressed in orthonormal polarization basis set $\mathbf{u}_1, \mathbf{u}_2$, as

$$\mathbf{E}_{\mathbf{u}_1,\mathbf{u}_2} = \mathbf{U}\mathbf{E}_{\mathbf{u}_x,\mathbf{u}_y}$$

where $\mathbf{U}$, from (1.37), is

$$\mathbf{U} = \begin{bmatrix} \langle \mathbf{u}_x, \mathbf{u}_1 \rangle & \langle \mathbf{u}_y, \mathbf{u}_1 \rangle \\ \langle \mathbf{u}_x, \mathbf{u}_2 \rangle & \langle \mathbf{u}_y, \mathbf{u}_2 \rangle \end{bmatrix} \tag{6.28}$$

The transformed coherency vector is

$$\begin{aligned} \hat{\mathbf{J}} &= \overline{\mathbf{E}_{\mathbf{u}_1,\mathbf{u}_2}(t) \otimes \mathbf{E}^*_{\mathbf{u}_1,\mathbf{u}_2}(t)} = \overline{\mathbf{U}\mathbf{E}_{\mathbf{u}_x,\mathbf{u}_y}(t) \otimes \mathbf{U}^*\mathbf{E}^*_{\mathbf{u}_x,\mathbf{u}_y}(t)} \\ &= (\mathbf{U} \otimes \mathbf{U}^*)\, \overline{\mathbf{E}_{\mathbf{u}_x,\mathbf{u}_y}(t) \otimes \mathbf{E}^*_{\mathbf{u}_x,\mathbf{u}_y}(t)} = (\mathbf{U} \otimes \mathbf{U}^*)\mathbf{J} \end{aligned} \tag{6.29}$$

where the identity (7.2) was used.

It is apparent from the manner in which $\hat{\mathbf{J}}$ is formed that, like $\mathbf{J}$, the first and fourth elements are real and the second and third are conjugates. Moreover, for an unpolarized wave, the first and fourth elements are equal and the second and third elements are zero.

If the transformation (6.29) is used in (6.27), it will be seen that the degree of polarization of a wave is given by the same equation in all orthonormal polarization bases,

$$R = \left[1 - \frac{4(\hat{J}_{11}\hat{J}_{22} - \hat{J}_{12}\hat{J}_{21})}{(\hat{J}_{11} + \hat{J}_{22})^2} \right]^{1/2}$$

A partially polarized wave can be treated as though it is the sum of two independent orthogonal completely polarized waves whose coherency vectors add (Mott, 1992, p. 382). As a special case, we can think of an unpolarized wave as being composed of two independent linearly polarized waves orthogonal to each other, each with equal power density. To see this, note that (6.18), the coherency vector of an unpolarized wave, can be split,

$$\mathbf{J} = Z_0\mathcal{P}\begin{bmatrix}1\\0\\0\\1\end{bmatrix} = Z_0\mathcal{P}\begin{bmatrix}1\\0\\0\\0\end{bmatrix} + Z_0\mathcal{P}\begin{bmatrix}0\\0\\0\\1\end{bmatrix}$$

Each vector in this form represents a linearly polarized wave. Just as readily, we could have written the coherency vector of an unpolarized wave as the sum of the coherency vectors of two independent circular waves of opposite rotation sense and equal power density.

Extension of the Coherency Vector Concept

The coherency vector of a wave was introduced in (6.11) with elements formed as time averages of products of the analytic signal electric field components. It is convenient in many developments to retain the concept. However, it can be seen from (6.38) that by choosing different receiving-antenna polarizations and measuring received power in a suitable integration period the coherency vector of a wave can be found without concern for the fields. By analogy with target matrices to be introduced later, we call this coherency vector the *directly measured* or *incoherently measured* vector, although it is power that is measured.

The Stokes Vector

A commonly used descriptive vector for a partially polarized wave is the Stokes vector introduced in Section 1.9 for completely polarized waves. It was defined for monochromatic waves as a transform of the coherency vector. We use the same definition for partially polarized waves but emphasize that now the definition is not restricted to single-frequency waves. The relationship is

$$\mathbf{G} = \mathbf{QJ} \tag{6.30}$$

where

$$\mathbf{Q} = \begin{bmatrix}1 & 0 & 0 & 1\\1 & 0 & 0 & -1\\0 & 1 & 1 & 0\\0 & j & -j & 0\end{bmatrix}$$

Multiplication and time averaging gives

$$\mathbf{G} = \begin{bmatrix} G_0 \\ G_1 \\ G_2 \\ G_3 \end{bmatrix} = \begin{bmatrix} J_{xx} + J_{yy} \\ J_{xx} - J_{yy} \\ J_{xy} + J_{yx} \\ j(J_{xy} - J_{yx}) \end{bmatrix} = \begin{bmatrix} \overline{|E_x(t)|^2 + |E_y(t)|^2} \\ \overline{|E_x(t)|^2 - |E_y(t)|^2} \\ 2\text{Re}\,\overline{E_x^*(t)E_y(t)} \\ 2\text{Im}\,\overline{E_x^*(t)E_y(t)} \end{bmatrix} \tag{6.31}$$

All elements of the vector are real.

Like the wave coherency vector, the Stokes vector can be found by selecting receiving antenna polarizations and measuring received power without determination of the electric field components that is implied by (6.31). The vector so determined will be called the directly measured or incoherently measured Stokes vector.

For quasimonochromatic waves, the Stokes vector, in terms of the amplitudes and relative phase of the analytic signal representation of the field components, is

$$\mathbf{G} = \begin{bmatrix} \overline{|a_x(t)|^2 + |a_y(t)|^2} \\ \overline{|a_x(t)|^2 - |a_y(t)|^2} \\ 2\,\overline{a_x(t)a_y(t)\cos[\Phi_y(t) - \Phi_x(t)]} \\ 2\,\overline{a_x(t)a_y(t)\sin[\Phi_y(t) - \Phi_x(t)]} \end{bmatrix}$$

We saw earlier that two quasimonochromatic waves with different bandwidths can have equal wave coherency vectors. It follows that two waves with different bandwidths can have the same Stokes vector.

The wave coherency vector in terms of the Stokes parameters, found by inverting (6.30), is

$$\mathbf{J} = \begin{bmatrix} J_{xx} \\ J_{xy} \\ J_{yx} \\ J_{yy} \end{bmatrix} = \frac{1}{2} \begin{bmatrix} G_0 + G_1 \\ G_2 - jG_3 \\ G_2 + jG_3 \\ G_0 - G_1 \end{bmatrix}$$

The determinant of the corresponding wave coherency matrix is

$$\det(\mathbf{J}_m) = \frac{1}{4}\left(G_0^2 - G_1^2 - G_2^2 - G_3^2\right) \tag{6.32}$$

We know from (6.12) that this determinant cannot be negative, and it follows that

$$G_0^2 \geq G_1^2 + G_2^2 + G_3^2$$

A Stokes vector for an antenna can be defined with the use of (6.16) and (6.30). It is

$$\mathbf{G}_A = \mathbf{Q}\mathbf{J}_A = \mathbf{Q}(\mathbf{h} \otimes \mathbf{h}^*) \tag{6.33}$$

Unpolarized Waves

For an unpolarized wave, $J_{xx} = J_{yy}$ and $J_{xy} = J_{yx} = 0$, where J_{xx} is Z_0 times the power density, $\mathcal{P}$, of the wave. It follows that for such a wave,

$$G_0 = 2Z_0\mathcal{P}$$

$$G_1 = G_2 = G_3 = 0$$

Complete Polarization

A monochromatic wave and a quasi-monochromatic wave that obeys the condition (6.19) and (6.20) both have a coherency matrix whose determinant is zero. It follows from (6.32) that

$$G_0^2 = G_1^2 + G_2^2 + G_3^2$$

Degree of Polarization

Earlier, we saw that the coherency vector elements of independent waves propagating in the same direction are additive. It follows that the Stokes vector elements of such waves can also be added. We saw also that the coherency vector of any wave can be uniquely expressed as the sum of coherency vectors for unpolarized and completely polarized waves. The Stokes vector of any wave can then be written as the Stokes vector of an unpolarized wave plus the Stokes vector of a completely polarized wave, and this representation is unique. We write

$$\mathbf{G} = \mathbf{G}^{(1)} + \mathbf{G}^{(2)}$$

where $\mathbf{G}^{(1)}$ represents an unpolarized wave and $\mathbf{G}^{(2)}$ represents a completely polarized wave. If the element constraints for these wave types are used, the elements are related by

$$G_0 = G_0^{(1)} + G_0^{(2)} \tag{6.34}$$

$$G_i = G_i^{(2)} \qquad i = 1, 2, 3 \tag{6.35}$$

Three elements of $\mathbf{G}^{(1)}$ are zero, while three elements of $\mathbf{G}^{(2)}$ may be found from these equations and a knowledge of $\mathbf{G}$. The remaining elements can be found from (6.34) and (6.35) and the fact that the determinant of the wave coherency matrix corresponding to $\mathbf{G}^{(2)}$ must be zero. This gives

$$G_0^{(2)} = \sqrt{G_1^2 + G_2^2 + G_3^2}$$

and

$$G_0^{(1)} = G_0 - \sqrt{G_1^2 + G_2^2 + G_3^2}$$

The degree of polarization was defined by (6.27) as the ratio of power densities of the polarized part and the total wave. But $G_0^{(2)}$ is proportional to the density of the polarized part, and G_0 is proportional, with the same proportionality constant, to the density of the total wave. Then the degree of polarization in terms of the Stokes parameters is

$$R = \frac{G_0^{(2)}}{G_0} = \frac{\sqrt{G_1^2 + G_2^2 + G_3^2}}{G_0} \tag{6.36}$$

The Poincaré Sphere

Equation (6.36) may be rewritten as

$$G_1^2 + G_2^2 + G_3^2 = (RG_0)^2 \tag{6.37}$$

and it is clear from this that the Stokes parameters may still be regarded as the rectangular coordinates of a point on the Poincaré sphere. The sphere radius for partially polarized waves is RG_0, and for unpolarized waves becomes zero. A polarization ellipse is not defined for the total wave, so the interpretation of latitude and azimuth angles of points on the sphere in terms of axial ratio and tilt angle must be limited to the polarized part of the wave.

If the wave is separated into unpolarized and completely polarized components, the Poincaré sphere for the total wave is also that for the completely polarized part, since

$$G_0^{(2)} = RG_0$$

If this is used in (6.37) and the relationships in (6.34) and (6.35) taken into account,

$$\left[G_1^{(2)}\right]^2 + \left[G_2^{(2)}\right]^2 + \left[G_3^{(2)}\right]^2 = \left[G_0^{(2)}\right]^2$$

Then, on the Poincaré sphere for the total wave, we may freely interpret points in terms of axial ratio and tilt angle of the completely polarized part of the wave.

6.3. RECEPTION OF PARTIALLY POLARIZED WAVES

When an incident wave $\mathbf{E}(t)$ falls on a receiving antenna with effective length $\mathbf{h}$, the received voltage is

$$V(t) = \mathbf{h}^T \mathbf{E}(t)$$

if the incident wave is completely polarized and if $\mathbf{E}$ and $\mathbf{h}$ are expressed in the same coordinates. With a partially polarized incident wave we measure, instead of the voltage, the received power, which is the time average

$$W = \frac{\overline{VV^*}}{8R_a} = \frac{1}{8R_a}\overline{\mathbf{h}^T \mathbf{E}(t) \mathbf{h}^\dagger \mathbf{E}^*(t)}$$

If the identities (7.1), (7.2), and (7.4) are used, the received power can be written as

$$W = \frac{1}{8R_a}(\mathbf{h} \otimes \mathbf{h}^*)^T \overline{\mathbf{E}(t) \otimes \mathbf{E}^*(t)} \tag{6.38}$$

If we wish to express the electric field in coordinates natural to a wave traveling in the ζ direction of Fig. 2.12, we write

$$\mathbf{E}(x, y, t) = \mathbf{diag}(-1, 1)\mathbf{E}(\xi, \eta, t)$$

Then, the power becomes

$$\begin{aligned} W &= \frac{1}{8R_a}\left[\mathbf{h}(x, y) \otimes \mathbf{h}^*(x, y)\right]^T \overline{\mathbf{diag}(-1, 1)\mathbf{E}(\xi, \eta, t) \otimes \mathbf{diag}(-1, 1)\mathbf{E}^*(\xi, \eta, t)} \\ &= \frac{1}{8R_a}\left[\mathbf{h}(x, y) \otimes \mathbf{h}^*(x, y)\right]^T \mathbf{diag}(1, -1, -1, 1)\overline{\mathbf{E}(\xi, \eta, t) \otimes \mathbf{E}^*(\xi, \eta, t)} \end{aligned} \tag{6.39}$$

The Kronecker product involving the electric field is the coherency vector of the incident wave, in its normal right-handed coordinate system. In (6.16), we defined a coherency vector for the antenna as

$$\mathbf{J}_A = \mathbf{h}(x, y) \otimes \mathbf{h}^*(x, y)$$

In terms of the coherency vectors of the receiving antenna and incident wave, the received power is

$$W = \frac{1}{8R_a}\mathbf{J}_A^T(x, y)\mathbf{diag}(1, -1, -1, 1)\mathbf{J}(\xi, \eta) \tag{6.40}$$

The coordinate notation can be omitted if it is remembered that the coherency vectors are to be expressed in right-handed coordinate systems whose z-axes are directed oppositely.

The received power can also be written, using $\mathbf{E}$ and $\mathbf{h}$ in the same coordinates, as

$$W = \frac{1}{8R_a}\mathbf{J}_A^T\overline{\mathbf{E}(t) \otimes \mathbf{E}^*(t)}$$

We do not call the Kronecker product of the field terms the coherency vector of the incident wave because the term "coherency vector" is used here to describe a wave in a right-handed coordinate system with the wave traveling in the direction of one of the axes.

If the Stokes vector of the receiving antenna, defined by (6.33), and the Stokes vector of the incident wave, given by (6.30), are used in (6.40), the received power can be written as

$$W = \frac{1}{16R_a}\mathbf{G}_A^T\mathbf{diag}(1, 1, -1, 1)\mathbf{G}$$

where $\mathbf{G}$ is the Stokes vector of the incident wave.

REFERENCES

M. J. Beran and G. B. Parrent, Jr., *Theory of Partial Coherence*, Prentice-Hall, Englewood Cliffs, NJ, 1964.

M. Born and E. Wolf, *Principles of Optics*, Pergamon Press, New York, 1965.

D. Gabor, "Theory of Communication", *J. Inst. Elect. Eng.*, **93** (III), 429–457, (1946).

H. Mott, *Antennas for Radar and Communication: A Polarimetric Approach*, Wiley-Interscience, New York, 1992.

P. Z. Peebles, *Probability, Random Variables, and Random Signal Principles*, McGraw-Hill, New York, 1980.

H. Urkowitz, *Signal Theory and Random Processes*, Artech House, Dedham, MA, 1983.

PROBLEMS

Note: A general-purpose mathematics program is desirable for solving some of the following problems.

6.1. Generate 100 random values of a Jones matrix using the rules,

$$\begin{aligned} T_{xx} &= -2 + 0.2C_1 \\ T_{xy} &= -T_{yx} = -(1 + 0.1C_2) - j(1 + 0.2C_3) \\ T_{yy} &= 4 + 0.2C_4 \end{aligned}$$

where $C_1 \ldots C_4$ are approximately Gaussian with zero mean and unit variance. If the magnitude of a C_i term is greater than 3σ, C_i is set to the 3σ magnitude.

6.2. (a) An amplitude-modulated signal is given by

$$V^r(t) = [1 + m(t)] \cos \overline{\omega} t$$

where $m(t)$ is much smaller than 1. Write the analytic signal representation of V^r in the strict form (6.6). (b) An angle-modulated signal is given by

$$V^r(t) = A \sin(\overline{\omega} t + m \sin \omega_m t)$$

Write the analytic signal representation of V^r in the form (6.9).

6.3. The Stokes vector of a wave is

$$\mathbf{G} = [\,6 \;\; 1 \;\; 2 \;\; 4\,]^T$$

Find the degree of polarization of the wave. Write $\mathbf{G}$ as the sum of the Stokes vectors for an unpolarized wave and a completely polarized wave. Find the elements of the two vectors.

6.4. Amplitude-modulated wave components are

$$\tilde{E}_x(z, t) = [1 + a \cos \omega t]e^{j(\overline{\omega}t - kz)}$$
$$\tilde{E}_y(z, t) = [1 + b \cos \omega t]e^{j(\overline{\omega}t - kz + c)}$$

where $\omega \ll \overline{\omega}$. Find the coherency vector of the wave and its degree of polarization.

6.5. A wave with the Stokes vector of Problem 6.3 is incident on an antenna that is impedance matched to its receiver and has effective length $\mathbf{h} = \mathbf{u}_x + (1 + j)\mathbf{u}_y$ and radiation resistance 40 Ω. $\mathbf{G}$ is in $\xi\eta\zeta$ coordinates of Fig. 2.12 and $\mathbf{h}$ in xyz. Find the received power.

6.6. A change of polarization basis for the electric field was discussed in Section 1.7, using transformation matrix $\mathbf{U}$. Develop the transformation for the coherency vector of a wave using the same transformation matrix.

6.7. Show that the determinant of a Hermitian positive semidefinite matrix is nonnegative.

CHAPTER 7

SCATTERING BY DEPOLARIZING TARGETS

If the scattering properties of a target vary with time because of changes in the target, such as movement of foliage or wave motion on bodies of water or changes in the intervening medium caused perhaps by precipitation, or if the target is examined from different aspect angles by multiple radar pulses, it is necessary to go beyond the Sinclair matrix description of the target. We will use the phrase "time-varying target" to describe one whose measured properties change for any of these reasons.

The wave scattered from a time-varying target is partially coherent and partially polarized; it is described by a vector whose elements are correlated complex random variables. The target information of interest is normally the received power or power density of the scattered wave, degree of polarization of the scattered wave, and the time variation of the target's scattering properties.

This chapter is concerned with the development of matrices characterizing a target in equations for received power and degree of polarization of a scattered wave. The chapter also includes brief discussions of variance and autocorrelation function of the received power. In all cases, the wave incident on the target is a single-frequency, completely polarized wave.

A large number of matrices have been used to describe a depolarizing target. These depend on the form desired for the received-power equation, the coordinate system adopted for the scattered wave, and the method of determining the matrix elements. The method of determining the matrix elements is critical, and some of the matrices cannot be compared to others because the elements are found in a different manner. The most widely used matrices to characterize a depolarizing target are the Mueller and Kennaugh matrices, which have real elements. The incident and scattered waves are described by Stokes vectors, also with real elements. Both matrices are discussed in detail in this chapter. Much of the work in the chapter is carried out, however, with complex target matrices, because they are more readily related to averages of products of a target's Sinclair or Jones matrix. The developments in all cases can be translated to forms using different matrices.

7.1. TARGETS

The simplest radar-target configuration has a stationary target, with no surface movement, which is examined by a fixed radar. It is sometimes called a *deterministic target* or a *point target*, but we will call it by the more descriptive name of *coherently scattering target*. Its scattered wave is monochromatic and completely polarized. The target can be represented by a Sinclair or Jones matrix. A coherent radar was defined in Chapter 4 as one in which the phase of the received pulse is retained and used in pulse integration. The description of a target as "coherently scattering" is consistent with the previous use, because the phase relationships of the Sinclair matrix elements of a coherently scattering target are the same from one pulse to the next, and coherent integration of received voltages is feasible. In Section 6.2, the degree of polarization of a wave was used to describe the scattering from a target. If the incident wave is completely polarized and if the degree of polarization of the scattered wave is 1, the target is coherently scattering.

If the target is in motion or if the radar-target relationship is changing, its electromagnetic description is less simple. The target's scattering surfaces may be moving relative to each other, as would be the situation with waves on water or with windblown foliage. The target may be moving, as would be an aircraft. The target may be stationary while the radar moves, as would be the situation with a terrain resolution cell that is examined by a synthetic aperture radar flying over the terrain. These targets are *incoherently scattering* or *depolarizing* targets. As with the coherently scattering target, other names are sometimes used, one being *distributed target*. The emphasis of this chapter is on the depolarized nature of the scattered waves, so we will call such target "depolarizing" to emphasize that characteristic. For greater precision, we should refer to a "depolarizing radar-target configuration", since the target itself might be a coherently scattering target and the radar receive a completely polarized wave if there were no changes in the radar-target relationship, but we use "depolarizing target" for convenience.

If the scattering-surface motions of the target are slow, or if the relative radar-target motion is slow, the target can be represented by a time-varying matrix having the same form as the Sinclair matrix. The Sinclair matrix description of a target is based on steady-state behavior of the illuminating and scattered waves. If the matrix is measured by a pulsed radar, the pulses must be sufficiently long to contain many cycles of the radar wave. On the other hand, each pulse must be short compared to the time required for significant target motions. These criteria help to establish which targets can be represented by $\mathbf{S}(t)$.

Let us suppose that multiple measurements are made of a time-varying target or of a target with a time-varying radar-target orientation to give $\mathbf{S}(n\Delta t)$. If $\mathbf{S}$ changes only by a small amount while the measurements are made, it is conceivable that its averaged value is useful; if the change is greater, the average value of $\mathbf{S}$ is of lesser use. The power, $W(n\Delta t)$, can be found in two ways, one of which is by measurement of the Sinclair matrix and the computation of power. The power found at each measurement is averaged. A preliminary step can be that of using each measurement of the Sinclair matrix to create another matrix from which the power is found.

A matrix characterizing the target in the received-power equation can be obtained directly by power measurements using different polarizations of transmitting and receiving antennas without the intervening step of measuring the Sinclair matrix. This measurement process is sometimes referred to as *incoherent* measurement of the matrix, in contrast to the coherent measurement of the Sinclair matrix. Phase angles are not measured in the incoherent method, while in the coherent measurement process they are.

A *completely depolarizing target* is one that scatters a wave whose degree of polarization is zero for all polarizations of the incident wave. The power to a radar receiver from a wave scattered from such a target will vary randomly over time, but the average power is constant, independent of the polarization characteristics of transmitting and receiving antennas. This definition does not explicitly include the time behavior of the target; if it is to be completely depolarizing for all incident waves, the scattering properties must vary during the radar pulse time, no matter how small, and from one pulse to the next, no matter how closely spaced in time the pulses are.

The distinction between a coherently scattering target and a depolarizing target depends on target characteristics, frequency of the illuminating wave, time required for a single radar measurement, integration time of the radar, polarization characteristics of the radar antennas, the uses to which the information will be put, and the desired accuracy. The same factors are involved in a decision to measure $\mathbf{S}(t)$ or to measure directly the elements of a 4×4 power matrix for the target. For example, a target that strongly scatters a horizontally polarized wave may appear to be a coherent scatterer with such illumination and a depolarizing scatterer when illuminated with a vertical wave. A target with small surface motions may appear depolarizing at high frequencies and coherently scattering at low. The matrices that describe the target at high and low frequencies may be unrelated.

If a target is described here as coherently scattering, the decision has been made that the Sinclair matrix is an applicable target descriptor; if the target is called depolarizing, a Sinclair matrix cannot adequately describe the target.

The matrices discussed here represent a depolarizing target in giving the received power and, for some matrices, the Stokes vector or coherency vector of the scattered wave. The time behavior of the target is also an important characteristic, and a discussion of it is given.

A matrix describing a coherently scattering target has 7 independent real parameters, or 5 for backscattering, while a matrix representing a depolarizing target has 16 independent real parameters for bistatic scattering and 10 for backscattering.

It is important to recognize that the physical target cannot be determined from a target matrix, even for a coherently scattering target, since more than one physical target can have the same scattering matrix. Even if knowledge of the target matrix is coupled with knowledge of the target's scale of time variation, the physical target causing the scattering cannot be determined.

Measurement Considerations

A target matrix is obtained by radiating waves of appropriate polarization toward the target and measuring the scattered waves. Sequential transmission of two orthogonally polarized pulses and measurement of two orthogonally polarized return signals for each transmission is sufficient for one measurement of the Sinclair matrix. N pairs of pulses are transmitted, either from the same point or from different locations, and the received signals are summed to form an overall matrix. For a constant target and unchanging radar-target aspect, all measurements of the Sinclair matrix, or other matrix, are the same, and the target matrix is independent of the radar pulse length, pulse period, and number of pulses summed. The resulting matrix is *radar independent*.

If the scattering properties of the target vary with time or if the radar-target geometry changes with time, the received signals are correlated random processes. A careful balance of factors is necessary if the time average of one of these signals is to represent a target adequately. Transmitted pulses must be short compared to a time in which significant target motions take place; otherwise some averaging will take place during a pulse. They must, however, be long enough when compared to a cycle of the transmitted wave for the time-invariant Maxwell equations to be valid, if the Sinclair matrix is to be utilized. The pulse repetition time of the radar must meet the requirements of the sampling theorem. This requires a short repetition time if the target return varies rapidly. On the other hand, since the average will be of a finite number of samples, it will approximate the true time average more closely if the time between measurements is greater than the signal decorrelation time; that is, if the measurements are independent. These conflicting requirements can be reconciled to some degree by using many samples and a long overall measurement time. If all these requirements cannot be met with a given target and radar, the resulting target matrix is *target and radar dependent*.

Subsequently, it will be noted that CW radar measurements can describe a target adequately and are radar independent.

Notation: Power

In this chapter, equations for received power are given that use target matrices formed by disparate methods. The powers cannot in general be equated. There are exceptions that will be apparent from subsequent developments.

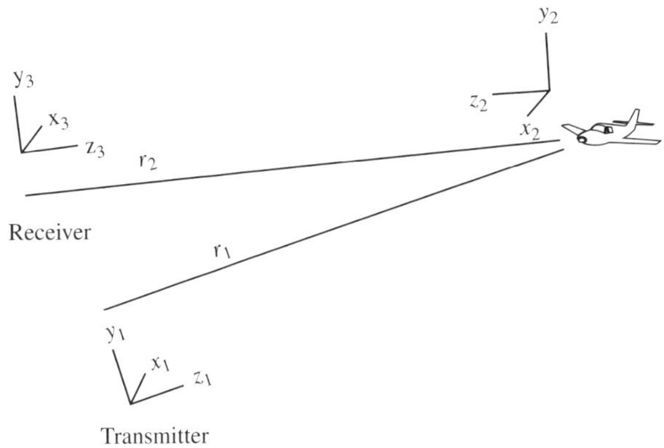

Fig. 7.1. Coordinate systems for scattering.

We will use as symbols for power:

W_c: Coherently-scattering target.

W_{av}: Depolarizing target with averaged matrix formed from **S** or **T**.

W_d: Depolarizing target using power measurements to determine matrix elements.

These distinctions will be clarified subsequently. Equations for power with the same subscripts can be compared, while those with different subscripts cannot be. Corresponding subscripts will be used for coherency and Stokes vectors.

Coordinate Systems

We use the three right-handed coordinate systems shown in Fig. 7.1. For brevity, they are referred to as:

System 1: $x_1y_1z_1$

System 2: $x_2y_2z_2$

System 3: $x_3y_3z_3$

7.2. AVERAGING THE SINCLAIR MATRIX

If the scattered wave has a sufficiently high degree of polarization, the target can be represented with small error by an averaged Sinclair or Jones matrix. If N measurements of the Sinclair matrix are coherently averaged, with $S_{xx,n}$ chosen as a phase reference, the averaged value is

$$\mathbf{S}_{\mathrm{av}} = \frac{1}{N}\sum_{1}^{N}\begin{bmatrix} |S_{xx,n}| & |S_{xy,n}|e^{j\beta_{xy,n}} \\ |S_{yx,n}|e^{j\beta_{yx,n}} & |S_{yy,n}|e^{j\beta_{yy,n}} \end{bmatrix}$$

where the phase angles are with respect to that of $S_{xx,n}$. The coordinate systems used are 1 and 3. For targets whose scattered waves have a high degree of polarization, the Sinclair matrix elements are strongly correlated in amplitude and phase from one pulse to the next, and coherent summing and averaging gives a Sinclair matrix that represents the target well. The matrix elements can be normalized, and various normalizing methods have been suggested (Krogager, 1993).

7.3. THE KRONECKER-PRODUCT MATRICES

The Kronecker product or direct product was defined by (1.41) for two-element vectors. The Kronecker product of 2×2 matrices is needed for subsequent developments. It is

$$\mathbf{A} \otimes \mathbf{B} = \begin{bmatrix} A_{11} & A_{12} \\ A_{21} & A_{22} \end{bmatrix} \otimes \begin{bmatrix} B_{11} & B_{12} \\ B_{21} & B_{22} \end{bmatrix} = \begin{bmatrix} A_{11}\mathbf{B} & A_{12}\mathbf{B} \\ A_{21}\mathbf{B} & A_{22}\mathbf{B} \end{bmatrix}$$

$$= \begin{bmatrix} A_{11}B_{11} & A_{11}B_{12} & A_{12}B_{11} & A_{12}B_{12} \\ A_{11}B_{21} & A_{11}B_{22} & A_{12}B_{21} & A_{12}B_{22} \\ A_{21}B_{11} & A_{21}B_{12} & A_{22}B_{11} & A_{22}B_{12} \\ A_{21}B_{21} & A_{21}B_{22} & A_{22}B_{21} & A_{22}B_{22} \end{bmatrix}$$

Useful identities, the first for scalars and the second for vectors or matrices, are (Pease, 1965, p. 324)

$$A \otimes B = AB \tag{7.1}$$

$$(\mathbf{A} \otimes \mathbf{B})(\mathbf{C} \otimes \mathbf{D}) = \mathbf{AC} \otimes \mathbf{BD} \tag{7.2}$$

The products must be meaningful.

Coherently Scattering Target Matrix: Sinclair Matrix Elements

Let the scattered wave whose electric field $\mathbf{E}^s$ is given by (3.4) fall on the receiving antenna of Fig. 7.1. Field $\mathbf{E}^s$ is in x_3 and y_3 components, and we think of the scattered wave as traveling in the opposite direction from z_3 in the right-handed coordinate system $x_3y_3z_3$. The received voltage is

$$V = \mathbf{h}_r^T \mathbf{E}^s = \frac{1}{\sqrt{4\pi}\, r_2} e^{-jkr_2} \mathbf{h}_r^T \mathbf{S} \mathbf{E}^i$$

with the receiving antenna effective length $\mathbf{h}_r$ also in system 3. The power to a receiver with resistance R_a is

$$W_c = \frac{1}{32\pi R_a r_2^2} (\mathbf{h}_r^T \mathbf{S} \mathbf{E}^i)(\mathbf{h}_r^T \mathbf{S} \mathbf{E}^i)^*$$

The incident wave in terms of the effective length of the transmitting antenna, taken from (3.3), is

$$\mathbf{E}^i = \frac{jZ_0 I}{2\lambda r_1} e^{-jkr_1}\mathbf{h}_t$$

With its use, the power equation becomes

$$W_c = \frac{1}{32\pi R_a r_2^2}(\mathbf{h}_r^T \mathbf{S}\mathbf{E}^i)(\mathbf{h}_r^T \mathbf{S}\mathbf{E}^i)^* = \frac{Z_0^2 I^2}{128\pi R_a \lambda^2 r_1^2 r_2^2}(\mathbf{h}_r^T \mathbf{S}\mathbf{h}_t) \otimes (\mathbf{h}_r^T \mathbf{S}\mathbf{h}_t)^* \tag{7.3}$$

It can be shown that

$$\mathbf{A}^T \otimes \mathbf{B}^T = (\mathbf{A} \otimes \mathbf{B})^T \tag{7.4}$$

If this identity and (7.2) are used, (7.3) becomes

$$W_c = \frac{Z_0^2 I^2}{128\pi R_a \lambda^2 r_1^2 r_2^2}(\mathbf{h}_r \otimes \mathbf{h}_r^*)^T (\mathbf{S} \otimes \mathbf{S}^*)(\mathbf{h}_t \otimes \mathbf{h}_t^*) \tag{7.5}$$

The Kronecker products in this form have complex elements, but W_c is real.

There is an alternative form for received power. If the transpose of the second scalar in (7.3) is taken, the equation becomes

$$\begin{aligned} W_c &= \frac{Z_0^2 I^2}{128\pi R_a \lambda^2 r_1^2 r_2^2}(\mathbf{h}_r^T \mathbf{S}\mathbf{h}_t) \otimes (\mathbf{h}_t^\dagger \mathbf{S}^\dagger \mathbf{h}_r^*) \\ &= \frac{Z_0^2 I^2}{128\pi R_a \lambda^2 r_1^2 r_2^2}(\mathbf{h}_r \otimes \mathbf{h}_t^*)^T (\mathbf{S} \otimes \mathbf{S}^\dagger)(\mathbf{h}_t \otimes \mathbf{h}_r^*) \end{aligned}$$

This form does not keep transmitting and receiving antenna effective lengths separate and is of less utility than the first form.

We use (7.5) and define the *Kronecker-product target matrix* by

$$\boldsymbol{\kappa}_S = \mathbf{S} \otimes \mathbf{S}^* = \begin{bmatrix} |S_{xx}|^2 & S_{xx}S_{xy}^* & S_{xx}^*S_{xy} & |S_{xy}|^2 \\ S_{xx}S_{yx}^* & S_{xx}S_{yy}^* & S_{xy}S_{yx}^* & S_{xy}S_{yy}^* \\ S_{xx}^*S_{yx} & S_{xy}^*S_{yx} & S_{xx}^*S_{yy} & S_{xy}^*S_{yy} \\ |S_{yx}|^2 & S_{yx}S_{yy}^* & S_{yx}^*S_{yy} & |S_{yy}|^2 \end{bmatrix} \tag{7.6}$$

With this definition, the received power becomes

$$W_c = \frac{Z_0^2 I^2}{128\pi R_a \lambda^2 r_1^2 r_2^2}(\mathbf{h}_r \otimes \mathbf{h}_r^*)^T \boldsymbol{\kappa}_S (\mathbf{h}_t \otimes \mathbf{h}_t^*) \tag{7.7}$$

Note that $\boldsymbol{\kappa}_s$ must have the same number of independent real parameters as **S**, seven. Matrix $\boldsymbol{\kappa}_s$ is symmetric for backscattering.

Basis Transform of κ_S

It was shown in Section 3.10 that the Sinclair matrix transforms under a change of basis as

$$\mathbf{S}' = \mathbf{U}^*\mathbf{S}\mathbf{U}^\dagger$$

where $\mathbf{U}$ is a unitary matrix given by (1.37). The Kronecker-product target matrix can be transformed to a new basis by

$$\begin{aligned}\boldsymbol{\kappa}'_S &= \mathbf{S}' \otimes \mathbf{S}'^* = (\mathbf{U}^*\mathbf{S}\mathbf{U}^\dagger) \otimes (\mathbf{U}\mathbf{S}^*\mathbf{U}^T) = (\mathbf{U}^* \otimes \mathbf{U})(\mathbf{S} \otimes \mathbf{S}^*)(\mathbf{U}^* \otimes \mathbf{U})^T \\ &= (\mathbf{U}^* \otimes \mathbf{U})\boldsymbol{\kappa}_S\, (\mathbf{U}^* \otimes \mathbf{U})^T\end{aligned}$$

This is a consimiliarity transform. The same transformation is applicable to matrices $\mathbf{D}_s$ and $\langle \boldsymbol{\kappa}_S \rangle$ defined in Sections 7.4 and 7.5.

Coherently Scattering Target Matrix: Jones Matrix Elements

The matrix $\boldsymbol{\kappa}_S$ is formed from the Sinclair matrix $\mathbf{S}$, which uses coordinate system 3 to represent the scattered wave. This is contrary to the normal manner of describing a wave, which utilizes an axis pointing in the direction of wave travel. For that reason it is desirable, if the characteristics of the scattered wave are to be considered, to use the Jones matrix $\mathbf{T}$ for a coherently scattering target rather than $\mathbf{S}$. The scattered field at the receiver in a Jones matrix formulation is

$$\mathbf{E}^s = \frac{1}{\sqrt{4\pi}\, r_2} e^{-jkr_2}\mathbf{T}\mathbf{E}^i = \frac{jZ_0 I}{2\sqrt{4\pi}\,\lambda r_1 r_2} e^{-jk(r_1+r_2)}\mathbf{T}\mathbf{h}_t \tag{7.8}$$

$\mathbf{E}^s$ has x_2 and y_2 components. The received voltage is

$$V = \left[\mathbf{diag}(-1, 1)\mathbf{h}_r\right]^T \frac{jZ_0 I}{2\sqrt{4\pi}\,\lambda r_1 r_2} e^{-jk(r_1+r_2)}\mathbf{T}\mathbf{h}_t \tag{7.9}$$

If a Kronecker product target matrix based on the Jones matrix is defined as

$$\boldsymbol{\kappa}_T = \mathbf{T} \otimes \mathbf{T}^* \tag{7.10}$$

the received power from a coherently scattering target, found by using (7.9), is

$$W_c = \frac{VV^*}{8R_a} = \frac{Z_0^2 I^2}{128\pi R_a \lambda^2 r_1^2 r_2^2}(\mathbf{h}_r \otimes \mathbf{h}_r^*)^T \mathbf{diag}(1, -1, -1, 1)\boldsymbol{\kappa}_T\, (\mathbf{h}_t \otimes \mathbf{h}_t^*) \tag{7.11}$$

The relationship between $\mathbf{S}$ and $\mathbf{T}$ can be used to obtain

$$\boldsymbol{\kappa}_T = \mathbf{diag}(1, -1, -1, 1)\boldsymbol{\kappa}_S \tag{7.12}$$

$\boldsymbol{\kappa}_T$ has seven independent real parameters.

7.4. MATRICES FOR A DEPOLARIZING TARGET: COHERENT MEASUREMENT

Consider a target whose scattered wave varies with time, either because the target properties vary or because the target is viewed from a different aspect angle with successive radar pulses. Measurement of the target with a radar pulse yields a Sinclair matrix **S**, and $\boldsymbol{\kappa}_S$ can be formed from each measurement of **S**.

The Averaged Kronecker Product Matrix

A useful matrix, the averaged Kronecker-product matrix $\langle\boldsymbol{\kappa}_S\rangle$ for a depolarizing target can be formed by averaging multiple measurements of $\boldsymbol{\kappa}_S$ made at intervals as the target changes or the radar-target aspect changes. It follows from (7.7) that the received power from a depolarizing target is

$$W_{\text{av}} = \frac{Z_0^2 I^2}{128\pi R_a \lambda^2 r_1^2 r_2^2} (\mathbf{h}_r \otimes \mathbf{h}_r^*)^T \langle \boldsymbol{\kappa}_S \rangle (\mathbf{h}_t \otimes \mathbf{h}_t^*) \tag{7.13}$$

where

$$\langle \boldsymbol{\kappa}_S \rangle = \frac{1}{N} \sum_{1}^{N} \boldsymbol{\kappa}_{Sn}$$

The angle brackets are used here and subsequently to symbolize pulse averaging.

If $\boldsymbol{\kappa}_T$ is averaged, the power using it is, from (7.11),

$$W_{\text{av}} = \frac{Z_0^2 I^2}{128\pi R_a \lambda^2 r_1^2 r_2^2} (\mathbf{h}_r \otimes \mathbf{h}_r^*)^T \mathbf{diag}(1, -1, -1, 1) \langle \boldsymbol{\kappa}_T \rangle (\mathbf{h}_t \otimes \mathbf{h}_t^*) \tag{7.14}$$

The target matrices formed by measurements of **S** or **T** followed by the formation of $\langle\boldsymbol{\kappa}_S\rangle$ or $\langle\boldsymbol{\kappa}_T\rangle$ are *coherently measured* matrices. They may also be referred to in this work as *indirectly measured* target matrices.

The Kennaugh Matrix

A target can be represented by a matrix having only real elements and the antennas by real vectors. If **Q** defined by (1.43) is used, (7.13) can be written as

$$W_{\text{av}} = \frac{Z_0^2 I^2}{256\pi R_a \lambda^2 r_1^2 r_2^2} \left[\mathbf{Q}(\mathbf{h}_r \otimes \mathbf{h}_r^*)\right]^T \mathbf{Q}^* \langle \boldsymbol{\kappa}_S \rangle \mathbf{Q}^{-1} \mathbf{Q}(\mathbf{h}_t \otimes \mathbf{h}_t^*)$$

The product,

$$\mathbf{G}_{Aj} = \mathbf{Q}(\mathbf{h}_j \otimes \mathbf{h}_j^*) = \begin{bmatrix} |h_{jx}|^2 + |h_{jy}|^2 \\ |h_{jx}|^2 - |h_{jy}|^2 \\ 2\text{Re}(h_{jx}^* h_{jy}) \\ 2\text{Im}(h_{jx}^* h_{jy}) \end{bmatrix} \qquad j = t, r \tag{7.15}$$

is the Stokes vector of the antenna, as defined in Section 6.2. If we define the Kennaugh matrix of the target as

$$\mathbf{K} = \mathbf{Q}^* \langle \kappa_S \rangle \mathbf{Q}^{-1} \tag{7.16}$$

the received power is

$$W_{\text{av}} = \frac{Z_0^2 I^2}{256\pi R_a \lambda^2 r_1^2 r_2^2} \mathbf{G}_{Ar}^T \mathbf{K} \mathbf{G}_{At} \tag{7.17}$$

$\mathbf{K}$ is symmetric for backscattering.

The Kennaugh matrix is perhaps the most widely used matrix in radar polarimetry for describing a depolarizing target. It is sometimes called the "Stokes reflection matrix", and a factor of one-half may also be included in the definition. All elements of the Kennaugh matrix are real. They are given in terms of the Sinclair matrix elements in Appendix C.

An Alternative Matrix Form

We began the development of the Kennaugh matrix form for power with the averaged Kronecker-product matrix $\langle \boldsymbol{\kappa}_S \rangle$. A parallel development using $\langle \boldsymbol{\kappa}_T \rangle$, defined in coordinate systems 1 and 2 of Fig. 7.1, leads to an alternative formulation, also using real vectors and a real matrix. In (7.14), use the antenna Stokes vectors of (7.15). This gives

$$W_{\text{av}} = \frac{Z_0^2 I^2}{128\pi R_a \lambda^2 r_1^2 r_2^2} \mathbf{G}_{Ar}^T \,\mathbf{diag}(1, 1, -1, 1) \mathbf{M}_{\text{av}} \mathbf{G}_{At} \tag{7.18}$$

where

$$\mathbf{M}_{\text{av}} = \mathbf{Q} \langle \kappa_T \rangle \mathbf{Q}^{-1} \tag{7.19}$$

We will see in Section 7.5 that the equation for received power using the target's *Mueller matrix* is similar to (7.18). Because of the manner in which it is formed and because it obeys a power equation like that obeyed by the Mueller matrix, we refer to $\mathbf{M}_{\text{av}}$ as the *averaged Mueller matrix* or as the *radar Mueller matrix*.

7.5. INCOHERENTLY MEASURED TARGET MATRICES

The measurement procedure outlined for finding $\langle \boldsymbol{\kappa}_S \rangle$ and $\langle \boldsymbol{\kappa}_T \rangle$ of a depolarizing target was referred to previously as a coherent or indirect measurement. In this section, we discuss an *incoherent* or *direct* method for obtaining matrices analogous to $\langle \boldsymbol{\kappa}_S \rangle$ and $\langle \boldsymbol{\kappa}_T \rangle$.

A Direct Determination of the Matrix Elements

Let the power received from a wave scattered by a depolarizing target be

$$W_d = \frac{Z_0^2 I^2}{128\pi R_a \lambda^2 r_1^2 r_2^2} (\mathbf{h}_r \otimes \mathbf{h}_r^*)^T \mathbf{D}_S (\mathbf{h}_t \otimes \mathbf{h}_t^*) \tag{7.20}$$

This equation is the same as (7.13) with $\mathbf{D}_S$ substituted for $\langle \boldsymbol{\kappa}_S \rangle$. However, the equation is intended to stand on its own, and W_d is not necessarily equal to that given by (7.13). The symbol $\mathbf{D}$ for the matrix is a reminder that the matrix elements are determined by direct measurement. Subscript S indicates that $\mathbf{D}_S$ uses the same coordinate systems, 1 and 3 of Fig. 7.1, as the Sinclair matrix. It does not imply that a relationship exists between $\mathbf{D}_S$ and $\mathbf{S}$.

To find the elements of $\mathbf{D}_S$, appropriate pairs of transmitting and receiving antennas are selected and the received power measured for each pair. The measurement may be done by using CW illumination. The received power is a random process, and W_d in (7.20) is a time average of the random process over a time interval long enough to allow a full range of radar-target motions to take place. Matrix $\mathbf{D}_S$ therefore inherently involves time averages of random processes. Ergodicity and stationarity are assumed. Note that not all elements of $\mathbf{D}_S$ are formed from the same sample waveform, since multiple measurements with different polarizations of the transmitted wave are needed. The measurement can also be carried out by sampling the random waveforms, giving

$$W_d = \frac{Z_0^2 I^2}{128\pi R_a \lambda^2 r_1^2 r_2^2} (\mathbf{h}_r \otimes \mathbf{h}_r^*)^T \overline{\mathbf{d}_S(t)} (\mathbf{h}_t \otimes \mathbf{h}_t^*)$$

where $\mathbf{d}_S$ is the matrix of sampled values and the overline indicates time averaging. The measurement procedure must meet criteria given in Section 7.1.

We note again that the scattered wave has a frequency spread, and the separation of target matrix and antenna effective lengths in the power equation requires that the frequency spread be small enough to allow the receiving antenna effective length $\mathbf{h}_r$ to be constant.

We can also write an equation for power like that of (7.14) with the substitution of $\mathbf{D}_T$ for the averaged value $\langle \boldsymbol{\kappa}_T \rangle$. It is

$$W_d = \frac{Z_0^2 I^2}{128\pi R_a \lambda^2 r_1^2 r_2^2} (\mathbf{h}_r \otimes \mathbf{h}_r^*)^T \mathbf{diag}(1, -1, -1, 1) \mathbf{D}_T (\mathbf{h}_t \otimes \mathbf{h}_t^*) \tag{7.21}$$

Subscript T is a reminder that the same coordinate systems are used for $\mathbf{D}_T$ and $\mathbf{T}$. Other than as an indication of the coordinate systems used, the subscripts T and S should not be construed as having significance.

It has only been conjectured at this point that $\mathbf{D}_T$ and $\mathbf{D}_S$ can represent a depolarizing target in equations for power. We must show that they can be found as constant matrices that give the correct power for any polarization of transmitting and receiving antennas. Moreover, the power equation must be valid for any target, coherently scattering, depolarizing, or completely depolarizing. It has also only been conjectured that $\langle \boldsymbol{\kappa}_T \rangle$ and $\langle \boldsymbol{\kappa}_S \rangle$ can represent a depolarizing target. Their ability to do so must also be examined.

Effect of Pulse Length

In determining the target matrix elements of (7.13), assume that the transmitted pulse is not short when compared to the time required for significant changes in the target scattering properties. With this assumption, the measured value of the Sinclair matrix is an average value over the pulse, $\mathbf{S}_{\mathrm{av}}$. The desired matrix is $\langle \boldsymbol{\kappa}_S \rangle = \langle \mathbf{S} \otimes \mathbf{S}^* \rangle$, but what is measured is $\langle \mathbf{S}_{\mathrm{av}} \otimes \mathbf{S}_{\mathrm{av}}^* \rangle$. They are not equal, and it follows that for a long pulse and a target that depolarizes an incident wave, the target matrix that is measured is dependent on the radar used.

Measurement Simultaneity

The formation of the Sinclair matrix requires that the elements be measured simultaneously. The transmission of multiple pulses and time-interpolation can effectively reduce the time intervals between measurements to a very small value. Simultaneous measurements of the Sinclair matrix can be made if orthogonally coded pulses are transmitted simultaneously from two orthogonally polarized antennas and if the received signals are separated by means of their coded nature. The required signal processing is extensive.

Element Relationships

The elements of $\mathbf{D}_S$ and $\mathbf{D}_T$ are not all independent. We can introduce element relationships in $\mathbf{D}_S$ by requiring that received power be real and nonnegative. Received power will be real if we require in (7.20) that

$$\mathbf{D}_S = \mathbf{q}\mathbf{D}_S^*\mathbf{q} \tag{7.22}$$

where

$$\mathbf{q} = \begin{bmatrix} 1 & 0 & 0 & 0 \\ 0 & 0 & 1 & 0 \\ 0 & 1 & 0 & 0 \\ 0 & 0 & 0 & 1 \end{bmatrix}$$

Expansion of (7.22) gives

$$\mathbf{D}_S = \begin{bmatrix} D_{S11} & D_{S12} & D_{S13} & D_{S14} \\ D_{S21} & D_{S22} & D_{S23} & D_{S24} \\ D_{S31} & D_{S32} & D_{S33} & D_{S34} \\ D_{S41} & D_{S42} & D_{S43} & D_{S44} \end{bmatrix} = \begin{bmatrix} D_{S11}^* & D_{S13}^* & D_{S12}^* & D_{S14}^* \\ D_{S31}^* & D_{S33}^* & D_{S32}^* & D_{S34}^* \\ D_{S21}^* & D_{S23}^* & D_{S22}^* & D_{S24}^* \\ D_{S41}^* & D_{S43}^* & D_{S42}^* & D_{S44}^* \end{bmatrix} \tag{7.23}$$

It can be seen from (7.23) that the 4 corner elements of $\mathbf{D}_S$ are real and nonnegative. Six of the noncorner elements are conjugates of the remaining six. The Mueller matrix, discussed subsequently in this section, is known to have 16 independent real elements in general (Azzam and Bashara, 1977, p. 149). The Mueller matrix is a transform of $\mathbf{D}_S$, which, therefore, must also have 16 independent real parameters. It follows that the noncorner elements of $\mathbf{D}_S$ are, in general, complex, and 6 of the noncorner elements are independent. Then $\mathbf{D}_S$ has 4 independent real elements and 6 independent complex elements.

For backscattering, there are additional relationships among the matrix elements. Note that $\langle \boldsymbol{\kappa}_S \rangle$ is symmetric for backscattering. Equation 7.13 then predicts that received power is not altered if transmitting and receiving antennas are interchanged, and the prediction is verified by power measurements, direct or indirect. Then $\mathbf{D}_S$ is symmetric for backscattering, and the following relationships are valid, in addition to those that apply for bistatic scattering:

$$D_{S41} = D_{S14}$$

$$D_{S21} = D_{S12}$$

$$D_{S42} = D_{S24}$$

$$D_{S23} \text{ is real}$$

One real and two complex elements are no longer independent, and a complex element for bistatic scattering is real for backscattering. The number of independent real parameters in $\mathbf{D}_S$ is reduced from 16 to 10; they appear as 3 real corner elements, and 3 complex and 1 real noncorner elements.

It is important to note that $\mathbf{D}_S$ is not a Kronecker product, either for bistatic scattering or backscattering. The elements of the Kronecker product $\boldsymbol{\kappa}_S$ in (7.6) obey the relationships found above for $\mathbf{D}_S$, but $\mathbf{D}_S$ does not obey all the constraints of $\boldsymbol{\kappa}_S$.

A comparison of the power equations shows that $\mathbf{D}_T$ is related to $\mathbf{D}_S$ by

$$\mathbf{D}_T = \mathbf{diag}(1, -1, -1, 1)\mathbf{D}_S \tag{7.24}$$

If this relationship is used in (7.22) and the indicated multiplication carried out, it will be found that the elements of $\mathbf{D}_T$ obey the same relationships as those of $\mathbf{D}_S$, or

$$\mathbf{D}_T = \begin{bmatrix} D_{T11} & D_{T12} & D_{T13} & D_{T14} \\ D_{T21} & D_{T22} & D_{T23} & D_{T24} \\ D_{T31} & D_{T32} & D_{T33} & D_{T34} \\ D_{T41} & D_{T42} & D_{T43} & D_{T44} \end{bmatrix} = \begin{bmatrix} D^*_{T11} & D^*_{T13} & D^*_{T12} & D^*_{T14} \\ D^*_{T31} & D^*_{T33} & D^*_{T32} & D^*_{T34} \\ D^*_{T21} & D^*_{T23} & D^*_{T22} & D^*_{T24} \\ D^*_{T41} & D^*_{T43} & D^*_{T42} & D^*_{T44} \end{bmatrix} \tag{7.25}$$

We saw that the corner elements of $\mathbf{D}_S$ must be real and nonnegative and that it has 4 independent real elements and 6 complex ones. Both findings are true for $\mathbf{D}_T$.

If the symmetry of $\mathbf{D}_S$ for backscattering and the relationship between $\mathbf{D}_S$ and $\mathbf{D}_T$ are used, it will be found that the elements of $\mathbf{D}_T$ must obey, in addition to the element relationships for bistatic scattering,

$$D_{T41} = D_{T14} \tag{7.26}$$

$$D_{T21} = -D_{T12} \tag{7.27}$$

$$D_{T42} = -D_{T24} \tag{7.28}$$

$$D_{T23} \text{ is real} \tag{7.29}$$

For backscattering, $\mathbf{D}_T$ has 10 independent real parameters: 3 real corner elements, and 3 complex and 1 real noncorner elements.

Coherency and Stokes Vectors of the Scattered Wave

The coherency vector of a single-frequency plane wave was defined in Section 1.9 as the Kronecker product of the time-invariant electric field vector with its conjugate. If the electric field is that of a pulsed system, an averaged coherency vector is

$$\mathbf{J}_{\text{av}} = \langle \mathbf{E} \otimes \mathbf{E}^* \rangle \tag{7.30}$$

If the relationship (3.2) between incident and scattered waves is used, together with (7.10), the averaged coherency vector of the scattered wave is

$$\mathbf{J}^s_{\text{av}} = \langle \mathbf{E}^s \otimes \mathbf{E}^{s*} \rangle = \frac{1}{4\pi r^2} \langle \kappa_T \rangle (\mathbf{E}^i \otimes \mathbf{E}^{i*})$$

The coherency vector of a partially polarized wave was defined by (6.11) as the time average $\overline{\mathbf{E}(t) \otimes \mathbf{E}^*(t)}$, where $\mathbf{E}(t)$ is the analytic signal representation of the field. The time-varying field in (6.11) can be measured by sampling if the criteria of Section 7.1 are observed. Note that the relative phases must be measured, so coherent measurement is required. Coherency vector $\mathbf{J}$ of (6.11) will not be equal to $\mathbf{J}_{\text{av}}$ as determined by measurements and use in (7.30) because the electric fields are measured in a different manner.

If the received power of (7.21) is compared to power given by (6.39), with the interpretation that $\mathbf{E}(t)$ in (6.39) is that of the scattered wave at the receiver, it is apparent that the coherency vector of the scattered wave is

$$\mathbf{J}^s = \frac{Z_0^2 I^2}{16\pi\lambda^2 r_1^2 r_2^2} \mathbf{D}_T (\mathbf{h}_t \otimes \mathbf{h}_t^*) = \frac{1}{4\pi r^2} \mathbf{D}_T (\mathbf{E}^i \otimes \mathbf{E}^{i*}) \tag{7.31}$$

It can be seen that $\mathbf{J}^s$ and $\mathbf{J}^s_{\text{av}}$ are corresponding vectors describing the scattered wave. Vector $\mathbf{J}^s_{\text{av}}$ is appropriate for use if the target matrix is formed coherently, and $\mathbf{J}^s$ is to be used if the incoherently measured matrix $\mathbf{D}_T$ is the target descriptor. The coherency vectors can be written in terms of target matrices $\mathbf{D}_S$ and $\langle \boldsymbol{\kappa}_S \rangle$ with the relationships (7.24) and (7.37).

Stokes vectors of the scattered wave are readily determined with the transform $\mathbf{G} = \mathbf{QJ}$. Doing so yields

$$\mathbf{G}^s = \frac{Z_0^2 I^2}{16\pi \lambda^2 r_1^2 r_2^2} \mathbf{M}\mathbf{G}_{At}$$

$$\mathbf{G}_{\text{av}}^s = \frac{Z_0^2 I^2}{16\pi \lambda^2 r_1^2 r_2^2} \mathbf{M}_{\text{av}}\mathbf{G}_{At}$$

The degree of polarization of a wave in terms of its coherency vector was given by (6.27) and in terms of its Stokes vector by (6.36). The vectors $\mathbf{J}_{\text{av}}^s$ and $\mathbf{G}_{\text{av}}^s$ can also be used to find a degree of polarization for the scattered wave, using the same equations.

Validity of $\mathbf{D}_T$ and $\mathbf{D}_S$

A coherently scattering target can be represented by a Sinclair or Jones matrix, both invariant to the polarization of receiving and transmitting antennas. Since $\boldsymbol{\kappa}_S$ and $\boldsymbol{\kappa}_T$ are Kronecker products of $\mathbf{S}$ and $\mathbf{T}$, they also can represent a coherently scattering target and give the correct received power for all antenna polarizations.

Both $\mathbf{D}_S$ and $\mathbf{D}_T$ are determined by selecting a finite number of antenna polarizations and measuring the received power. It is of interest to consider if the resulting matrices are valid representations of a depolarizing target for antenna polarizations other than the pairs used to create the matrices. In Section 7.6, we will consider the same question for $\langle \boldsymbol{\kappa}_S \rangle$ and $\langle \boldsymbol{\kappa}_T \rangle$. First, note that if $\mathbf{D}_S$ and $\mathbf{D}_T$ are found by direct measurement of the power received from a coherently scattering target, they will be the same as $\boldsymbol{\kappa}_S$ and $\boldsymbol{\kappa}_T$. It follows that $\mathbf{D}_S$ and $\mathbf{D}_T$ are independent of antenna polarizations if the target is a coherent scatterer.

Consider an idealized completely depolarizing target. It was defined in Section 7.1 as one whose scattered wave is unpolarized and for which received power is independent of the polarizations of transmitting and receiving antennas. Equations 7.20 and 7.21 predict constant received power if and only if

$$\mathbf{D}_S = \mathbf{D}_T = A \begin{bmatrix} 1 & 0 & 0 & 1 \\ 0 & 0 & 0 & 0 \\ 0 & 0 & 0 & 0 \\ 1 & 0 & 0 & 1 \end{bmatrix} \tag{7.32}$$

We conclude that this matrix correctly represents a completely depolarizing target for all antenna polarizations.

The coherency vector of a partially polarized wave can be separated to be the sum of the coherency vectors of an unpolarized wave and a completely polarized wave (Born and Wolf, 1965, p. 551). We may then write (7.31) as

$$\mathbf{J}^s = \mathbf{J}^{s(1)} + \mathbf{J}^{s(2)} = \frac{Z_0^2 I^2}{16\pi \lambda^2 r_1^2 r_2^2} \mathbf{D}_T (\mathbf{h}_t \otimes \mathbf{h}_t^*)$$

$$= \frac{Z_0^2 I^2}{16\pi \lambda^2 r_1^2 r_2^2} \left(\mathbf{D}_T^{(1)} + \mathbf{D}_T^{(2)} \right) (\mathbf{h}_t \otimes \mathbf{h}_t^*) \tag{7.33}$$

where superscripts (1) and (2) for the coherency vectors refer respectively to unpolarized and completely polarized states of the scattered wave. We may think of $\mathbf{D}_T^{(1)}$ as being that part of the target matrix that produces the unpolarized wave represented by $\mathbf{J}^{s(1)}$, but the separation is not unique. This is discussed in greater detail in Chapter 9.

Consider the class of scatterers for which a member can be thought of as a coherently scattering target combined with a completely depolarizing target, and for which $\mathbf{D}_T^{(1)}$ is independent of antenna polarizations and given by the matrix of (7.32). The signals from the coherently scattering part of the target and the completely depolarizing part are uncorrelated, and their coherency vectors are additive. With this choice, $\mathbf{D}_T^{(1)}$ and $\mathbf{D}_T^{(2)}$ are independent of the polarization of $\mathbf{h}_t$. The separation of (7.33) is valid for all antenna polarizations, and $\mathbf{D}_T^{(2)}$ is a matrix representing a coherently scattering target. Matrix $\mathbf{D}_T$ is independent of the antenna polarization. We conclude that a target that is the combination of a coherently scattering target and an independent completely depolarizing target can be represented by matrix $\mathbf{D}_T$, or $\mathbf{D}_S$, which is independent of the polarization of transmitting and receiving antennas. We cannot, at this point, say that a general depolarizing target can be so represented.

The Mueller Matrix

The matrices $\mathbf{D}_T$ and $\langle \boldsymbol{\kappa}_T \rangle$ correspond in the sense that they obey similar equations for received power. The real target matrix $\mathbf{M}_{\text{av}}$ was formed from $\langle \boldsymbol{\kappa}_T \rangle$, and a corresponding matrix can be formed from $\mathbf{D}_T$. Using the transform (7.19), we define the real matrix

$$\mathbf{M} = \mathbf{Q}\mathbf{D}_T\mathbf{Q}^{-1} \tag{7.34}$$

$\mathbf{M}$ is the Mueller matrix of the target; it is widely used in optics. It corresponds to the radar Mueller matrix $\mathbf{M}_{\text{av}}$ in the same sense that $\mathbf{D}_T$ corresponds to $\langle \boldsymbol{\kappa}_S \rangle$. It is similar in form to $\mathbf{M}_{\text{av}}$ and obeys the same equation for received power. From (7.21) and (7.34), received power is

$$W_d = \frac{Z_0^2 I^2}{256\pi R_a \lambda^2 r_1^2 r_2^2} \mathbf{G}_{Ar}^T \mathbf{diag}(1, 1, -1, 1) \mathbf{M} \mathbf{G}_{At} \tag{7.35}$$

where $\mathbf{G}_{Ar}$ and $\mathbf{G}_{At}$ are the Stokes vectors of the antennas. Like $\mathbf{D}_T$, of which it is a transform, $\mathbf{M}$ has 16 independent real parameters for bistatic scattering and 10 for backscattering.

The Mueller matrix is important in optical studies, where the concept of antenna effective length is not used. Instead of the Kronecker products of antenna effective lengths, the real Stokes vectors of transmitting and receiving devices, which can be measured optically, are used.

We showed that a target which is a combination of a coherent scatterer and a completely depolarizing scatterer, with the scattered waves from the two scatterers uncorrelated, can be represented by $\mathbf{D}_T$, but not that the representation is valid for any target. It is important, however, to recognize that the Mueller matrix is known to represent any depolarizing or coherently scattering target correctly. It follows that $\mathbf{D}_S$ and $\mathbf{D}_T$ can represent any depolarizing or coherently scattering target.

Alternative Forms for the Mueller Matrix

The Mueller matrix and Stokes vector forms given here are widely used but not universal. All the forms in use may be obtained by selecting matrix $\mathbf{Q}$, one choice of which is given by (1.43), to give the desired form of the Stokes vector. The Stokes vector, for any choice of $\mathbf{Q}$, is related to the wave coherency vector by $\mathbf{G} = \mathbf{QJ}$, and the Mueller matrix must satisfy (7.34). The elements of $\mathbf{G}$ and $\mathbf{M}$ should be real, and the elements of the Stokes vector

$$\mathbf{G} = [\ G_0 \ \ G_1 \ \ G_2 \ \ G_3\ ]$$

are commonly accepted as meeting these conditions (Mott, 1992, p. 393)

$$G_0^2 = G_1^2 + G_2^2 + G_3^2 \qquad \text{Conpletely polarized wave}$$

$$G_1 = G_2 = G_3 = 0 \qquad \text{Unpolarized wave}$$

G_0 is commonly taken as proportional to the received power of the wave. It is desirable to choose $\mathbf{Q}$ so that these conditions are met.

An Alternative Incoherently Measured Target Matrix

The Mueller matrix was defined as a transform of $\mathbf{D}_T$ using the transformation that related $\mathbf{M}_{\text{av}}$ to $\langle \boldsymbol{\kappa}_T \rangle$. A target matrix $\mathbf{K}_D$ can similarly be defined as a transform of $\mathbf{D}_S$ by using (7.16) relating $\mathbf{K}$ to $\langle \boldsymbol{\kappa}_S \rangle$. Then, we define

$$\mathbf{K}_D = \mathbf{Q}^* \mathbf{D}_S \mathbf{Q}^{-1}$$

If this transform is used in (7.20), the received power becomes

$$W_d = \frac{Z_0^2 I^2}{256\pi R_a \lambda^2 r_1^2 r_2^2} \mathbf{G}_{Ar}^T \mathbf{K}_D \mathbf{G}_{At} \tag{7.36}$$

$\mathbf{K}_D$ is an incoherently measured matrix corresponding to the Kennaugh matrix $\mathbf{K}$, which is determined by coherent measurement. It can be referred to as the *incoherently* or *directly measured Kennaugh matrix*. It can also be called the *optical Kennaugh matrix*. It differs from the Mueller matrix only in the coordinate systems used for its determination.

Since $\mathbf{K}_D$ is a transform of $\mathbf{D}_S$, it has the same number of real independent parameters, 16 for bistatic scattering and 10 for backscattering. It is symmetric for backscattering.

7.6. MATRIX PROPERTIES AND RELATIONSHIPS

Averaging does not affect the relationship expressed by (7.12), so

$$\langle \kappa_T \rangle = \mathbf{diag}(1, -1, -1, 1)\langle \kappa_S \rangle \tag{7.37}$$

We noted with (7.24) that the incoherently measured equivalent matrices obey the same relationship. If (7.17) and (7.18) for received power are compared, it will be seen that

$$\mathbf{K} = \mathbf{diag}(1, 1, -1, 1)\mathbf{M}_{av} \tag{7.38}$$

Similarly, if, (7.35) and (7.36) are compared, it will be seen that

$$\mathbf{K}_D = \mathbf{diag}(1, 1, -1, 1)\mathbf{M}$$

A coherency vector using $\mathbf{D}_T$ and a Stokes vector using $\mathbf{M}$ were given previously for the scattered wave. Corresponding forms were not given for the matrices $\mathbf{D}_S$ and $\mathbf{K}_D$. When the latter matrices are used, the scattered field components call for the use of coordinate system 3, which is awkward for presenting the scattered wave's characteristics. However, the relationships above can be used to give the scattered wave description in terms of $\mathbf{D}_S$ and $\mathbf{K}_D$,

$$\mathbf{J}^s = \frac{1}{4\pi r^2}\mathbf{D}_T\mathbf{J}^i = \frac{1}{4\pi r^2}\mathbf{diag}(1, -1, -1, 1)\mathbf{D}_S\mathbf{J}^i$$

$$\mathbf{G}^s = \frac{1}{4\pi r^2}\mathbf{M}\mathbf{G}^i = \frac{1}{4\pi r^2}\mathbf{diag}(1, 1, -1, 1)\mathbf{K}_D\mathbf{G}^i$$

Matrix Equality

We have indicated correspondences between $\mathbf{D}_S$ and $\langle \kappa_S \rangle$, $\mathbf{D}_T$ and $\langle \kappa_T \rangle$, $\mathbf{M}$ and $\mathbf{M}_{\mathrm{av}}$, and $\mathbf{K}_D$ and $\mathbf{K}$. We ask now if any of these matrix pairs are equal.

In determining $\langle \kappa_S \rangle$, it is assumed that the measurement procedure and time used to measure $\mathbf{S}$ depend on the radar and on requirements for use of the time-invariant Maxwell equations and the Lorentz reciprocity theorem, and not solely on the target. It is of interest to ask if this averaged matrix yields for a depolarizing target an equivalent description to that of the $\mathbf{D}_S$ matrix; in other words, is $\langle \boldsymbol{\kappa}_S \rangle$ equal to $\mathbf{D}_S$?

If the relationships developed previously are used, $\mathbf{D}_S$, with only independent elements shown, is

$$\mathbf{D}_S = \begin{bmatrix} D_{S11} & D_{S12} & \cdot & D_{S14} \\ D_{S21} & D_{S22} & D_{S23} & D_{S24} \\ \cdot & \cdot & \cdot & \cdot \\ D_{S41} & D_{S42} & \cdot & D_{S44} \end{bmatrix}$$

An arbitrary choice was made of which independent elements to show.

The averaged bistatic matrix $\langle \boldsymbol{\kappa}_S \rangle$, with the linearly dependent or conjugate elements not shown, has these elements remaining,

$$\langle \boldsymbol{\kappa}_S \rangle = \langle \begin{bmatrix} |S_{xx}|^2 & S_{xx}S_{xy}^* & \cdot & |S_{xy}|^2 \\ S_{xx}S_{yx}^* & S_{xx}S_{yy}^* & S_{xy}S_{yx}^* & S_{xy}S_{yy}^* \\ \cdot & \cdot & \cdot & \cdot \\ |S_{yx}|^2 & S_{yx}S_{yy}^* & \cdot & |S_{yy}|^2 \end{bmatrix} \rangle \tag{7.39}$$

There are nonlinear relationships among the elements of the unaveraged matrix for a coherently scattering target that make some of them nonindependent until averaged. There is, however, no linear or conjugate relationship among the terms shown. Therefore, when the average value of the matrix is taken, all the elements in (7.39) will be independent, and no additional ones will be independent. Recall that both $\mathbf{D}_S$ and $\langle \boldsymbol{\kappa}_S \rangle$ have six complex and four real independent elements.

For backscattering, the additional requirements on element relationships of $\mathbf{D}_S$ reduce the number of independent real elements to four real and three complex,

$$\mathbf{D}_{Sb} = \begin{bmatrix} D_{Sb11} & D_{Sb12} & \cdot & D_{Sb14} \\ \cdot & D_{Sb22} & D_{Sb23} & D_{Sb24} \\ \cdot & \cdot & \cdot & \cdot \\ \cdot & \cdot & \cdot & D_{Sb44} \end{bmatrix}$$

The average Kronecker product matrix of a coherently scattering target, specialized to backscattering, with the linearly dependent and conjugate elements not shown, is

$$\langle \boldsymbol{\kappa}_{Sb} \rangle = \langle \begin{bmatrix} |S_{xx}|^2 & S_{xx}S_{xy}^* & \cdot & |S_{xy}|^2 \\ \cdot & S_{xx}S_{yy}^* & \cdot & S_{xy}S_{yy}^* \\ \cdot & \cdot & \cdot & \cdot \\ \cdot & \cdot & \cdot & |S_{yy}|^2 \end{bmatrix} \rangle$$

There is a nonlinear relationship among the unaveraged elements of this matrix, and element (22) can be found from elements (12), (14), and (24). This nonlinear relationship will not survive the averaging process, so we do not utilize it. The averaged matrix $\langle \boldsymbol{\kappa}_{Sb} \rangle$ has 3 real and 3 complex elements, giving 9 independent real parameters, as contrasted to the 10 independent real parameters of $\mathbf{D}_{Sb}$. This does not allow the averaged matrix to represent a depolarizing target in the same sense as does $\mathbf{D}_{Sb}$.

Examination shows that $\langle \boldsymbol{\kappa}_{Sb} \rangle$ has one independent element fewer than $\mathbf{D}_{Sb}$ because in forming $\langle \boldsymbol{\kappa}_{Sb} \rangle$ we used the symmetry of the backscattering Sinclair matrix. To form $\mathbf{D}_S$ we required only that it give a real, nonnegative power and that it be symmetric for backscattering. This causes elements (23) and (32) of $\mathbf{D}_{Sb}$ to be real and equal but does not require that they be equal to (14) and (41), in which case $\mathbf{D}_{Sb}$ and $\langle \boldsymbol{\kappa}_{Sb} \rangle$ would be identical. In the following subsection, it is shown that $\langle \boldsymbol{\kappa}_{Sb} \rangle$ can represent only approximately a completely depolarizing target. On the other hand, it was shown in the discussion relative to (7.32) that $\mathbf{D}_{Sb}$ can represent such a target. We conclude that for backscattering, incoherently measured matrix $\mathbf{D}_S$ and coherently measured matrix $\langle \boldsymbol{\kappa}_S \rangle$ are not equal for depolarizing targets. It is reasonable to assume that they are unequal for bistatic scattering, also. This inequality holds also for the matrix pairs $\mathbf{D}_T$ and $\langle \boldsymbol{\kappa}_T \rangle$, $\mathbf{M}$ and $\mathbf{M}_{\text{av}}$, and $\mathbf{K}_D$ and $\mathbf{K}$.

Utility of the Averaged Target Matrices

Matrix $\langle \boldsymbol{\kappa}_S \rangle$ has the same form as $\mathbf{D}_S$, although not the same element values. However, measurements of $\langle \boldsymbol{\kappa}_S \rangle$ can be repeated for a target, and the matrix can, therefore, characterize the target in the equation for received power and in the determination of the degree of polarization of the scattered wave. In general, the received power and the degree of polarization will be different for matrices $\mathbf{D}_S$ and $\langle \boldsymbol{\kappa}_S \rangle$, but repeatability is a significant factor and makes $\langle \boldsymbol{\kappa}_S \rangle$ a useful target matrix. For backscattering, the use of $\langle \boldsymbol{\kappa}_{Sb} \rangle$ for a depolarizing target can only be considered to give an approximate power for some polarizations, as the following development will suggest.

The averaged Kronecker-product matrix specialized to backscattering is

$$\langle \kappa_{Sb} \rangle = \left\langle \begin{bmatrix} |S_{xx}|^2 & S_{xx}S_{xy}^* & S_{xx}^*S_{xy} & |S_{xy}|^2 \\ S_{xx}S_{xy}^* & S_{xx}S_{yy}^* & |S_{xy}|^2 & S_{xy}S_{yy}^* \\ S_{xx}^*S_{xy} & |S_{xy}|^2 & S_{xx}^*S_{yy} & S_{xy}^*S_{yy} \\ |S_{xy}|^2 & S_{xy}S_{yy}^* & S_{xy}^*S_{yy} & |S_{yy}|^2 \end{bmatrix} \right\rangle$$

Thought will suggest that for a target that completely depolarizes all incident waves,

$$\langle |S_{xx}|^2 \rangle = \langle |S_{xy}|^2 \rangle = \langle |S_{yy}|^2 \rangle \tag{7.40}$$

In addition to this requirement, the three Sinclair matrix elements are uncorrelated, so that for a completely depolarizing target,

$$\langle \kappa_{Sb} \rangle_u = C \begin{bmatrix} 1 & 0 & 0 & 1 \\ 0 & 0 & 1 & 0 \\ 0 & 1 & 0 & 0 \\ 1 & 0 & 0 & 1 \end{bmatrix} \tag{7.41}$$

This matrix differs from $\mathbf{D}_S$ for a completely depolarizing target, given by (7.32), with only the corner elements nonzero.

Study of $\langle \boldsymbol{\kappa}_{Sb} \rangle$ will show that for horizontal and vertical linear polarizations the received power is the same. However, for other polarizations, the power differs from that received for horizontal and vertical linear. Thus the matrix does not meet the requirement that for a completely depolarizing target the received power must be independent of antenna polarizations. It follows that $\langle \boldsymbol{\kappa}_{Sb} \rangle$ cannot be an accurate representation of a depolarizing target. Further study will show, however, that the received power from the completely depolarizing target, found by using $\langle \boldsymbol{\kappa}_{Sb} \rangle$, does not depart greatly from a constant value for any antenna polarization. Then, $\langle \boldsymbol{\kappa}_{Sb} \rangle$ can be considered approximately valid as a representation of a depolarizing target in the power equation. If only horizontal- or vertical-linear polarizations of the incident wave are used, the representation need not be considered approximate.

If received power using the Kennaugh matrix in (7.17) is to be independent of the polarization of transmitting and receiving antennas, only the (11) element of $\mathbf{K}$ can have value. If the Kennaugh matrix for a completely depolarizing target is found from the average Kronecker-product matrix (7.41), it will be found to be

$$\mathbf{K}_{bu} = C \begin{bmatrix} 2 & 0 & 0 & 0 \\ 0 & 0 & 0 & 0 \\ 0 & 0 & 1 & 0 \\ 0 & 0 & 0 & 1 \end{bmatrix} \tag{7.42}$$

This Kennaugh matrix does not accurately represent a completely depolarizing target for backscattering.

7.7. MODIFIED MATRICES

We noted that $\langle \boldsymbol{\kappa}_S \rangle$, but not $\langle \boldsymbol{\kappa}_{Sb} \rangle$, can characterize a depolarizing target in the equation for received power and in determining the degree of polarization of a scattered wave. We can, however, modify $\langle \boldsymbol{\kappa}_{Sb} \rangle$ so that it can represent a completely depolarizing target. We do this by constructing a modified or associated matrix, first using the independent elements of $\langle \boldsymbol{\kappa}_{Sb} \rangle$ and then completing the matrix in accordance with the element relationships in $\mathbf{D}_{Sb}$. Those relationships require only that the (23) and (32) elements be real and equal, not that they be equal to the (14) and (41) elements as in $\langle \boldsymbol{\kappa}_{Sb} \rangle$. The modified, or associated, matrix is

$$\langle \boldsymbol{\kappa}_{Sb} \rangle_{\text{mod}} = \left\langle \begin{bmatrix} |S_{xx}|^2 & S_{xx}S_{xy}^* & S_{xx}^*S_{xy} & |S_{xy}|^2 \\ S_{xx}S_{xy}^* & S_{xx}S_{yy}^* & a & S_{xy}S_{yy}^* \\ S_{xx}^*S_{xy} & a & S_{xx}^*S_{yy} & S_{xy}^*S_{yy} \\ |S_{xy}|^2 & S_{xy}S_{yy}^* & S_{xy}^*S_{yy} & |S_{yy}|^2 \end{bmatrix} \right\rangle$$

where a is real.

In Section 7.5, the concept of separating a 4×4 power matrix for a depolarizing target into matrices for a completely depolarizing target and a coherently-scattering target was introduced. The procedure is developed more fully in Section 9.9. The separation is not independent of the polarization of the transmitting antenna for a general depolarizing target. For targets that can be considered a combination of a coherently scattering target and an independent completely depolarizing target, with scattered waves uncorrelated, the separation is valid for all antenna polarizations. For targets not having this property, the separation must be considered an approximation.

If $\langle \boldsymbol{\kappa}_{Sb} \rangle_{\text{mod}}$ is separated into matrices for a completely depolarizing target and a coherently scattering target, it is

$$\langle \boldsymbol{\kappa}_{Sb} \rangle_{\text{mod}} = \left\langle \begin{bmatrix} |S_{xx}|^2 - b & S_{xx}S_{xy}^* & S_{xx}^*S_{xy} & |S_{xy}|^2 - b \\ S_{xx}S_{xy}^* & S_{xx}S_{yy}^* & a & S_{xy}S_{yy}^* \\ S_{xx}^*S_{xy} & a & S_{xx}^*S_{yy} & S_{xy}^*S_{yy} \\ |S_{xy}|^2 - b & S_{xy}S_{yy}^* & S_{xy}^*S_{yy} & |S_{yy}|^2 - b \end{bmatrix} \right\rangle + b \begin{bmatrix} 1 & 0 & 0 & 1 \\ 0 & 0 & 0 & 0 \\ 0 & 0 & 0 & 0 \\ 1 & 0 & 0 & 1 \end{bmatrix}$$

If we consider a linear x-directed incident electric field, b can be found from (9.36),

$$\begin{aligned} b &= \frac{1}{2}\langle |S_{xx}|^2 + |S_{xy}|^2 \rangle \\ &\quad - \frac{1}{2}\left[\langle |S_{xx}|^2 + |S_{xy}|^2 \rangle^2 - 4\left(\langle |S_{xx}|^2\rangle\langle |S_{xy}|^2\rangle - \langle S_{xx}S_{xy}^*\rangle\langle S_{xx}^*S_{xy}\rangle\right)\right]^{1/2} \end{aligned}$$

In the equation for the modified matrix, elements (23) and (32) have not been specified. Let us now require that the matrix for the coherently scattering target be that for backscattering. This specifies the undetermined elements as

$$a = \langle |S_{xy}|^2 \rangle - b$$

if b is smaller than $|S_{xx}|^2$, $|S_{xy}|^2$, and $|S_{yy}|^2$. It can be shown that b is small if the degree of polarization of the scattered wave is high. If the separated matrices are recombined, the modified matrix becomes

$$\langle \boldsymbol{\kappa}_{Sb} \rangle = \left\langle \begin{bmatrix} |S_{xx}|^2 & S_{xx}S_{xy}^* & S_{xx}^*S_{xy} & |S_{xy}|^2 \\ S_{xx}S_{xy}^* & S_{xx}S_{yy}^* & |S_{xy}|^2 - b & S_{xy}S_{yy}^* \\ S_{xx}^*S_{xy} & |S_{xy}|^2 - b & S_{xx}^*S_{yy} & S_{xy}^*S_{yy} \\ |S_{xy}|^2 & S_{xy}S_{yy}^* & S_{xy}^*S_{yy} & |S_{yy}|^2 \end{bmatrix} \right\rangle$$

This matrix is uniquely determined by the elements of the averaged backscattering Kronecker product matrix and the additional requirement that the part of the matrix representing a coherently scattering target describe a backscattering target. Therefore, this modified matrix can be used to determine the degree of polarization of the scattered wave, given a polarization of the transmitting antenna.

A different modified matrix can be developed by assuming that the incident wave is y polarized. The procedures for doing so are given in Section 9.9. The two forms may differ, but either can be used to find the degree of polarization of the scattered wave.

In order to find the degree of polarization of the scattered wave, it is desirable to have the scattered wave in the coordinates of system 2 of Fig. 7.1. This is readily accomplished by using the averaged Kronecker product matrix based on the Jones matrix, so a modified matrix based on the Jones matrix is

$$\langle \boldsymbol{\kappa}_{Tb} \rangle_{\text{mod}} = \mathbf{diag}(1, -1, -1, 1)\langle \boldsymbol{\kappa}_{Sb} \rangle_{\text{mod}}$$

The Kennaugh matrix $\mathbf{K}$ is a transform of $\langle \boldsymbol{\kappa}_S \rangle$, and the average Mueller matrix $\mathbf{M}_{\text{av}}$ is a transform of $\langle \boldsymbol{\kappa}_T \rangle$. For backscattering, $\mathbf{K}$ and $\mathbf{M}_{\text{av}}$ have the same defect as $\langle \boldsymbol{\kappa}_{Sb} \rangle$ and $\langle \boldsymbol{\kappa}_{Tb} \rangle$. However, the defining transforms can also be applied to the modified averaged Kronecker product matrices to obtain modified Kennaugh and average Mueller matrices that are polarimetrically valid for backscattering.

7.8. NAMES

Mueller matrix $\mathbf{M}$ is a transform of $\mathbf{D}_T$, whose elements are incoherently measured. Historically, the Mueller matrix arose in optics and its elements are measured incoherently, so the treatment here is in accord with past and present practice. The averaged Mueller matrix $\mathbf{M}_{\text{av}}$ is based on measurement of the Jones matrix and averages of its element products. This is commonly done in radar, so the averaged Mueller matrix is also called here a radar Mueller matrix.

The Kennaugh matrix $\mathbf{K}$ was developed for microwave radar and is found by averaging the products of elements of a Sinclair matrix. Matrix $\mathbf{K}_D$ has the same form as the standard Kennaugh matrix, but its elements are incoherently measured, as is common in optical studies. For that reason it has been called here the directly measured or optical Kennaugh matrix. There is a transform relationship between the optical Kennaugh matrix and the Mueller matrix, and one between the radar Mueller matrix and the Kennaugh matrix. For depolarizing targets, there is no relationship between the Mueller matrix and the radar Mueller matrix, nor is there one between the Kennaugh matrix and the optical Kennaugh matrix. As defined here, there is no relationship between the Mueller matrix and the Kennaugh matrix except for coherently scattering targets.

7.9. ADDITIONAL TARGET INFORMATION

The target matrices discussed here provide no information about the time scale of target motions. That information can be measured or found from the autocorrelation function for received power,

$$R_w(\tau) = \overline{w(t)w(t+\tau)}$$

where $w(t)$ can be measured or found from one of the power equations, such as (7.20). Matrix $\mathbf{D}_S$ in that equation can be found by measuring the average power during a time span sufficient to allow a representative range of target motions, or by sampling. To find the autocorrelation function, however, sampling is appropriate.

7.10. TARGET COVARIANCE AND COHERENCY MATRICES

The decision to form the four Sinclair elements in a 2×2 matrix is arbitrary and one could order them as a target vector just as readily. A vector can also be formed from sums and differences of the elements.

The Bistatic Covariance Matrix

We first form a target vector with the Sinclair elements,

$$\mathbf{X}_4 = \begin{bmatrix} S_{xx} & S_{xy} & S_{yx} & S_{yy} \end{bmatrix}^T$$

A matrix to represent an incoherently scattering target can be defined by an average of $\mathbf{X}_4\mathbf{X}_4^\dagger$,

$$\mathbf{C}_4 = \langle \mathbf{X}_4\mathbf{X}_4^\dagger \rangle = \langle \begin{bmatrix} |S_{xx}|^2 & S_{xx}S_{xy}^* & S_{xx}S_{yx}^* & S_{xx}S_{yy}^* \\ S_{xy}S_{xx}^* & |S_{xy}|^2 & S_{xy}S_{yx}^* & S_{xy}S_{yy}^* \\ S_{yx}S_{xx}^* & S_{yx}S_{xy}^* & |S_{yx}|^2 & S_{yx}S_{yy}^* \\ S_{yy}S_{xx}^* & S_{yy}S_{xy}^* & S_{yy}S_{yx}^* & |S_{yy}|^2 \end{bmatrix} \rangle \tag{7.43}$$

This matrix is the *target covariance matrix* (van Zyl and Zebker, 1990, p. 282).

The voltage received from a coherently scattering target is

$$V = \frac{jZ_0 I}{\sqrt{4\pi}(2\lambda r_1 r_2)} e^{-jk(r_1+r_2)} \mathbf{h}_r^T \mathbf{S} \mathbf{h}_t$$

Let us assume that the voltage can also be given by

$$V = \frac{jZ_0 I}{\sqrt{4\pi}(2\lambda r_1 r_2)} e^{-jk(r_1+r_2)} \mathbf{H}_4^T \mathbf{X}_4$$

where $\mathbf{H}_4$ is to be determined. If these forms for voltage are equated, it is readily seen that

$$\mathbf{H}_4 = \mathbf{h}_r \otimes \mathbf{h}_t \tag{7.44}$$

Received power from a depolarizing target is given by

$$W_{\text{av}} = \frac{\langle VV^* \rangle}{8R_a} = \frac{Z_0^2 I^2}{128\pi R_a \lambda^2 r_1^2 r_2^2} \langle (\mathbf{H}_4^T \mathbf{X}_4)(\mathbf{H}_4^\dagger \mathbf{X}_4^*) \rangle$$

$$= \frac{Z_0^2 I^2}{128\pi R_a \lambda^2 r_1^2 r_2^2} \mathbf{H}_4^T \mathbf{C}_4 \mathbf{H}_4^* \tag{7.45}$$

This formulation for received voltage and power does not permit the separation of the effects of transmitting and receiving antennas. Covariance matrix $\mathbf{C}_4$ is Hermitian positive semidefinite.

The elements of $\mathbf{C}_4$ are the same as those of $\langle \boldsymbol{\kappa}_S \rangle$; just as $\langle \boldsymbol{\kappa}_S \rangle$ had a counterpart matrix $\mathbf{D}_S$ whose elements could be directly measured if desired, so too does $\mathbf{C}_4$ have a corresponding matrix, obeying the same received-power equation, whose elements can be directly measured.

The covariance matrix measurements are repeatable if it is formed by averaging products of the Sinclair matrix, and the matrix is therefore a useful one for describing targets.

If the covariance matrix is specialized to backscattering, it will have the same characteristics as the averaged Kronecker product matrix specialized to backscattering; that is, it does not represent a depolarizing target in the same sense as does the $\mathbf{D}_{Sb}$ matrix or the Mueller matrix. A modified covariance matrix for backscattering can be defined in a manner similar to the definitions of the modified matrices of Section 7.7.

The Backscattering Covariance Matrix

For backscattering, the voltage received from a coherently scattering target is

$$V = \frac{jZ_0 I}{\sqrt{4\pi}(2\lambda r^2)} e^{-j2kr} \mathbf{h}_r^T \mathbf{S} \mathbf{h}_t$$

where we allow for separate, but co-located, antennas. In a widely used matrix formulation for backscattering with a symmetric Sinclair matrix, a target vector is defined as

$$\mathbf{X}_3 = [\, S_{xx} \quad \sqrt{2} S_{xy} \quad S_{yy} \,]^T$$

and a target matrix as

$$\mathbf{C}_3 = \langle \mathbf{X}_3 \mathbf{X}_3^\dagger \rangle = \left\langle \begin{bmatrix} |S_{xx}|^2 & \sqrt{2} S_{xx} S_{xy}^* & S_{xx} S_{yy}^* \\ \sqrt{2} S_{xx}^* S_{xy} & 2|S_{xy}|^2 & \sqrt{2} S_{xy} S_{yy}^* \\ S_{xx}^* S_{yy} & \sqrt{2} S_{xy}^* S_{yy} & |S_{yy}|^2 \end{bmatrix} \right\rangle \tag{7.46}$$

The square of the Euclidean norm of $\mathbf{X}_3\mathbf{X}_3^\dagger$ is equal to the trace of $\mathbf{S}^*\mathbf{S}$. If $\mathbf{S}$ is transformed to a new polarization basis by a consimiliarity transform

$$\hat{\mathbf{S}} = \mathbf{U}^*\mathbf{S}\mathbf{U}^\dagger$$

where $\mathbf{U}$ is a unitary matrix given by (3.13), it can be shown that the trace of $\mathbf{S}^*\mathbf{S}$ is preserved. It follows that the Euclidean norm of $\mathbf{X}_3$ is invariant to a change of polarization basis. The 3×3 or *backscattering covariance matrix*, $\mathbf{C}_3$, is Hermitian positive semidefinite. The (13) element for the unaveraged matrix can be found from a nonlinear operation on elements (12), (23), and (22), but after averaging the (13) element is independent.

The received voltage can be written as

$$V = \frac{jZ_0 I}{\sqrt{4\pi}(2\lambda r^2)} e^{-j2kr}\mathbf{H}_3^T\mathbf{X}_3$$

where $\mathbf{H}_3$ is a joint antenna vector,

$$\mathbf{H}_3 = \begin{bmatrix} h_{rx}h_{tx} \\ \frac{1}{\sqrt{2}}(h_{rx}h_{ty} + h_{ry}h_{tx}) \\ h_{ry}h_{ty} \end{bmatrix} \tag{7.47}$$

The magnitude of $\mathbf{H}_3$ is not constant if $\mathbf{h}_r$ and $\mathbf{h}_t$ are varied while keeping their magnitudes constant. If receiving and transmitting antennas are equal, the magnitude of $\mathbf{H}_3$ does remain constant as the effective antenna length is varied.

Power received from a depolarizing target can be written as

$$\begin{aligned} W_{\text{av}} &= \frac{\langle VV^*\rangle}{8R_a} = \frac{Z_0^2 I^2}{128\pi R_a \lambda^2 r^4}\langle(\mathbf{H}_3^T\mathbf{X}_3)(\mathbf{H}_3^T\mathbf{X}_3)^\dagger\rangle \\ &= \frac{Z_0^2 I^2}{128\pi R_a \lambda^2 r^4}\mathbf{H}_3^T\mathbf{C}_3\mathbf{H}_3^* \end{aligned} \tag{7.48}$$

The backscattering covariance matrix has three real and three complex independent elements. A comparison shows that it has the same independent elements as the 4×4 covariance matrix specialized to backscattering or the average Kronecker product matrix similarly specialized. The matrix is based on the symmetry of the Sinclair matrix. It shares the failing of other matrices so based: It cannot represent a depolarizing target with complete accuracy.

The 4 x 4 covariance matrix discussed here has a corresponding matrix that can be determined by direct measurement of received power for chosen antenna polarizations and its use in an equation of the form (7.45). We wish to determine if $\mathbf{C}_3$ has such a counterpart. Can $\mathbf{C}_G$ in

$$W = \frac{Z_0^2 I^2}{128\pi R_a \lambda^2 r^4}\mathbf{H}_3^T\mathbf{C}_G\mathbf{H}_3^*$$

be found for a depolarizing target by selecting $\mathbf{H}_3$ and measuring W?

If the equation is to be valid for a depolarizing target it must be valid for a completely depolarizing target, for which the received power is independent of the polarizations of transmitting and receiving antennas. We, therefore, require for the joint antenna vector that $\mathbf{H}_3^T \mathbf{C}_G \mathbf{H}_3^*$ be constant. If various simple antenna pairs are chosen to determine the matrix elements, the resulting equations are inconsistent. No matrix can be found to satisfy this constraint.

The power equation given above can be satisfied approximately if we choose

$$\mathbf{C}_G = C \begin{bmatrix} 1 & 0 & 0 \\ 0 & 2 & 0 \\ 0 & 0 & 1 \end{bmatrix}$$

If this matrix is used in the equation for received power, it will be found that the same power is received for horizontal and vertical linear antenna polarizations, but not for other polarizations.

Since no 3×3 matrix can be found that gives the correct power for a completely depolarizing target, it follows that the backscattering matrix $\mathbf{C}_3$ is only an approximation to the correct representation of a depolarizing target.

Target Coherency Matrix

The target vector used to form the covariance matrix is not the only one that can be formed from the elements of $\mathbf{S}$. A target vector that is sometimes used is

$$\mathbf{k}_4 = [\, k_0 \;\; k_1 \;\; k_2 \;\; k_3 \,]^T$$

where $k_i = \mathrm{Tr}(\mathbf{S}\boldsymbol{\sigma}_i)$ and the $\boldsymbol{\sigma}_i$ are the Pauli spin matrices,

$$\boldsymbol{\sigma}_0 = \begin{bmatrix} 1 & 0 \\ 0 & 1 \end{bmatrix} \quad \boldsymbol{\sigma}_1 = \begin{bmatrix} 1 & 0 \\ 0 & -1 \end{bmatrix} \quad \boldsymbol{\sigma}_2 = \begin{bmatrix} 0 & 1 \\ 1 & 0 \end{bmatrix} \quad \boldsymbol{\sigma}_3 = \begin{bmatrix} 0 & -j \\ j & 0 \end{bmatrix}$$

From the definitions of target vectors $\mathbf{X}_4$ and $\mathbf{k}_4$, the received power is readily converted from the covariance matrix form to

$$W_{\mathrm{av}} = \frac{Z_0^2 I^2}{128\pi R_a \lambda^2 r_1^2 r_2^2} \mathbf{N}_4^T \mathbf{L}_4 \mathbf{N}_4^*$$

where $\mathbf{N}_4$ is a joint antenna vector defined by

$$\mathbf{N}_4 = \begin{bmatrix} h_{rx}h_{tx} + h_{ry}h_{ty} \\ h_{rx}h_{tx} - h_{ry}h_{ty} \\ h_{rx}h_{ty} + h_{ry}h_{tx} \\ -j(h_{rx}h_{ty} - h_{ry}h_{tx}) \end{bmatrix}$$

and

$$\mathbf{L}_4 = \langle \mathbf{k}_4 \mathbf{k}_4^\dagger \rangle$$

$\mathbf{L}_4$ is sometimes called the "coherency matrix." The matrix $\overline{\mathbf{E}(t) \otimes \mathbf{E}^*(t)}$ is also called a coherency matrix (Born and Wolf, 1965, p. 545), and for that reason we use "coherency matrix" for a target only when preceded by the word "target."

For backscattering, the fourth element of the target vector $\mathbf{k}_4$ is zero, and a three-element target vector $\mathbf{L}_3$ can be formed by taking its first three elements. If a joint antenna vector $\mathbf{N}_3$ is formed by taking the first three elements of $\mathbf{N}_4$, the received power becomes

$$W_{\text{av}} = \frac{Z_0^2 I^2}{128\pi R_a \lambda^2 r_1^2 r_2^2} \mathbf{N}_3^T \mathbf{L}_3 \mathbf{N}_3^*$$

where

$$\mathbf{L}_3 = \langle \mathbf{k}_3 \mathbf{k}_3^\dagger \rangle$$

7.11. A SCATTERING MATRIX WITH CIRCULAR COMPONENTS

In circular-component form, the scattered field of a coherently scattering target is

$$\mathbf{E}^s(L, R) = \frac{1}{\sqrt{4\pi} r_2} \mathbf{A} \mathbf{E}^i(L, R) e^{-jkr_2}$$

where $\mathbf{A}$ is

$$\mathbf{A} = \mathbf{U}_c^{-1} \mathbf{T} \mathbf{U}_c$$

with $\mathbf{T}$ the target's Jones matrix and $\mathbf{U}_c$ given by (3.9). If the relationship between the target's Jones and Sinclair matrices is used, $\mathbf{A}$ becomes

$$\mathbf{A} = \mathbf{U}_c^{-1} \mathbf{diag}(-1, 1) \mathbf{S} \mathbf{U}_c = -\mathbf{U}_c^T \mathbf{S} \mathbf{U}_c$$

The latter form shows $\mathbf{A}$ and $\mathbf{S}$ to be related by a unitary consimilarity or unitary congruence transform.

The received power given by (3.11) can be written in Kronecker-product form,

$$W_c = \frac{Z_0^2 I^2}{128\pi R_a \lambda^2 r_1^2 r_2^2} (\mathbf{h}_r \otimes \mathbf{h}_r^*)^T (\mathbf{A} \otimes \mathbf{A}^*)(\mathbf{h}_t \otimes \mathbf{h}_t^*) \tag{7.49}$$

We can proceed in two ways to form a matrix representing a depolarizing target. In the first way, a matrix $\mathbf{D}_{\text{circ}}$ can be defined to give the received power in an equation of this form,

$$W_d = \frac{Z_0^2 I^2}{128\pi R_a \lambda^2 r_1^2 r_2^2} (\mathbf{h}_r \otimes \mathbf{h}_r^*)^T \mathbf{D}_{\text{circ}} (\mathbf{h}_t^T \otimes \mathbf{h}_t^*)$$

The elements of $\mathbf{D}_{\text{circ}}$ can be directly measured. In an alternative procedure, the elements of $\mathbf{A}$ are measured, and the averaged value $\langle \mathbf{A} \otimes \mathbf{A}^* \rangle$ used in an equation of the form (7.49)

7.12. THE GRAVES POWER DENSITY MATRIX

The power density of the scattered wave from a time-varying target is given by the use of the *Graves polarization power scattering matrix* (Graves, 1956). To develop the form for the polarization power scattering matrix, or power density matrix, we begin with the relationship (3.4) between incident and scattered fields, given in coordinate systems 1 and 3 of Fig. 7.1. The power density of the scattered wave from a coherently scattering target, found from (3.4), is

$$\mathcal{P}_c^s = \frac{1}{2Z_0}\mathbf{E}^{s\dagger}\mathbf{E}^s = \frac{1}{8\pi Z_0 r^2}\mathbf{E}^{i\dagger}\mathbf{S}^\dagger\mathbf{S}\mathbf{E}^i$$

If multiple measurements are made of the scattering from a depolarizing target or of a coherently scattering target at different aspect angles, an average power density is

$$\mathcal{P}_{\text{av}}^s = \frac{1}{2Z_0}\langle\mathbf{E}^{s\dagger}\mathbf{E}^s\rangle = \frac{1}{8\pi Z_0 r^2}\langle\mathbf{E}^{i\dagger}\mathbf{S}^\dagger\mathbf{S}\mathbf{E}^i\rangle = \frac{1}{8\pi Z_0 r^2}\mathbf{E}^{i\dagger}\boldsymbol{\sigma}_{\text{av}}\mathbf{E}^i \tag{7.50}$$

where

$$\boldsymbol{\sigma}_{\text{av}} = \langle\mathbf{S}^\dagger\mathbf{S}\rangle \tag{7.51}$$

Multiplication gives the matrix,

$$\boldsymbol{\sigma}_{\text{av}} = \langle\begin{bmatrix} |S_{xx}|^2 + |S_{yx}|^2 & S_{xx}^* S_{xy} + S_{yx}^* S_{yy} \\ S_{xx}S_{xy}^* + S_{yx}S_{yy}^* & |S_{xy}|^2 + |S_{yy}|^2 \end{bmatrix}\rangle \tag{7.52}$$

which is valid for bistatic scattering. Matrix $\boldsymbol{\sigma}_{\text{av}}$ is the Graves matrix. It is Hermitian positive semidefinite. The equation for power density provides no information about the polarization of the scattered wave, and none can be obtained from the Graves matrix.

If (7.52) is compared to the elements of the Kennaugh matrix given in Appendix C, it will be seen that

$$\sigma_{\text{av}} = \begin{bmatrix} K_{11} + K_{12} & K_{13} - jK_{14} \\ K_{13} + jK_{14} & K_{11} - K_{12} \end{bmatrix} \tag{7.53}$$

The voltage induced in a horizontally polarized receiving antenna at distance r is

$$V_H = \mathbf{h}_r^T\mathbf{E}^s = \frac{1}{\sqrt{4\pi}r}\begin{bmatrix} h_{rx} & 0 \end{bmatrix}\mathbf{S}\mathbf{E}^i = \frac{h_{rx}}{\sqrt{4\pi}r}(S_{xx}E_x^i + S_{xy}E_y^i)$$

and the received power is

$$W_H = \frac{|h_{rx}|^2}{32\pi R_a r_2^2}\mathbf{E}^{i\dagger}\langle\begin{bmatrix} |S_{xx}|^2 & S_{xx}^* S_{xy} \\ S_{xx}S_{xy}^* & |S_{xy}|^2 \end{bmatrix}\rangle\mathbf{E}^i = \frac{|h_{rx}|^2}{32\pi R_a r_2^2}\mathbf{E}^{i\dagger}\boldsymbol{\sigma}_H\mathbf{E}^i$$

Similarly, the power received by a vertically polarized antenna is

$$W_V = \frac{|h_{ry}|^2}{32\pi R_a r_2^2}\mathbf{E}^{i\dagger}\langle\begin{bmatrix} |S_{yx}|^2 & S_{yx}^* S_{yy} \\ S_{yx} S_{yy}^* & |S_{xy}|^2 \end{bmatrix}\rangle\mathbf{E}^i = \frac{|h_{ry}|^2}{32\pi R_a r_2^2}\mathbf{E}^{i\dagger}\sigma_V\mathbf{E}^i$$

where $\boldsymbol{\sigma}_H$ and $\boldsymbol{\sigma}_V$ are defined by the equations. It may be seen that

$$\boldsymbol{\sigma}_{\text{av}} = \boldsymbol{\sigma}_H + \boldsymbol{\sigma}_V$$

The Graves matrix has thus been separated into parts that yield the powers received respectively by linear horizontally and vertically polarized antennas. Both $\boldsymbol{\sigma}_H$ and $\boldsymbol{\sigma}_V$ are Hermitian. None of the matrices is restricted to backscattering. The average scattered power density can now be written as

$$\mathcal{P}_{\text{av}}^s = \frac{1}{8\pi Z_0 r^2}\mathbf{E}^{i\dagger}(\boldsymbol{\sigma}_H + \boldsymbol{\sigma}_H)\mathbf{E}^i$$

The power-density scattering matrix need not be formed by measuring the Sinclair matrix and averaging over multiple measurements. It can also be found directly from the equations for received power to horizontally and vertically polarized receiving antennas,

$$W_{Hd} = \frac{Z_0^2 I^2 |h_{rx}|^2}{128\pi R_a \lambda^2 r_1^2 r_2^2}\mathbf{h}_t^\dagger \boldsymbol{\sigma}_H \mathbf{h}_t$$

$$W_{Vd} = \frac{Z_0^2 I^2 |h_{ry}|^2}{128\pi R_a \lambda^2 r_1^2 r_2^2}\mathbf{h}_t^\dagger \boldsymbol{\sigma}_V \mathbf{h}_t$$

To find $\boldsymbol{\sigma}_H$ and $\boldsymbol{\sigma}_V$ directly, values of $\mathbf{h}_t$ are selected and the power measured for each value. We have not distinguished by our notation the resulting matrices from those found by averaging products of the Sinclair matrix.

Power density, from (7.50), is proportional to $\hat{\mathbf{h}}_t^\dagger \boldsymbol{\sigma}_{\text{av}} \hat{\mathbf{h}}_t$, which, by the Rayleigh-Ritz theorem and recognition that $\boldsymbol{\sigma}_{\text{av}}$ is positive semidefinite, lies between λ_1 and 0, inclusive, where λ_1 is the largest eigenvalue of $\boldsymbol{\sigma}_{av}$.

We examine the incoherently measured Graves matrix, denoted now as $\boldsymbol{\sigma}$, as we examined other matrices, to determine whether it can represent a depolarizing target or not: If it can represent a completely depolarizing target, it can represent a depolarizing target. A completely depolarizing target should scatter a wave with the same power density, independent of the polarization of the transmitting antenna. Then it is required that $\hat{\mathbf{h}}_t^\dagger \boldsymbol{\sigma} \hat{\mathbf{h}}_t$ be constant for all polarizations of the transmitting antenna. This can be true only if the eigenvalues of $\boldsymbol{\sigma}$ are equal. They are equal only if

$$\boldsymbol{\sigma} = \sigma_{xx}\begin{bmatrix} 1 & 0 \\ 0 & 1 \end{bmatrix}$$

This matrix can represent a completely-depolarizing target, but it is also the Graves matrix for a flat plate at normal incidence. As with the other target matrices discussed here, a physical target cannot be inferred from the Graves matrix.

7.13. MEASUREMENT CONSIDERATIONS

The target matrices discussed here can be grouped into two classes. The first contains matrices that are found by measurement with a continuous wave or by sampling of the received power with short pulses if care is taken to meet the constraints discussed in Section 7.1. They include $\mathbf{D}_S$, $\mathbf{D}_T$, $\mathbf{M}$, and $\mathbf{K}_D$, and are called here **D**-type matrices. The second class contains matrices formed as an average of products of the Sinclair or Jones matrix elements. It includes the averaged forms of the Kronecker products, $\langle \boldsymbol{\kappa}_S \rangle$ and $\langle \boldsymbol{\kappa}_T \rangle$, the averaged forms of the Mueller and Kennaugh matrices, $\mathbf{M}_{\text{av}}$ and $\mathbf{K}$, and the modified forms of these matrices discussed in Section 7.7. It also includes the 4×4 and 3×3 covariance matrices and the target coherency matrices. The Graves polarization power scattering matrix can be considered to have elements that are directly measured or to have elements formed from the Sinclair matrix. Both forms are useful, but they are not interchangeable. Both matrix classes can be used to find, in a repeatable manner, received power and degree of polarization of the scattered wave, although the values so determined will not in general be the same for corresponding matrices from the two classes. The two matrix classes are inherently not comparable. Both are based on measurements, and they are based on measurements of different quantities.

The target matrices can also be grouped according to the coordinate systems used in defining them. Matrices using coordinate systems 1 and 2 of Fig. 7.1 are $\mathbf{D}_T$ and its transform $\mathbf{M}$, and $\langle \boldsymbol{\kappa}_T \rangle$ and its transform $\mathbf{M}_{\text{av}}$. Those using systems 1 and 3 are $\mathbf{D}_S$ and its transform $\mathbf{K}_D$, and $\langle \boldsymbol{\kappa}_S \rangle$ and its transform $\mathbf{M}_{\text{av}}$. Also using systems 1 and 3 are the covariance matrices $\mathbf{C}_4$ and $\mathbf{C}_3$, the target coherency matrices $\mathbf{L}_4$ and $\mathbf{L}_3$, and the Graves matrix $\boldsymbol{\sigma}$.

We have noted throughout this discussion that matrices using coordinate systems 1 and 2 give simpler equations for the scattered wave and are more convenient than those using systems 1 and 3 for finding the degree of polarization of the scattered wave, while matrices using systems 1 and 3 give simpler forms for received power.

Coordinate systems 1 and 2 coincide for direct transmission through an optical device. Moreover, many optical measurements are made with continuous waves. Then $\mathbf{D}_T$ and $\mathbf{M}$ are useful. It is also sometimes desirable in optics to determine the real Stokes vector of a wave, and $\mathbf{M}$ leads naturally to this determination.

Most radar measurements are for backscattering, for which it is convenient to use coordinate systems 1 and 3. The requirement of measuring distance leads to the use of a pulsed transmission, and the coherent addition of received pulses makes it desirable to use multiple measurements of $\mathbf{S}$. Appropriate measurements are of $\langle \boldsymbol{\kappa}_S \rangle$ or **K**. It is noted specifically that a synthetic aperture radar uses the pulsed nature of the transmitted wave to identify resolution cells on extended terrain, and CW measurements cannot be used. If the degree of polarization of the scattered wave is to be found, the averaged matrices must be modified in the manner described in Section 7.7. Moreover, these matrices and others from the second class described above represent a depolarizing target for backscattering only approximately.

7.14. DEGREE OF POLARIZATION AND POLARIMETRIC ENTROPY

The coherency vector of a wave scattered from a target is given by

$$\mathbf{J}^s = \left[\, J_{11}^s \;\; J_{12}^s \;\; J_{21}^s \;\; J_{22}^s \,\right]^T = \frac{1}{4\pi r^2}\mathbf{D}_T(\mathbf{E}^i \otimes \mathbf{E}^{i*}) \approx \frac{1}{4\pi r^2}\langle \boldsymbol{\kappa}_T \rangle (\mathbf{E}^i \otimes \mathbf{E}^{i*})$$

and the degree of polarization of the scattered wave is

$$R = \left[1 - \frac{(J_{11}^s J_{22}^s - J_{12}^s J_{21}^s)}{(J_{11}^s + J_{22}^s)^2}\right]^{1/2}$$

There are two important antenna polarizations for a target in a backscattering configuration. They are given by the eigenvectors of the Hermitian positive semidefinite Graves matrix of the target. One of the eigenvectors gives maximum scattered wave power density if it is used as the effective length of the transmitting antenna; the other gives a submaximum power density. The degrees of polarization, R_1 and R_2, that correspond to maximum and submaximum power densities clearly depend only on the target. Here, we will find a coefficient that is approximately related to the two degrees of polarization and provides information about the depolarizing nature of a target.

As an example, form the covariance matrix for backscattering by

$$\mathbf{X}_4 = [\, S_{xx} \;\; S_{xy} \;\; S_{xy} \;\; S_{yy} \,]^T$$
$$\mathbf{C}_4 = \mathbf{X}_4 \mathbf{X}_4^\dagger + a\mathbf{I}$$

The eigenvalues of $\mathbf{C}_4$ are

$$\lambda_1 = \lambda_0 + a \qquad \lambda_2 = \lambda_3 = \lambda_4 = a$$

If $a = 0$, $\mathbf{C}_4$ has rank 1 and only one nonzero eigenvalue, λ_0.

For a target with a relatively small unpolarized scattering component, the eigenvalues differ greatly, while for a large unpolarized component they are more nearly equal. Cloude and Pottier (1997) devised a measure of the equality of the eigenvalues and recognized that it is related to the scattered wave's degree of polarization. In analogy to the concept of entropy in information theory, they defined a *polarimetric entropy*.

If x_n is an event that occurs with probability $P(x_n)$,

$$I(x_n) = -\log_b P(x_n)$$

units of information are contained in the statement that x_n has occurred (Ziemer and Tranter, 1995, p. 668). The logarithmic base is arbitrary. The average information associated with the outcome of an experiment is defined as the entropy,

$$H = E[I(x_n)] = -\sum_{1}^{N} P(x_n)\left(\log_b P(x_n)\right)$$

where N is the number of possible outcomes. Entropy can be considered an average uncertainty and is maximum when each outcome is equally likely.

Using the four real eigenvalues of the covariance matrix, define

$$P_n = \frac{\lambda_n}{\sum_1^4 \lambda_j}$$

as a probability. One can justify this because of the observed behavior that a large λ_n, compared to the average value of the four eigenvalues, corresponds to a high degree of scattered wave polarization for the corresponding value of the antenna polarization. We can then interpret the greatest value of P_n fairly broadly as the probability that the wave is completely polarized. In this discussion, logarithmic base 4 was used to make the maximum entropy 1.

Examples show an approximate relationship between the polarimetric entropy and the characteristic degrees of polarization of the scattered wave from the target,

$$H \approx 1 - \sqrt{1 - R_1 R_2}$$

The individual values of the degree of polarization provide the greatest amount of information about the depolarizing behavior of a target, but it can be seen from this relationship that polarimetric entropy is a useful target descriptor.

7.15. VARIANCE OF POWER

Depolarization of a scattered wave causes the received power for a radar pulse to differ from the power for the next pulse. A strongly depolarizing target will show a greater variability in sample values than will a weakly depolarizing target. Then a measure of the ability of the target to depolarize an incident wave is the variance of the power. The received power for each pulse, W_n, is a sampled value of a continuous random variable. The normalized variance of W_n can be determined. Factors other than depolarization contribute to the variance, and it cannot be compared directly to the degree of polarization of the scattered wave. It must be used with care to characterize a target; nonetheless, it is a useful target descriptor.

7.16. SUMMARY OF POWER EQUATIONS AND MATRIX RELATIONSHIPS

Summarized here are some of the more important equations for scattered fields and power received from a depolarizing target.

Coherent scattering: Fields:

$$\mathbf{E}^s = \frac{1}{\sqrt{4\pi}r}e^{-jkr}\mathbf{S}\mathbf{E}^i = \frac{1}{\sqrt{4\pi}r}e^{-jkr}\mathbf{diag}(-1, 1)\mathbf{T}\mathbf{E}^i$$

Coherent scattering: Received voltage:

$$V = \frac{jZ_0 I}{2\sqrt{4\pi}\lambda r_1 r_2}e^{-jk(r_1+r_2)}\mathbf{h}_r^T\mathbf{S}\mathbf{h}_t = \frac{jZ_0 I}{2\sqrt{4\pi}\lambda r_1 r_2}e^{-jk(r_1+r_2)}\mathbf{h}_r^T\mathbf{diag}(-1, 1)\mathbf{T}\mathbf{h}_t$$

Coherent scattering: Received power:

$$W_c = \frac{Z_0^2 I^2}{128\pi R_a \lambda^2 r_1^2 r_2^2}(\mathbf{h}_r^T\mathbf{S}\mathbf{h}_t)(\mathbf{h}_r^T\mathbf{S}\mathbf{h}_t)^*$$

$$W_c = \frac{Z_0^2 I^2}{128\pi R_a \lambda^2 r_1^2 r_2^2}(\mathbf{h}_r \otimes \mathbf{h}_r^*)^T \left\{ \begin{array}{c} \boldsymbol{\kappa}_S \\ \mathbf{diag}(1, -1, -1, 1)\boldsymbol{\kappa}_T \end{array} \right\} (\mathbf{h}_t \otimes \mathbf{h}_t^*)$$

Depolarized scattering: Scattered waves with incoherently measured matrices:

$$\mathbf{J}^s = \frac{1}{4\pi r^2}\mathbf{D}_T\mathbf{J}^i = \frac{Z_0^2 I^2}{16\pi\lambda^2 r_1^2 r_2^2}\mathbf{D}_T(\mathbf{h}_t \otimes \mathbf{h}_t^*)$$

$$\mathbf{G}^s = \frac{1}{4\pi r^2}\mathbf{M}\mathbf{G}^i = \frac{Z_0^2 I^2}{16\pi\lambda^2 r_1^2 r_2^2}\mathbf{M}\mathbf{G}_{At}$$

Depolarized scattering: Scattered waves with averaged matrices:

$$\mathbf{J}_{\text{av}}^s = \frac{1}{4\pi r^2}\langle\boldsymbol{\kappa}_T\rangle\mathbf{J}^i = \frac{Z_0^2 I^2}{16\pi\lambda^2 r_1^2 r_2^2}\langle\boldsymbol{\kappa}_T\rangle(\mathbf{h}_t \otimes \mathbf{h}_t^*)$$

$$\mathbf{G}_{\text{av}}^s = \frac{1}{4\pi r^2}\mathbf{M}_{\text{av}}\mathbf{G}^i = \frac{Z_0^2 I^2}{16\pi\lambda^2 r_1^2 r_2^2}\mathbf{M}_{\text{av}}\mathbf{G}_{At}$$

Depolarized scattering: Received power with incoherently measured matrices:

$$W_d = \frac{Z_0^2 I^2}{128\pi R_a \lambda^2 r_1^2 r_2^2} (\mathbf{h} \otimes \mathbf{h}_r^*)^T \left\{ \begin{array}{c} \mathbf{diag}(1, -1, -1, 1)\mathbf{D}_T \\ \mathbf{D}_S \end{array} \right\} (\mathbf{h}_t \otimes \mathbf{h}_t^*)$$

$$W_d = \frac{Z_0^2 I^2}{128\pi R_a \lambda^2 r_1^2 r_2^2} \mathbf{G}_{Ar}^T \left\{ \begin{array}{c} \mathbf{diag}(1, 1, -1, 1)\mathbf{M} \\ \mathbf{K}_D \end{array} \right\} \mathbf{G}_{At}$$

Depolarized scattering: Received power with averaged matrices:

$$W_{\text{av}} = \frac{Z_0^2 I^2}{128\pi R_a \lambda^2 r_1^2 r_2^2} (\mathbf{h} \otimes \mathbf{h}_r^*)^T \left\{ \begin{array}{c} \mathbf{diag}(1, -1, -1, 1)\langle \boldsymbol{\kappa}_T \rangle \\ \langle \boldsymbol{\kappa}_S \rangle \end{array} \right\} (\mathbf{h}_t \otimes \mathbf{h}_t^*)$$

$$W_{\text{av}} = \frac{Z_0^2 I^2}{128\pi R_a \lambda^2 r_1^2 r_2^2} \mathbf{G}_{Ar}^T \left\{ \begin{array}{c} \mathbf{diag}(1, 1, -1, 1)\mathbf{M}_{\text{av}} \\ \mathbf{K} \end{array} \right\} \mathbf{G}_{At}$$

Power density as an average value:

$$\mathcal{P}_{\text{av}}^s = \frac{1}{8\pi Z_0 r^2} \mathbf{E}^{i\dagger} \langle \mathbf{S}^\dagger \mathbf{S} \rangle \mathbf{E}^i$$

Power density with incoherently measured Graves matrix:

$$\mathcal{P}_{\text{av}}^s = \frac{1}{8\pi Z_0 r^2} \mathbf{E}^{i\dagger} \boldsymbol{\sigma} \mathbf{E}^i$$

Matrix relationships:

$$\begin{aligned}
\mathbf{D}_T &= \mathbf{diag}(1, -1, -1, 1)\mathbf{D}_S \\
\mathbf{J} &= \overline{\mathbf{E}(t) \otimes \mathbf{E}^*(t)} \\
\mathbf{J}_{\text{av}} &= \langle \mathbf{E} \otimes \mathbf{E}^* \rangle \\
\mathbf{G} &= \mathbf{QJ} \\
\mathbf{G}_{\text{av}} &= \mathbf{QJ}_{\text{av}} \\
\mathbf{M} &= \mathbf{QD}_T\mathbf{Q}^{-1} \\
\mathbf{M}_{\text{av}} &= \mathbf{Q}\langle \boldsymbol{\kappa}_T \rangle \mathbf{Q}^{-1} \\
\mathbf{K}_D &= \mathbf{Q}^*\mathbf{D}_S\mathbf{Q}^{-1} \\
\mathbf{K} &= \mathbf{Q}^*\langle \boldsymbol{\kappa}_S \rangle \mathbf{Q}^{-1} \\
\mathbf{M} &= \mathbf{diag}(1, 1, -1, 1)\mathbf{K}_D \\
\mathbf{M}_{\text{av}} &= \mathbf{diag}(1, 1, -1, 1)\mathbf{K}
\end{aligned}$$

REFERENCES

R. M. A. Azzam and N. M, Bashara, *Ellipsometry and Polarized Light*, North-Holland, New York, 1977.

M. Born and E. Wolf, 1965, *Principles of Optics*, 3rd ed. Pergamon Press, New York, 1965.

S. R. Cloude and E. Pottier, "An Entropy Based Classification Scheme for Land Applications of Polarimetric SAR," *IEEE Trans. GRS*, **35**(1), pp. 68–78 (January, 1997).

C. D. Graves, "Radar Polarization Power Scattering Matrix," *Proc. IRE*, **44**(2), pp. 248–252, 1956.

E. Krogager, *Aspects of Polarimetric Radar Imaging*, Ph. D. thesis, Technical University of Denmark, 1993.

H. Mott, *Antennas for Radar and Communications: A Polarimetric Approach*, Wiley-Interscience, New York, 1992.

M. C. Pease, *Methods of Matrix Algebra*, Academic Press, New York, 1965.

J. J. van Zyl and H. A. Zebker, "Imaging Radar Polarimetry", Chapt. 5 in *PIER 3, Progress in Electromagnetics Research*, J. A. Kong, ed., Elsevier, New York, 1990.

R. E. Ziemer and W. H. Tranter, *Principles of Communications*, 4th ed., Wiley, New York, 1995.

PROBLEMS

Note: A general-purpose mathematics program is desirable for solving some of the following problems.

7.1. The backscattering Kronecker-product matrix for a coherently scattering target is

$$\kappa_S = \begin{bmatrix} 4 & 2e^{-j\pi/4} & 2e^{j\pi/4} & 1 \\ 2e^{-j\pi/4} & 8 & 1 & 4e^{j\pi/4} \\ 2e^{j\pi/4} & 1 & 8 & 4e^{-j\pi/4} \\ 1 & 4e^{j\pi/4} & 4e^{-j\pi/4} & 16 \end{bmatrix}$$

Find the Sinclair matrix of the target. Find the antenna polarization for transmitting and receiving antenna that maximizes received power.

7.2. In terms of the Jones matrix elements, find the average Mueller matrix of a target.

7.3. The modified Kronecker-product matrix for backscattering of Section 7.7 was developed for an incident x-polarized wave. Develop a modified matrix if the incident wave is y polarized. State the contraints on derived matrix elements.

7.4. The covariance matrix of a target, specialized to backscattering, is

$$\mathbf{C}_4 = \begin{bmatrix} 5 & \sqrt{2}(1-j) & \sqrt{2}(1-j) & 8 \\ \sqrt{2}(1+j) & 2 & 1 & 2\sqrt{2}(1+j) \\ \sqrt{2}(1+j) & 1 & 2 & 2\sqrt{2}(1+j) \\ 8 & 2\sqrt{2}(1-j) & 2\sqrt{2}(1-j) & 17 \end{bmatrix}$$

Find the polarimetric entropy of the target. From the elements of $\mathbf{C}_4$ find the Graves matrix and the characteristic degrees of polarization of the wave scattered from the targets. Verify that $H \approx 1 - \sqrt{1 - R_1 R_2}$.

7.5. Find the average Kronecker-product matrix $\langle \boldsymbol{\kappa}_T \rangle$ of the target of Problem 7.1.

7.6. What are the SI units of the incoherently measured target matrix $\mathbf{D}_S$? The Graves matrix $\boldsymbol{\sigma}$?

7.7. The incoherently measured matrix of a depolarizing target is

$$\mathbf{D}_T = \begin{bmatrix} 5.5 & \sqrt{2}(1-j) & \sqrt{2}(1+j) & 2 \\ -\sqrt{2}(1-j) & -8.25 & -1 & -2\sqrt{2}(1+j) \\ -\sqrt{2}(1+j) & -1 & -8.25 & -2\sqrt{2}(1-j) \\ 2 & 2\sqrt{2}(1+j) & 2\sqrt{2}(1-j) & 17.125 \end{bmatrix}$$

If the incident wave is linear with a y directed electric field, find the degree of polarization of the scattered wave.

CHAPTER 8

OPTIMAL POLARIZATIONS FOR RADAR

We saw in Chapter 3 that antenna polarizations can be chosen to maximize the power received from a coherently scattering target. In this chapter, we recognize that maximizing received power may not be the appropriate criterion for a specific application and extend the discussion to depolarizing targets.

It was noted in Section 3.11 that it is not necessary to construct an antenna with an effective length chosen to optimize the received power from a target. Instead, information that would be obtained by use of the optimum antenna could be synthesized from that obtained by the use of two orthogonally polarized antennas. A similar conclusion is valid for the optimally polarized antennas discussed in this chapter, although the synthesis may be less simple.

8.1. ANTENNA SELECTION CRITERIA

Radar clutter is an unwanted return that may be completely or partially polarized. Noise is an undesired electrical phenomenon that adds to the desired radar return signal. It may arise inside the receiver or from external sources. If it is external, it may have a degree of polarization ranging from zero to one; if it is internal, it may affect one radar channel more than another and thus act as if it were partially polarized. A distinction between clutter and noise is that clutter power is a function of the transmitted signal power and noise is not. If we are to distinguish polarimetrically between a desired target and clutter or noise, we must know the polarimetric properties of the target and the clutter. We then treat clutter and, with

Remote Sensing with Polarimetric Radar, by Harold Mott

some reservations, noise as targets and choose antenna polarizations to optimize the relationship between the desired signal and the undesired.

If the signal from a desired target is degraded by unpolarized noise or clutter, an antenna polarization that maximizes received power from the target gives a maximum target–noise or target–clutter ratio, but a different approach is necessary if it is desired, for example, to locate roads or other manmade structures in forested areas. An antenna polarization that maximizes the return from a road may give a strong return from the forest. The scattering properties of the forest must therefore be taken into account, and the antenna polarization should be chosen to maximize the contrast between the scatterers rather than received power from the desired target. A polarization chosen to maximize the contrast may not maximize the return from the desired scatterer or minimize that from the undesired one. The received signal may be degraded by noise. If so, a polarization that enhances contrast may give an unacceptable signal–noise ratio, and the conflicting requirements of contrast enhancement and signal–noise improvement must be balanced.

8.2. LAGRANGE MULTIPLIERS

Consider the maximization of a function of two variables, $f(x, y)$, subject to the constraint

$$y = y_a(x)$$

A straightforward approach is to differentiate and find the solution to

$$\frac{d}{dx} f[x, y_a(x)] = \frac{\partial f}{\partial x} + \frac{\partial f}{\partial y}\frac{d}{dx}[y_a(x)] = 0 \tag{8.1}$$

The maximum value of $f(x, y)$ occurs at the solution x_1 of this equation and the corresponding value $y_a(x_1)$.

An alternative method can be used to obtain a solution. Suppose we wish to maximize $f(x, y)$ with the constraint

$$g(x, y) = 0$$

We introduce a new function $F = f + \lambda g$, with the requirement that $g = 0$. In this function, λ is a Lagrange multiplier. The new function is maximized by requiring

$$\frac{\partial f}{\partial x} + \lambda \frac{\partial g}{\partial x} = 0 \tag{8.2}$$

$$\frac{\partial f}{\partial y} + \lambda \frac{\partial g}{\partial y} = 0 \tag{8.3}$$

In the differentiation, x and y are taken as independent variables. To show that a solution to these equations maximizes $f(x, y)$, use

$$g(x, y) = y_a(x) - y = 0$$

With this use, the equations of (8.2) and (8.3) become

$$\frac{\partial F}{\partial x} = \frac{\partial f}{\partial x} + \lambda \frac{dy_a}{dx} = 0$$

$$\frac{\partial F}{\partial y} = \frac{\partial f}{\partial y} - \lambda = 0$$

If these equations are combined, (8.1) results.

Other constraints can be introduced by adding additional functions like λg, with different Lagrange multipliers, to the function f (Morse and Feshbach, 1953, p. 278).

A. COHERENTLY SCATTERING TARGETS

We noted previously that antennas selected to maximize received power from a coherently scattering target may not be optimum, and we now consider choosing antennas to satisfy more general criteria.

8.3. MAXIMUM POWER

The procedure for selecting an antenna to maximize backscattered received power from a coherently scattering target was discussed in Section 3.11 and will be noted only briefly here. The selection of antennas for bistatic scattering is given in greater detail.

Backscattering

For backscattering, transmitting and receiving antennas can have the same effective length, **h**, which must satisfy

$$\mathbf{Sh} = \gamma \mathbf{h}^* \tag{8.4}$$

$$(\mathbf{S}^*\mathbf{S} - |\gamma|^2\mathbf{I})\mathbf{h} = \mathbf{0} \tag{8.5}$$

where **S** is the Sinclair matrix and **I** the identity matrix. If the two eigenvalues of the second equation are distinct, it is straightforward to find the eigenvectors. One of these, when used as the normalized effective length of transmitting and receiving antennas, gives a maximum received power and the other a submaximum power. If the eigenvalues are the same, any vector is an eigenvector. For this case, there is no coneigenvector solution for (8.4), although it may be possible for specific Sinclair matrices to find an effective length that satisfies the equation.

Bistatic Scattering

The power density of a scattered wave, adapted from (7.50), is

$$\mathcal{P}^s = \frac{Z_0^2 I^2}{32\pi\lambda^2 r_1^2 r_2^2}\mathbf{h}_t^\dagger \boldsymbol{\sigma} \mathbf{h}_t \tag{8.6}$$

where $\boldsymbol{\sigma}$ is the Graves matrix for a coherently-scattering target, and r_1 and r_2 are transmitter–target and target–receiver distances. The effective length of the transmitting antenna is $\mathbf{h}_t$. Coordinate systems 1 and 3 of Fig. 7.1 are used.

A transmitting antenna polarization to maximize power density can be found by the method of Lagrange multipliers. The antenna effective length in (8.6) must be held constant. This constraint can be written as

$$\mathbf{h}_t^\dagger \mathbf{I} \mathbf{h}_t = \text{constant} = c$$

Define

$$F = \mathbf{h}_t^\dagger \boldsymbol{\sigma} \mathbf{h}_t + \lambda(c - \mathbf{h}_t^\dagger \mathbf{I} \mathbf{h}_t)$$

F can be maximized by finding its partial derivatives with respect to the complex elements of $\mathbf{h}_t^*$, forming a gradient vector with the derivatives, and setting the vector to zero. Doing so leads to

$$(\boldsymbol{\sigma} - \lambda\mathbf{I})\mathbf{h}_t = \mathbf{0} \tag{8.7}$$

It can be seen from this equation that the transmitting antenna effective length which maximizes power density at the receiver, either for backscattering or bistatic scattering, is an eigenvector of the Graves matrix.

This process maximizes the ratio

$$R = \frac{\mathbf{h}_t^\dagger \boldsymbol{\sigma} \mathbf{h}_t}{\mathbf{h}_t^\dagger \mathbf{I} \mathbf{h}_t} \tag{8.8}$$

If (8.7) is substituted into (8.8), it will be seen that $R = \lambda$ if $\mathbf{h}_t$ is properly chosen. To maximize received power density, the eigenvector corresponding to the greater eigenvalue of (8.7) should be selected. If the two eigenvalues are equal, any vector satisfies (8.7) and the power density of the scattered wave is independent of the polarization of the incident wave.

The electric field of the scattered wave in coordinate system 3 of Fig. 7.1 is

$$\mathbf{E}^s = \mathbf{S}\mathbf{h}_t$$

where unnecessary constants are omitted.

We saw in Section 2.13 that maximum power is received by an antenna whose effective length is proportional to the conjugate of the electric field incident on the antenna, and the effective length and the field are given in the same coordinate system. Then for maximum received power,

$$\mathbf{h}_r = |\mathbf{h}_r| \frac{\mathbf{S}^* \mathbf{h}_t^*}{|\mathbf{S}^* \mathbf{h}_t^*|}$$

Maximum power is received if $\mathbf{h}_t$ is selected as the eigenvector corresponding to the maximum eigenvalue of (8.7) and $\mathbf{h}_r$ is chosen according to this equation, and the effective lengths are substituted into (3.15).

8.4. POWER CONTRAST: BACKSCATTERING

Consider two targets with Sinclair matrices $\mathbf{S}_a$ and $\mathbf{S}_b$ whose scattered waves are completely polarized. It is straightforward to maximize the power ratio of target a to target b. For complete polarization, the antenna with effective length $\mathbf{h}$ used for transmitting and receiving can be selected to give zero copolarized power from target b. While this may not give maximum received power from target a, the contrast is maximum.

Received power from a target is zero if $\mathbf{h}^T \mathbf{S} \mathbf{h} = 0$. If antenna polarization ratio P is used, this equation becomes either

$$S_{yy} P^2 + 2 S_{xy} P + S_{xx} = 0$$

or

$$S_{xx} \left(\frac{1}{P} \right)^2 + 2 S_{xy} \left(\frac{1}{P} \right) + S_{yy} = 0$$

If $S_{yy} \neq 0$, P can be found from the first equation, and if $S_{xx} \neq 0$, $1/P$ can be found from the second. The first form is used here, but the results would be unaltered if the second form were used. The two roots of the quadratic are

$$P_3, P_4 = \frac{\sqrt{S_{xy}^2 - S_{xx} S_{yy}} \mp S_{xy}}{S_{yy}}$$

If either polarization is chosen, the power received from the target is zero. If target a is the desired target and one of these polarizations is chosen using the Sinclair matrix elements of target b, the ratio of powers from the two targets is infinite. The chosen polarization will not in general maximize the received power from target a and may not give a satisfactory signal–noise ratio.

B. DEPOLARIZING TARGETS

A notable attempt to find optimum antennas for depolarizing targets was that of Ioannidis and Hammers (1979) using a Lagrange multiplier approach. Their procedure is relatively complex and involves extensive numerical computations for some targets. The iterative approach of the following section also requires numerical computations but is more efficient than the Ioannidis and Hammers method.

8.5. ITERATIVE PROCEDURE FOR MAXIMIZING POWER CONTRAST

Yang (1999; Yang et al., 2000) developed an efficient method for maximizing the power ratio between two depolarizing targets and discusses the computer implementation of the process. Received power using the Kennaugh matrix and antenna Stokes vectors, and neglecting unnecessary constants, is

$$W = \mathbf{Y}^T \mathbf{K} \mathbf{X} = \mathbf{Y}^T [K_{ij}] \mathbf{X} \qquad i, j = 0, 1, 2, 3$$

$\mathbf{X}$ and $\mathbf{Y}$ are Stokes vectors of transmitting and receiving antennas, respectively. Let $\mathbf{K}$ be symmetric. This equation can be written

$$W = \left[\begin{array}{c|c} 1 & \mathbf{Y}_1^T \end{array}\right] \left[\begin{array}{c|c} K_{00} & \mathbf{a}^T \\ \hline \mathbf{a} & \mathbf{U} \end{array}\right] \left[\begin{array}{c} 1 \\ \hline \mathbf{X}_1 \end{array}\right] \tag{8.9}$$

where $\mathbf{X}_1$, $\mathbf{Y}_1$, $\mathbf{a}$, and $\mathbf{U}$ are formed in an obvious manner from elements of the Stokes vectors and Kennaugh matrix. It is clear that $\mathbf{X}_1^T \mathbf{X}_1 = \mathbf{Y}_1^T \mathbf{Y}_1 = 1$. Multiplication in (8.9) gives

$$W = K_{00} + \mathbf{a}^T \mathbf{X}_1 + \mathbf{a}^T \mathbf{Y}_1 + \mathbf{Y}_1^T \mathbf{U} \mathbf{X}_1$$

which can be expanded and written as

$$W = F_0 + F_1 x_1 + F_2 x_2 + F_3 x_3$$

where the F_i depend on $\mathbf{K}$ and $\mathbf{Y}_1$. To determine the antenna polarization that maximizes the received power contrast between a desired and an undesired target represented by their Kennaugh matrices $\mathbf{A}$ and $\mathbf{B}$, the ratio

$$R_{ab} = \frac{\mathbf{Y}^T \mathbf{A} \mathbf{X}}{\mathbf{Y}^T \mathbf{B} \mathbf{X}} = \frac{A_0 + A_1 x_1 + A_2 x_2 + A_3 x_3}{B_0 + B_1 x_1 + B_2 x_2 + B_3 x_3} \tag{8.10}$$

is maximized. The A_i are determined straightforwardly from the elements of $\mathbf{A}$ and the elements of $\mathbf{Y}_1$. The B_i are found from $\mathbf{B}$ and $\mathbf{Y}_1$.

Maximization of $\mathbf{R}_{ab}$

For specified values of A_i and B_i, (8.10) can be maximized by a proper choice of the x_i. Let this value be $R_m(A_i, B_i) = \max(R_{ab})$. For convenience, we write R_m without the functional notation. In the iterative process of maximizing (8.10), R_m takes on successive values, $R_m^{(n)}$.

We write

$$R_m = \frac{A_0 + A_1x_1 + A_2x_2 + A_3x_3}{B_0 + B_1x_1 + B_2x_2 + B_3x_3} \tag{8.11}$$

and understand that we have substituted R_m for R_{ab} by choosing optimal values for the x_i. The equation corresponds to

$$(A_0 - R_mB_0) + (A_1 - R_mB_1)x_1 + (A_2 - R_mB_2)x_2 + (A_3 - R_mB_3)x_3 = 0 \tag{8.12}$$

If we choose

$$x_i = \frac{A_i - R_mB_i}{\sqrt{\sum_{i=1}^{3}(A_i - R_mB_i)^2}} \tag{8.13}$$

the equality (8.12) allows R_m to be found. Substituting (8.13) into (8.12) yields

$$(A_0 - R_mB_0)^2 = \sum_{i=1}^{3}(A_i - R_mB_i)^2 \tag{8.14}$$

This is a quadratic equation in R_m, which can be found from

$$R_m = \frac{z_{12} \pm \sqrt{z_{12}^2 - z_1z_2}}{z_2} \tag{8.15}$$

where

$$z_1 = A_0^2 - \sum_{1}^{3} A_i^2 \tag{8.16}$$

$$z_2 = B_0^2 - \sum_{1}^{3} B_i^2 \tag{8.17}$$

$$z_{12} = A_0B_0 - \sum_{1}^{3} A_iB_i \tag{8.18}$$

After R_m is found, $\mathbf{X}_1$ can be found from (8.13).

The Optimization Process

The determination of **X** and **Y** giving the maximum value of R_{ab} is straightforward:

1. Choose $\mathbf{Y}_1$ arbitrarily, with the constraint that its magnitude be 1. Let this value be $\mathbf{Y}_1^{(1)}$.
2. Find A_i and B_i with the use of $\mathbf{Y}_1^{(1)}$ and matrices **A** and **B**. These values are $A_i^{(1)}$ and $B_i^{(1)}$.
3. Use $A_i^{(1)}$ and $B_i^{(1)}$ in (8.15)–(8.18) to find $R_m^{(1)}$. Find $\mathbf{X}_1^{(1)}$ from (8.13). It has unit magnitude.
4. Since numerator and denominator of R_{ab} are real, and since **A** and **B** are symmetric, R_{ab} can also be written as

$$R_{ab} = \frac{\mathbf{X}^T\mathbf{A}\mathbf{Y}}{\mathbf{X}^T\mathbf{B}\mathbf{Y}} = \frac{A_0' + A_1'y_1 + A_2'y_2 + A_3'y_3}{B_0' + B_1'y_1 + B_2'y_2 + B_3'y_3}$$

   Using $\mathbf{X}_1^{(1)}$ and matrices **A** and **B**, find the A_i' and B_i'. They are $A_i'^{(1)}$ and $B_i'^{(1)}$.
5. Find $\mathbf{Y}_1$ from (8.13) and (8.15)–(8.18). This is $\mathbf{Y}_1^{(2)}$.
6. Examine the values of $\mathbf{X}_1$ and $\mathbf{Y}_1$ to ascertain that the iterative process is or is not complete. See below.
7. Use the most recent value of $\mathbf{Y}_1$ to determine a new value of $\mathbf{X}_1$, beginning with Step 2. As outlined here, $\mathbf{Y}_1^{(n)}$ is used to find $\mathbf{X}_1^{(n)}$, and $\mathbf{X}_1^{(n)}$ is used to find $\mathbf{Y}_1^{(n+1)}$.
   If

$$\sum_{i=1}^{3} |x_i^{(n+1)} - x_i^n| \le \epsilon \qquad \text{and} \qquad \sum_{i=1}^{3} |y_i^{(n+1)} - y_i^n| \le \epsilon$$

   the optimization process is complete. When it is complete, **Y** and **X** are equal, reflecting the equality of transmitting and receiving antennas. The final value of R_m is a maximum of the power ratio (8.10).

The power ratio may have more than one local maximum. The solution found will, in general, depend on the initial choice of $\mathbf{Y}_1$. It has been suggested (Yang et al., 2000) that several starting points be chosen and the resulting power ratios compared to find the optimum solution. In Section 8.9, it will be shown that if unpolarized clutter is added to the return from a coherently scattering target, the transmitting antenna polarization that maximizes the scattered power density also maximizes the received copolarized power. This polarization may offer a good starting point for the numerical optimization discussed here.

The use of **X** and **Y** in (8.10) does not imply that this optimization is restricted to rectangular coordinates. The Stokes vectors and Kennaugh matrix may be transformed to another polarization basis before the iterative optimization process is carried out.

Power Contrast for Bistatic Scattering

The numerical procedure for selecting an antenna to maximize the received copolarized power for symmetric Kennaugh matrices can be readily adapted to the determination of optimal antennas for bistatic scattering. The major modifications are the use of separate vectors in (8.9) rather than vector **a** only, and the use of the transposes of **A** and **B** to find $\mathbf{Y}_1$ in Steps 4 and 5 of the process. The optimal values of **X** and **Y** may not be equal.

Maximizing the Cross-Polarized Power Ratio

For backscattering, **A** and **B** are symmetric; when R_{ab} is maximum, antenna Stokes vectors **X** and **Y** are equal. The power received from the desired target, **A**, is the copolarized power. Vector **X** which maximizes (8.10) does not in general maximize received power from target **A**. If additional clutter is present, maximizing the contrast in the cross-polarized received power, rather than the contrast in the copolarized power, may be useful. The cross-polarized power is that power received if transmitting and receiving antennas are orthogonal.

Two antennas with effective lengths $\mathbf{h}_1$ and $\mathbf{h}_2$ are orthogonal if their inner product is zero. If this relationship is used to find the Stokes vectors of orthogonal antennas, it will be seen that they are related by

$$\mathbf{G}_2 = \mathbf{diag}(1, -1, -1, -1)\mathbf{G}_1 \tag{8.19}$$

To maximize the cross-polarized power ratio, replace **Y** in (8.10) by its orthogonal vector, using the relationship (8.19), to obtain (Yang, 1999; Yang et al., 1997),

$$R_{\text{cross}} = \frac{\mathbf{Y}_{\text{ortho}}^T\mathbf{AX}}{\mathbf{Y}_{\text{ortho}}^T\mathbf{BX}} = \frac{\mathbf{Y}^T\mathbf{diag}(1, -1, -1, -1)\mathbf{AX}}{\mathbf{Y}^T\mathbf{diag}(1, -1, -1, -1)\mathbf{BX}} = \frac{\mathbf{Y}^T\hat{\mathbf{A}}\mathbf{X}}{\mathbf{Y}^T\hat{\mathbf{B}}\mathbf{X}}$$

Matrices $\hat{\mathbf{A}}$ and $\hat{\mathbf{B}}$ are not symmetric. The numerical optimization outlined in this section must therefore be carried out in the manner outlined for bistatic scattering. Stokes vectors **X** and **Y** will not be equal to each other.

8.6. THE BACKSCATTERING COVARIANCE MATRIX

If (7.48) is used, the ratio of the powers received from two scatterers represented by their covariance matrices is

$$R_{ab} = \frac{\mathbf{H}_3^T\mathbf{C}_{3a}\mathbf{H}_3^*}{\mathbf{H}_3^T\mathbf{C}_{3b}\mathbf{H}_3^*} \tag{8.20}$$

where the covariance matrices are given by (7.46), and the joint antenna vector by (7.47). The numerator in (8.20) represents the target of interest. If Lagrange multipliers are used to maximize the ratio of (8.20), the developments of Section 8.3 suggest that the joint antenna vector $\mathbf{H}_3$ can be found by solution of

$$(\mathbf{C}_{3a} - \lambda \mathbf{C}_{3b})\mathbf{H}_3^* = \mathbf{0} \tag{8.21}$$

Two problems arise if this is done. First, if the same antenna is used for transmitting and receiving, (7.47) shows that the elements of $\mathbf{H}_3$ are not independent. The generalized eigenvector of $\mathbf{C}_{3a}$ and $\mathbf{C}_{3b}$ cannot then be used to find $\mathbf{H}_3$. Second, the Lagrange multiplier method maximizes the numerator of (8.20) while holding the denominator constant. In the search for the optimum contrast ratio, the magnitudes of $\mathbf{h}_r$ and $\mathbf{h}_t$ must be constant. This constraint was not utilized in obtaining the eigenvalue equation (8.21). Instead of explicitly requiring constant effective length magnitudes, let $\mathbf{C}_{3b}$ be the identity matrix, an allowable covariance matrix. For this case, the magnitude of $\mathbf{H}_3$ must be constant if the denominator of (8.20) is to be constant. We noted in Section 7.10, however, that this magnitude does not remain constant when transmitting and receiving antenna effective lengths are varied while keeping their magnitudes constant. In general, therefore, the denominator is not constant and $\mathbf{H}_3^*$ cannot be an eigenvector of (8.21).

It has been noted (Cloude, 1990) that the result of the optimization of (8.20) by solving the eigenvalue equation (8.21) gives some transmit and receive antenna pairs an amplitude weighting larger than for other pairs. The optimization attempt is biased toward these states and gives an erroneous result.

8.7. THE BISTATIC COVARIANCE MATRIX

If (7.45) is used, the ratio of powers received from targets a and b is

$$R_{ab} = \frac{\mathbf{H}_4^T \mathbf{C}_{4a} \mathbf{H}_4^*}{\mathbf{H}_4^T \mathbf{C}_{4b} \mathbf{H}_4^*}$$

where $\mathbf{C}_4$ and $\mathbf{H}_4$ are the target covariance matrix and the joint antenna vector. The elements of $\mathbf{H}_4$ are related by $\mathbf{H}_4^T \mathbf{q} \mathbf{H}_4 = 0$, where

$$\mathbf{q} = \begin{bmatrix} 0 & 1 \\ -1 & 0 \end{bmatrix} \otimes \begin{bmatrix} 0 & 1 \\ -1 & 0 \end{bmatrix}$$

R_{ab} can be maximized by a Lagrange multiplier approach that takes the given constraint on the joint antenna vector into account if we define

$$F = \mathbf{H}_4^T \mathbf{C}_{4a} \mathbf{H}_4^* + \lambda(c - \mathbf{H}_4^T \mathbf{C}_{4b} \mathbf{H}_4^*) - \mu \mathbf{H}_4^T \mathbf{q} \mathbf{H}_4$$

and set its gradient with respect to the components of $\mathbf{H}_4$ to zero. The procedure is difficult to implement.

8.8. MAXIMIZING POWER CONTRAST BY MATRIX DECOMPOSITION

In Section 9.9 it is shown that, for a specified incident-wave polarization, a matrix for a depolarizing target can be separated into a part that gives an unpolarized scattered wave and a part that gives a completely polarized scattered wave. The separation is based on the coherency vector of the scattered wave in coordinate system 2 of Fig. 7.1. It is therefore desirable for our discussion to use $\mathbf{D}_T$ or the modified average Kronecker product matrix $\langle \boldsymbol{\kappa}_{Tb} \rangle_{\text{mod}}$. We use $\mathbf{D}$ to represent either matrix. The decomposition in Section 9.9 is carried out with $\mathbf{D}$ in rectangular form and an x or y directed linear wave incident wave, but the incident waves can be selected arbitrarily, with $\mathbf{D}$ transformed to correspond to the chosen incident-wave polarization.

Maximum Signal–Noise Ratio with the D Matrix: Backscattering

If the target is a combination of a coherently-scattering target and an independent scatterer whose scattered wave is unpolarized for all incident waves, the decomposition of $\mathbf{D}$ is independent of the polarization basis and is

$$\mathbf{D} = \begin{bmatrix} A & 0 & 0 & A \\ 0 & 0 & 0 & 0 \\ 0 & 0 & 0 & 0 \\ A & 0 & 0 & A \end{bmatrix} + \begin{bmatrix} D_{11} - A & D_{12} & D_{12}^* & D_{14} - A \\ -D_{12} & D_{22} & A - D_{14} & D_{24} \\ -D_{12}^* & A - D_{14} & D_{22}^* & D_{24}^* \\ D_{14} - A & -D_{24} & D_{24}^* & D_{44} - A \end{bmatrix}$$

A can be found from (9.36) or (9.42). The Sinclair matrix can be found from the second matrix of (9.46), and the antenna polarization that maximizes coherent power can be found by methods discussed previously.

If the conditions leading to (9.46) are not met, the process is still useful if the degree of polarization of the scattered wave is high for all incident waves. The unpolarized part of the scattered wave is small and does not alter the optimal antenna polarization significantly from the value it would have for the coherently-scattering part of the target. However, its presence does not allow Sinclair matrix $\mathbf{S}$ to be found from $\mathbf{D}$ in (9.46). In that case, the decomposition (9.40) or (9.41), or the equivalent decomposition in another polarization basis, can be carried out to remove the completely depolarizing part of $\mathbf{D}$ for a specified polarization and leave a coherently scattering part from which $\mathbf{S}$ can be found. Sinclair matrix $\mathbf{S}$ is exact only for the specified polarization, but is an approximation to a Sinclair matrix for any incident wave.

Backscattered Power Contrast

Selecting antennas to maximize the contrast between two depolarizing targets is more difficult than selecting antennas to maximize the power from one target.

Minimizing Power from the Undesired (Clutter) Target. Assume that the degree of polarization of the scattered wave from the undesired target is high for any polarization of the incident wave. The conditions given previously for separation of the undesired target's **D** matrix into a part representing a coherently scattering target and one representing a completely depolarizing target are met, at least approximately. After the decomposition, a Sinclair matrix can be found for the coherently scattering part of the undesired target, and an antenna can be chosen to make the coherent return from the undesired target zero. There are two such polarizations; that giving the greatest power from the desired target should be chosen.

Maximizing Power from the Desired Target. Assume the degree of polarization of the scattered wave from the desired target is high for any polarization of the incident wave and that from the undesired target is low. Conditions are met, at least approximately, for decomposing the **D** matrix of the desired target according to the procedures given previously. After this is done, an antenna polarization can be selected to give maximum power from the coherently scattering part of the desired target.

8.9. OPTIMIZATION WITH THE GRAVES MATRIX

The Graves matrix yields the power density of the scattered wave, not the received power. It is useful in combination with other matrices, however, in the selection of antennas for maximizing received power or power contrast.

Maximizing Power Density

Power density can be maximized while keeping the magnitude of the transmitting antenna effective length constant by maximizing the ratio

$$R = \frac{\mathbf{h}_t^\dagger \sigma \mathbf{h}_t}{\mathbf{h}_t^\dagger \mathbf{I} \mathbf{h}_t}$$

In this equation, σ is used either for the average Graves matrix discussed in Section 7.12 or for the corresponding matrix with incoherently measured elements. Maximizing R leads to the eigenvalue equation,

$$(\sigma - \lambda \mathbf{I})\mathbf{h}_t = \mathbf{0}$$

The equation satisfies the constraint that the magnitude of $\mathbf{h}_t$ be constant. The eigenvalues of σ are real and positive and are equal to the ratio R. The eigenvector corresponding to the largest eigenvalue is chosen in order to maximize power density.

Power Density Contrast

We can maximize the ratio of the power densities from two targets if we use (8.6) for the scattered power density and choose the transmitting antenna effective length to maximize

$$R_{ab} = \frac{\mathbf{h}_t^\dagger \boldsymbol{\sigma}_a \mathbf{h}_t}{\mathbf{h}_t^\dagger \boldsymbol{\sigma}_b \mathbf{h}_t} \tag{8.22}$$

The ratio (8.22) can be maximized by creating

$$F = \mathbf{h}_t^\dagger \boldsymbol{\sigma}_a \mathbf{h}_t + \lambda(\text{constant} - \mathbf{h}_t^\dagger \boldsymbol{\sigma}_b \mathbf{h}_t)$$

and setting its gradient with respect to the elements of $\mathbf{h}_t^*$ to zero,

$$(\boldsymbol{\sigma}_a - \lambda \boldsymbol{\sigma}_b)\mathbf{h}_t = \mathbf{0} \tag{8.23}$$

The process of maximizing (8.22) by solving (8.23) does not explicitly utilize the constraint that the magnitude of $\mathbf{h}_t$ be constant. This constraint can be taken into account by converting the constrained maximization of (8.22) to an unconstrained maximization (Yang et al., 2000). Expanding $F = \mathbf{h}_t^\dagger \boldsymbol{\sigma} \mathbf{h}_t$, where $\boldsymbol{\sigma}$ and $\mathbf{h}_t$ are in x and y components, and requiring that the magnitude of $\mathbf{h}_t$ be 1, leads to $F = \mathbf{A}^T \mathbf{G}$, where

$$\mathbf{A} = \begin{bmatrix} A_0 \\ A_1 \\ A_2 \\ A_3 \end{bmatrix} = \frac{1}{2} \begin{bmatrix} \sigma_{11} + \sigma_{22} \\ \sigma_{11} - \sigma_{22} \\ 2\text{Re}(\sigma_{12}) \\ -2\text{Im}(\sigma_{12}) \end{bmatrix} \tag{8.24}$$

and $\mathbf{G}$ is the antenna Stokes vector,

$$\mathbf{G} = \begin{bmatrix} 1 \\ G_1 \\ G_2 \\ G_3 \end{bmatrix} = \frac{1}{2} \begin{bmatrix} 1 \\ |h_{tx}|^2 - |h_{ty}|^2 \\ 2\text{Re}(h_{tx}^* h_{ty}) \\ 2\text{Im}(h_{tx}^* h_{ty}) \end{bmatrix}$$

When written with vectors $\mathbf{A}$ and $\mathbf{G}$, (8.22) becomes

$$R_{ab} = \frac{\mathbf{A}^T \mathbf{G}}{\mathbf{B}^T \mathbf{G}} = \frac{A_0 + A_1 G_1 + A_2 G_2 + A_3 G_3}{B_0 + B_1 G_1 + B_2 G_2 + B_3 G_3} \tag{8.25}$$

where $\mathbf{A}$ is composed of the elements of $\boldsymbol{\sigma}_a$ according to (8.24) and $\mathbf{B}$ is formed from the corresponding elements of $\boldsymbol{\sigma}_b$.

The expanded form of (8.25) has the same form as that of (8.11) and can be solved in the same manner. In (8.25), however, the A_i are constants and an iterative solution is unnecessary. The Stokes vector of the transmitting antenna that maximizes the power density contrast can therefore be found, after (8.24) is used to obtain the A_i, from (8.13) through (8.18).

Maximum Received Power with Noise

Suppose a depolarizing target has a Kennaugh matrix that is the sum of the Kennaugh matrices of a coherently scattering target and unpolarized clutter,

$$\mathbf{K} = \mathbf{K}_{\text{coh}} + \mathbf{N} = \mathbf{K}_{\text{coh}} + N_0 \begin{bmatrix} 1 & 0 & 0 & 0 \\ 0 & 0 & 0 & 0 \\ 0 & 0 & 0 & 0 \\ 0 & 0 & 0 & 0 \end{bmatrix} \tag{8.26}$$

The Graves matrices corresponding to $\mathbf{K}$ and $\mathbf{K}_{\text{coh}}$ are $\boldsymbol{\sigma}$ and $\boldsymbol{\sigma}_{\text{coh}}$. They can be found from the Kennaugh matrices by (7.53). Doing so will show that

$$\boldsymbol{\sigma} = \boldsymbol{\sigma}_{\text{coh}} + \begin{bmatrix} N_0 & 0 \\ 0 & N_0 \end{bmatrix}$$

The eigenvalues of $\boldsymbol{\sigma}$ and $\boldsymbol{\sigma}_{\text{coh}}$ differ, but their eigenvectors are the same.

Received power for backscattering, when using the same antenna, with Stokes vector $\mathbf{G}$, for transmitting and receiving, can be written as

$$W = \mathbf{G}^T\mathbf{K}\mathbf{G} = \mathbf{G}^T\mathbf{K}_{\text{coh}}\mathbf{G} + \mathbf{G}^T\mathbf{N}\mathbf{G} \tag{8.27}$$

The antenna effective length that maximizes the first term of this sum is an eigenvector of $\boldsymbol{\sigma}$ and the second term is independent of $\mathbf{h}_t$. Therefore, the effective length that maximizes received power for a depolarizing target satisfying (8.26) is an eigenvector of Graves matrix $\boldsymbol{\sigma}$.

Many depolarizing targets do not satisfy the assumption that the Kennaugh matrix is the sum of matrices of a coherently scattering target and unpolarized clutter. Then $\mathbf{K}$ may not be separable; even if it is separable, $\mathbf{N}$ will not have the form used in (8.26), and the second term in the sum of (8.27) will not be independent of antenna polarization. The Graves matrix formed from elements of the Kennaugh matrix can still be used to find the antenna polarization that maximizes the power density of the scattered wave. This polarization may not give maximum power, but it will be a good approximation to the polarization that does.

After the transmitting antenna polarization is found, its Stokes vector $\mathbf{G}_{At}$ can be used with the target Kennaugh or Mueller matrix to find the Stokes vector $\mathbf{G}^s$ of the scattered wave.

The degree of polarization of the scattered wave, adapted from (6.36), is

$$R = \frac{\sqrt{(G_1^s)^2 + (G_2^s)^2 + (G_3^s)^2}}{G_0^s}$$

With the help of the degree of polarization, the Stokes vector of the scattered wave can be separated into the Stokes vectors of a completely polarized wave and an unpolarized wave,

$$\mathbf{G}^s = \mathbf{G}^{s(1)} + \mathbf{G}^{s(2)} = \begin{bmatrix} (1-R)G_0^s \\ 0 \\ 0 \\ 0 \end{bmatrix} + \begin{bmatrix} RG_0^s \\ G_1^s \\ G_2^s \\ G_3^s \end{bmatrix}$$

The received power, neglecting a multiplying constant, is given by the product of the wave Stokes vector with the Stokes vector of the receiving antenna,

$$\begin{aligned} W &= \mathbf{G}_{Ar}^T \mathbf{diag}(1, 1, -1, 1)\mathbf{G}^s \\ &= \mathbf{G}_{Ar}^T \mathbf{diag}(1, 1, -1, 1)\mathbf{G}^{s(1)} + \mathbf{G}_{Ar}^T \mathbf{diag}(1, 1, -1, 1)\mathbf{G}^{s(2)} \end{aligned}$$

where

$$\mathbf{G}_{Ar}^T = [1 \quad G_{Ar1} \quad G_{Ar2} \quad G_{Ar3}]$$

The power received from the unpolarized wave is independent of the receiving antenna polarization, and the power from the polarized wave will be a maximum if the antenna Stokes vector elements are related to the elements of the Stokes vector of the polarized part of the wave by

$$\begin{bmatrix} 1 \\ G_{Ar1} \\ G_{Ar2} \\ G_{Ar3} \end{bmatrix} = C \begin{bmatrix} RG_0^s \\ G_1^s \\ -G_2^s \\ G_3^s \end{bmatrix}$$

The receiving antenna effective length components are readily found from this vector.

This process is valid for all incident-wave polarizations.

Initial Vectors for Maximizing Contrast

In this section we saw that the Graves matrix eigenvector maximizes the received copolarized power for backscattering if the target return is that from a coherently scattering target plus unpolarized noise. If the additive noise is not unpolarized, the Graves matrix eigenvector is a good approximation to the optimal antenna effective length if the additive noise has a small degree of polarization. This effective length may in some cases serve as an initial vector for the power ratio maximization of Section 8.5. It neglects the properties of the undesired target but maximizes, at least approximately, the power received from the desired target. Another choice which might be useful is the vector, found from (8.25), which maximizes the power density ratio. It does not maximize the received power ratio but can serve as a starting point for power ratio maximization.

Maximizing Contrast with Graves and Mueller Matrices

Graves and Kennaugh (or Mueller) matrices can be used jointly to maximize, approximately, the contrast in received power from two depolarizing targets. Four approaches are possible: In the first, power density from target *a* is maximized with the Graves matrix for target *a* to give the polarization of the transmitting antenna. Then the receiving antenna is selected to maximize the received power from the polarized part of the scattered wave. This process ignores target *b* and maximizes the received power from target *a*. It is most useful if the signal from target *b* has a small degree of polarization.

In the second approach, power density from target *a* is maximized with the Graves matrix for target *a* to give the polarization of the transmitting antenna. Then the receiving antenna is selected to give zero received power from the polarized part of the wave scattered from target *b*. The power from target *b* is minimum if $\mathbf{G}_{Ar}$ is chosen to make the power from the polarized part of the Stokes vector from *b* equal to zero,

$$\mathbf{G}_{Ar} = C[RG_0^s \quad -G_1^s \quad G_2^s - G_3^s]^T$$

where C is a constant. The antenna effective length can be found from the receiving antenna Stokes vector.

The first method, choosing the receiving antenna to maximize power received from the polarized part of the scattered wave from target *a*, will result in better polarization estimates if target *b* scatters a wave with a small degree of polarization. The second method, choosing the receiving antenna to minimize power received from the polarized part of the scattered wave from target *b*, will give better estimates if the wave scattered by target *b* has a high degree of polarization.

A third approach to maximizing the received power contrast is to use the Graves matrices of targets *a* and *b* to maximize the contrast in power densities at the receiver. After the transmitter polarization is found in this way, the Mueller matrix of target *a* is used to find the scattered Stokes vector. This vector is separated into unpolarized and completely polarized parts. Finally, a receiving antenna is selected to maximize the power from the polarized wave from target *a*.

In a fourth approach, the transmitting antenna is chosen by the third method. Then the Mueller matrix of target *b* is used to find the Stokes vector of the scattered wave from target *b*. Finally, a receiving antenna is selected to minimize the power from the polarized part of this wave.

It is relatively easy to perform all four optimizations. The polarization pair that results in the greatest power ratio can then be selected.

REFERENCES

S. R. Cloude, "Polarimetric Optimisation Based on the Target Covariance Matrix", *Elec. Lett.*, **26**(20), 1670–1671 (1990).

G. A. Ioannidis and D. E. Hammers, "Optimum Antenna Polarizations for Target Discrimination in Clutter", *IEEE Trans. AP*, **27**(3), 357–363 (May 1979).

P. M. Morse and H. Feshbach, *Methods of Theoretical Physics*, Mc-Graw-Hill, New York, 1953.

J. Yang, Y. Yamaguchi, H. Yamada, and S. M. Lin, "The Formulae of the Characteristic Polarization States in the Co-pol Channel and the Optimal Polarization State for Contrast Enhancement", *IEICE Trans. Commun.*, **E80-B**(10), 1570–1575 (1997).

J. Yang, *On Theoretical Problems in Radar Polarimetry*, Ph. D. thesis, Niigata University, 1999.

J. Yang, Y. Yamaguchi, W.-M. Boerner, and S. Lin, "Numerical Methods for Solving the Optimal Problem of Contrast Enhancement", *IEEE Trans. GRS*, **38**(2) (March 2000).

PROBLEMS

Note: A general-purpose mathematics program is desirable for solving some of the following problems.

8.1. The Kennaugh matrix of a desired target for backscattering is

$$\mathbf{K}_a = \begin{bmatrix} 13.0625 & -5.8125 & 3\sqrt{2} & \sqrt{2} \\ -5.8125 & 9.3125 & -\sqrt{2} & -3\sqrt{2} \\ 3\sqrt{2} & -\sqrt{2} & 10.25 & 0 \\ \sqrt{2} & -3\sqrt{2} & 0 & -6.25 \end{bmatrix}$$

and that of an undesired target is

$$\mathbf{K}_b = \begin{bmatrix} 2.875 & 0.625 & 2.5 & 0.25 \\ 0.625 & 0.375 & 0.5 & 1.25 \\ 2.5 & 0.5 & 2.75 & 0 \\ 0.25 & 1.25 & 0 & -0.25 \end{bmatrix}$$

Use the iterative procedure of Section 8.5 to find the polarization of the antenna used for transmitting and receiving that maximizes the contrast in received powers between desired and undesired targets.

8.2. The undesired target of Problem 8.1 is coherently scattering. Find the Sinclair matrix corresponding to $\mathbf{K}_b$ and the antenna polarization that minimizes the received power from it. Compare this polarization to that found in Problem 8.1.

8.3. Find the Graves matrix for the desired target of Problem 8.1. Determine the polarization of the antenna used for transmitting and receiving that maximizes the power density of the wave scattered from the target. Compare the polarization to that found in Problem 8.1.

8.4. In the subsection of Section 8.9 discussing the maximizing of contrast with the Graves and Mueller matrices, four approaches are outlined. Apply all four approaches to matrix $\mathbf{K}_a$ of Problem 8.1 and the matrix

$$\mathbf{K}_b = \begin{bmatrix} 3.875 & 0.625 & 2.5 & 0.25 \\ 0.625 & 0.375 & 0.5 & 1.25 \\ 2.5 & 0.5 & 3.25 & 0 \\ 0.25 & 1.25 & 0 & 0.25 \end{bmatrix}$$

to maximize the contrast in received powers.

8.5. The coherency vector of a wave incident on an antenna is given by

$$\mathbf{J} = [\ 4 \ \ 1.5(\sqrt{3} - j) \ \ 1.5(\sqrt{3} + j) \ \ 2.25\ ]^T$$

Find the coherency vector of the receiving antenna that maximizes received power. Find the coherency vector of the receiving antenna that minimizes received power.

8.6. Write a vector/matrix equation to express the relationship between the elements of an antenna Stokes vector. *Note*: It may be necessary to define a matrix.

8.7. The ratio of received powers from desired and undesired targets is

$$R = \frac{\mathbf{Y}^T \mathbf{A} \mathbf{X}}{\mathbf{Y}^T \mathbf{B} \mathbf{X}}$$

where $\mathbf{A}$ and $\mathbf{B}$ are the Kennaugh matrices of desired and undesired targets, and $\mathbf{X}$ and $\mathbf{Y}$ are the Stokes vectors of transmitting and receiving antennas respectively. Ratio R is maximized in a Lagrange multiplier approach by maximizing the numerator while holding the denominator constant and obeying the antenna constraints developed in Problem 8.6. Develop the function F, to be maximized, in the manner in which it was developed in Section 8.2, including the constraints on the denominator and those on the antenna Stokes vectors.

8.8. Show that if f and g in Section 8.2 are functions of N variables, maximization of $F = f + \lambda g$ is carried out by setting the gradient of F, with respect to the variables, to zero.

CHAPTER 9

CLASSIFICATION OF TARGETS

The assignment of radar targets to classes on the basis of voltages received from the targets is an application of the principles of pattern recognition. The measured quantities in polarimetric remote sensing are the voltages from which the elements of the Sinclair matrix are found or the received powers from which the parameters of a power-type target matrix are found.

A. CLASSIFICATION CONCEPTS

Classification can be based directly on the measured signals, or the measured quantities can be used to form other parameters, or features, to represent the target. It is desirable that the number of features be smaller than the number of measurements and that the features be more suitable for classification than the original measurements. Mathematical techniques have been developed for selecting desirable feature sets (Therrien, 1989, p. 71), but sets can be formed intuitively from an understanding of a classification problem, and that is done in the latter part of this chapter. The ultimate test of the classification method, the number of classes, the measurements made of a target's scattered wave, and the features used to classify it must be usefulness and accuracy of classification. *Note*: In this discussion, we will use "features" to mean either the selected features that represent the target in the classification process or the basic measurements if they are used in classification.

Remote Sensing with Polarimetric Radar, by Harold Mott

9.1. REPRESENTATION AND CLASSIFICATION OF TARGETS

In the procedures to be considered, a target feature is plotted as a coordinate in a multidimensional *feature space*, or hyperspace, with a dimension equal to that of the number of features used. A set of features representing a target corresponds to a point in this *classification space*, and classification is based on the distance of this point from a reference point. The features should be of the same order numerically and represent target properties that are naturally comparable. If they do not meet these criteria, the concept of distance in the feature space lacks credibility. This discussion is about variables that are continuous, represent similar quantities, and are numerically comparable. The received voltages in radar are random, and the features representing the target are random variables. A *feature vector*, or *classification vector*, whose components are the coordinates of the target features, can be formed in the hyperspace. The vector components are restricted at this point to be real, and a complex quantity is treated as two real components of a vector. The plotted feature may be of a nonvisualizable quantity, such as a voltage, and the hyperspace may seem abstract. It is useful, nevertheless, to associate a hyperspace with the features.

It may be apparent when measurements are made that distinct groups exist among the measurements or the extracted features. Targets can be classified by placing them in the closest natural group in the classification hyperspace. The process in which class assignments depend on the features without the use of externally acquired information is called *unsupervised classification* or *clustering*. It is called "unsupervised" by analogy to an unsupervised learning process in which the learning system itself creates categories in the course of learning.

The focus of this discussion is the classification of targets about which externally acquired information is available. For airborne observation of the ground, for example, ground-based observers may visually identify areas as forested or farmland. Measurements made in these areas provide references to which unknown targets can be compared. The procedure in which a target is assigned to classes that have been identified in some manner other than by measurements used in the classification process is called *supervised classification*. This is by analogy to supervised learning in which corrections are provided by a teacher. There can be overlap of the classes. Terrain identified as farmland by ground observers may produce polarimetric measurements more characteristic of forest than of farmland. It must be noted also that the system supervisor is imperfect. If a terrain cell is to be classified as earth or water, a ground observer must arbitrarily decide if a swampy area is one or the other. Measurements of the signals of targets known from external observation to belong to specific classes will be referred to as verified or validated measurements, and the features derived from the measurements are validated features. In the classification process, the features of an unknown target will be compared to the validated features and class assignments made on the basis of the comparison.

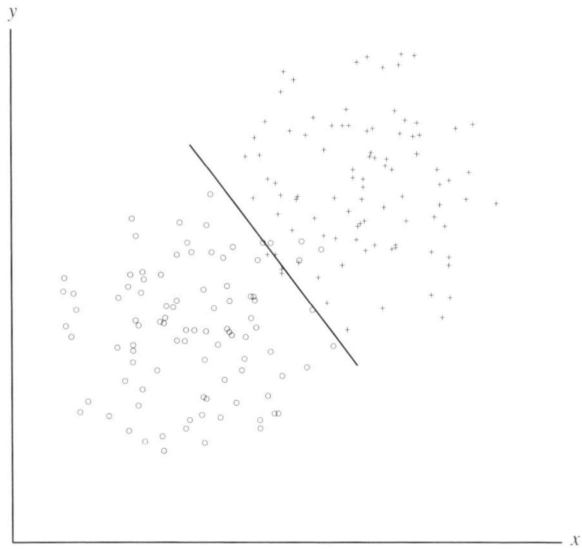

Fig. 9.1. Overlapping classes.

Overlapping Sets

Figure 9.1 shows a scatter plot for two validated classes with different means. The points for the classes form two clusters, and there is overlap of the clusters. Some members of class 1 appear to be associated more closely with class 2 members than with class 1, and vice versa. Classification of targets by comparison to validated sets that overlap is more subject to error than classification with disjoint classes. In the overlap region, it may be necessary to accept misclassification or to defer classification.

Distance in Classification Space

The measurements or features of a target are represented as a point in a hyperspace, and it is natural to think of a distance between two points in the space. The most commonly used distance is the straight-line or Euclidean distance, but there are other useful distance measures.

A distance function or *metric*, $d(x, y)$, is one that meets the following conditions (Therrien, 1989, p. 47):

$$d(\mathbf{x}, \mathbf{y}) \begin{cases} > 0 & \mathbf{x} \neq \mathbf{y} \\ = 0 & \mathbf{x} = \mathbf{y} \end{cases}$$

$$d(\mathbf{x}, \mathbf{y}) = d(\mathbf{y}, \mathbf{x})$$

$$d(\mathbf{x}, \mathbf{y}) + d(\mathbf{y}, \mathbf{z}) \geq d(\mathbf{x}, \mathbf{z})$$

The third condition is the triangular inequality and is familiar from plane analytic geometry.

The Euclidean distance is given by

$$d_E(\mathbf{x}, \mathbf{y}) = \left[\sum_1^N (x_n - y_n)^2 \right]^{1/2}$$

It is readily seen that it meets the three conditions given above. The squared *Mahalonobis distance* between the end points of real random vectors $\mathbf{X}$ and $\mathbf{Y}$ that take on values $\mathbf{x}$ and $\mathbf{y}$ is

$$d_M^2(\mathbf{x}, \mathbf{y}) = (\mathbf{x} - \mathbf{y})^T \mathbf{C}^{-1} (\mathbf{x} - \mathbf{y})$$

where $\mathbf{C}$ is the covariance matrix of $\mathbf{X}$. If $\mathbf{X}$ and $\mathbf{Y}$ belong to clusters with different covariance matrices, the Mahalonobis distance is approximate.

Classification Methods

Consider the assignment of a target represented by vector $\mathbf{x}$ to class ω_i. The first classification method to be considered is outlined in Section 9.2. We assume that the class-conditional probability density $p_{\mathbf{X}|\omega_i}(\mathbf{x}|\omega_i)$ is known. This information is unlikely to be available in remote sensing, but a form for the density can be assumed and the parameters needed to complete the definition of the densities found by measurements.

9.2. BAYES DECISION RULE

Let H_i be the hypothesis that the object belongs to class ω_i and consider the conditional probability $P(H_i|\mathbf{x})$ that, given $\mathbf{x}$, H_i is true. This probability is often written by using ω_i itself as the hypothesis H_i, or $P(\omega_i|\mathbf{x}) = P(H_i|\mathbf{x})$. The probability expressed by the this equation is the *posterior probability*. The probability $P(H_i) = P(\omega_i)$, where H_i is the hypothesis that vector $\mathbf{x}$ belongs to class ω_i, is the *prior probability*.

A reasonable classification scheme is to assign the object to the class for which the posterior probability,

$$P(\omega_i|\mathbf{x}) = \frac{p_{\mathbf{X}|\omega_i}(\mathbf{x}|\omega_i) P(\omega_i)}{p_{\mathbf{X}}(\mathbf{x})} \tag{9.1}$$

adapted from (B.27), is greatest. Since the denominator is the same for all classes, the object is assigned to the class with the greatest numerator. This classification method results in an assignment with the smallest probability of error. It is called the *Bayes minimum probability of error rule*. An appropriate classification method is: Choose ω_i if $P(\omega_i|\mathbf{x}) > P(\omega_j|\mathbf{x}) \quad$ for all $j \neq i$. The posterior probabilities are not known in the general case. We therefore rewrite this equation using (9.1). Doing so leads to the classification rule: To assign vector $\mathbf{x}$ to one of classes having known prior probabilities, substitute it into the known class-conditional densities, and choose ω_i if

$$\frac{p_{\mathbf{X}|\omega_i}(\mathbf{x}|\omega_i)}{p_{\mathbf{X}|\omega_j}(\mathbf{x}|\omega_j)} > \frac{P(\omega_j)}{P(\omega_i)} \qquad \text{for all } j \neq i \tag{9.2}$$

The class-conditional probability densities are sometimes called *likelihood functions*, and their ratio is the likelihood ratio. The decision rule (9.2) is the *likelihood ratio test.*

Error Probability

Let H be the hypothesis that an error in classification has been made. Then

$$P(H) = P(\text{error}) = \int_{-\infty}^{\infty} P(\text{error}|\mathbf{x})\, p_{\mathbf{X}}(\mathbf{x})\, d\mathbf{x}$$

The classification space can be divided into nonoverlapping regions R_1 and R_2 associated, respectively, with classes ω_1 and ω_2. The class conditional densities $p_{\mathbf{X}|\omega_1}(\mathbf{x}|\omega_1)$ and $p_{\mathbf{X}|\omega_2}(\mathbf{x}|\omega_2)$ overlap so that a vector belonging to the distribution with density $p_{\mathbf{X}|\omega_2}(\mathbf{x}|\omega_2)$ can be located with non-zero probability in R_1, and vice versa. The probabilities of the four possible decisions for assigning a vector are

$$P(\text{assign to } \omega_i \mid \text{belongs to } \omega_i) = \int_{R_i} p_{\mathbf{X}|\omega_i}(\mathbf{x}|\omega_i)\, d\mathbf{x} \tag{9.3}$$

$$P(\text{assign to } \omega_i \mid \text{belongs to } \omega_j) = \int_{R_i} p_{\mathbf{X}|\omega_j}(\mathbf{x}|\omega_j)\, d\mathbf{x} \tag{9.4}$$

for $i = 1, 2$ and $i \neq j$. The probabilities of the two incorrect decisions are the class error probabilities,

$$\epsilon_i = P(\text{assign to } \omega_j \mid \text{belongs to } \omega_i) = \int_{R_j} p_{\mathbf{X}|\omega_i}(\mathbf{x}|\omega_i)\, d\mathbf{x} \quad i, j = 1, 2 \quad i \neq j$$

The Bayes probability of error is the probability of error if a vector belongs to ω_1 multiplied by the probability that the vector belongs to that class, plus the probability of error if it belongs to ω_2 times the probability that it belongs to that class, or

$$\epsilon_B = P(\text{error}) = \epsilon_1 P(\omega_1) + \epsilon_2 P(\omega_2)$$

The Bayes error is the minimum error probability for classification (Therrien, 1989, p. 144). Its computation is difficult except in special cases (Fukunaga, 1990, p. 54).

Bayes Risk

A classification decision has a cost associated with it. Let H_i be the hypothesis that an object represented by $\mathbf{x}$ belongs to class ω_i, of two classes. The costs associated with each decision will be related to four terms:

$$c_{ii}P(\text{decide } H_i \mid H_i \text{ true}) = c_{ii}P(H_i|H_i) \qquad i = 1, 2$$

$$c_{ij}P(\text{decide } H_i \mid H_j \text{ true}) = c_{ij}P(H_i|H_j) \qquad i = 1, 2 \qquad i \neq j$$

where c_{ij} is the applicable cost of a decision. The probabilities in the cost terms are those of (9.3) and (9.4). The average cost of making a decision if $\mathbf{x}$ is from class ω_1 is

$$\begin{aligned} C(\text{decision} \mid H_1) &= c_{11}P(H_1 \mid H_1) + c_{21}P(H_2 \mid H_1) \\ &= c_{11}\int_{R_1} p_{\mathbf{X}|\omega_1}(\mathbf{x}|\omega_1)\,d\mathbf{x} + c_{21}\int_{R_2} p_{\mathbf{X}|\omega_1}(\mathbf{x}|\omega_1)\,d\mathbf{x} \\ &= c_{11}\int_{R_1} p_{\mathbf{X}|\omega_1}(\mathbf{x}|\omega_1)\,d\mathbf{x} + c_{21}\left[1 - \int_{R_1} p_{\mathbf{X}|\omega_1}(\mathbf{x}|\omega_1)\,d\mathbf{x}\right] \end{aligned} \tag{9.5}$$

The average cost of a decision when $\mathbf{x}$ is from ω_2 is

$$\begin{aligned} C(\text{decision} \mid H_2) &= c_{12}P(H_1 \mid H_2) + c_{22}P(H_2 \mid H_2) \\ &= c_{12}\int_{R_1} p_{\mathbf{X}|\omega_2}(\mathbf{x}|\omega_2)\,d\mathbf{x} + c_{22}\left[1 - \int_{R_1} p_{\mathbf{X}|\omega_2}(\mathbf{x}|\omega_2)\,d\mathbf{x}\right] \end{aligned} \tag{9.6}$$

The average cost of a decision, without regard to which hypothesis is true, is the average of these costs, taking into account the prior probabilities of $\mathbf{x}$ being from the two classes. It is

$$C = P(H_1)C(\text{decision} \mid H_1) + P(H_2)C(\text{decision} \mid H_2) \tag{9.7}$$

If (9.5) and (9.6) are substituted into (9.7), it can be arranged as

$$\begin{aligned} C &= P(H_1)c_{21} + P(H_2)c_{22} \\ &\quad + \int_{R_1} \left[P(H_2)(c_{12} - c_{22})p_{\mathbf{X}|\omega_2}(\mathbf{x}|\omega_2) - P(H_1)(c_{21} - c_{11})p_{\mathbf{X}|\omega_1}(\mathbf{x}|\omega_1)\right] d\mathbf{x} \\ &= C_F + \int_{R_1} f(\mathbf{x})\,d\mathbf{x} \end{aligned} \tag{9.8}$$

The average cost, not taking into account the class of $\mathbf{x}$, is the *Bayes risk*.

It is reasonable to assume all costs positive and that the cost of an incorrect decision is greater than that of a correct one. Both product terms in the integral

(9.8) are therefore positive. Cost C is positive, so to minimize it, the integral should be negative and have the largest possible magnitude.

To minimize the integral in (9.8), R_1 should be the region where the integrand is negative, since each $f(\mathbf{x})\,d\mathbf{x}$ then contributes a negative increment to the overall integral. Region R_1 found in this way may consist of unconnected regions. To put this another way, $\mathbf{x}$ should be placed in R_1 if it makes the integrand negative. This defines R_1. After R_1 and R_2 are determined, an object represented by $\mathbf{x}$ is assigned to ω_1 if it lies in R_1. The decision rule can be expressed by: If

$$\frac{p_{\mathbf{X}|\omega_2}(\mathbf{x}|\omega_2)}{p_{\mathbf{X}|\omega_1}(\mathbf{x}|\omega_1)} \begin{cases} > \frac{P(\omega_1)(c_{21}-c_{11})}{P(\omega_2)(c_{12}-c_{22})} & \text{choose } \omega_2 \\ < \frac{P(\omega_1)(c_{21}-c_{11})}{P(\omega_2)(c_{12}-c_{22})} & \text{choose } \omega_1 \end{cases}$$

The left side of the decision equation is the likelihood ratio, and the right side is the *threshold* of the test. This is the *Bayes minimum risk rule*. It does not give the minimum probability of error in general. If $c_{21} - c_{11} = c_{12} - c_{22}$, the rule becomes the likelihood ratio test and is the same as the minimum-probability-of-error rule.

The Bayes risk can be written using integration over both regions R_1 and R_2 and can be extended to more than two classes. The Bayes risk for any number of classes is (Hand, 1981, p. 6)

$$C = \sum_j \int_{R_j} \left[\sum_i c_{ji} P(\omega_i) p_{\mathbf{X}|\omega_i}(\mathbf{x}|\omega_i) \right] d\mathbf{x} \tag{9.9}$$

This multiclass Bayes risk will be minimized by defining the regions R_k such that $\mathbf{x}$ is assigned to ω_k when (Hand, 1981, p. 6)

$$\sum_i c_{ki} P(\omega_i) p_{\mathbf{X}|\omega_i}(\mathbf{x}|\omega_i) < \sum_i c_{ji} P(\omega_i) p_{\mathbf{X}|\omega_i}(\mathbf{x}|\omega_i) \qquad j \neq k$$

It has been noted (Skolnik, 1962, p. 380) that in the detection of targets in clutter there is no satisfactory basis for assigning costs, and situations can be envisioned in terrain classification for which assigning costs is difficult.

9.3. THE NEYMAN–PEARSON DECISION RULE

The Neyman–Pearson decision rule is widely used for the problem of deciding that a radar return is noise alone or signal plus noise. It has an application to the multiclass problem, since before an assignment of a radar return is made, a decision must be made that the return is not so corrupted with noise as to be useless. If it is incorrectly decided that a radar return is noise alone, a missed detection occurs, and if it is incorrectly decided that a signal from a target is present, a false alarm results.

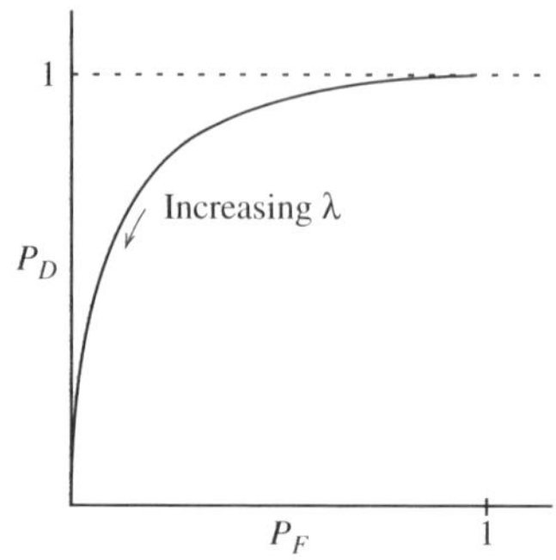

Fig. 9.2. Receiver operating characteristic.

The decision rule, which we will not develop here, is: If

$$\frac{p_{\mathbf{X}|\omega_1}(\mathbf{x}|\omega_1)}{p_{\mathbf{X}|\omega_2}(\mathbf{x}|\omega_2)} \begin{cases} > \lambda & \text{choose } \omega_1 \\ < \lambda & \text{choose } \omega_2 \end{cases}$$

where the left side of the equation is the likelihood ratio. The parameter λ can be found from the *receiver operating characteristic*, which is a plot, of which Fig. 9.2 is typical, of the probability of detection versus the probability of false alarm. It has been shown that λ is the slope of the operating characteristic (Van Trees, 1968, p. 36), which is usually obtained experimentally by applying the decision rule to a set of test samples and counting errors.

9.4. BAYES ERROR BOUNDS

It was noted in Section 9.2 that the Bayes error is the minimum error probability in classification. In Appendix D, a discussion of the upper bound on the Bayes error probability is given.

9.5. ESTIMATION OF PARAMETERS FROM DATA

In remote sensing it is unlikely that the class-conditional probability densities are known. If their form is assumed, the parameters that complete the definition of the probability densities can be determined by measurements. Classification can then proceed according to the procedures outlined in Section 9.2. If the correct form of the probability density functions is chosen and if the density parameters are estimated accurately, the error probability approaches the Bayes error. In this section, we consider the determination of the probability density parameters.

Let real random vector $\mathbf{X}$ of M elements represent one measurement of a target. "One measurement" includes multiple measurements needed to form a measurement vector. The parameters necessary to specify a pdf can be formed into vector $\mathbf{Y}$ with Q real elements. An estimate of $\mathbf{Y}$ will be made utilizing N measurements of vector $\mathbf{X}$ with measured values $\mathbf{x}$. This estimate can be written as $\hat{\mathbf{Y}}(\mathbf{Z})$, where $\mathbf{Z}$ is a real random vector of MN elements formed from the N measurements of M-element vector $\mathbf{X}$,

$$\mathbf{Z} = \left[\mathbf{X}_1^T \cdots \mathbf{X}_N^T\right]^T$$

$\hat{\mathbf{Y}}$ is a random vector. The circumflex indicates an estimate for vectors and matrices. We consider one class at a time and work with probability densities rather than class-conditional densities.

Properties of Estimates

An estimate of a parameter vector is *unbiased* if the expected value of the estimate is equal to the true vector, thus, $E\left[\hat{\mathbf{Y}}(\mathbf{Z})\right] = \mathbf{Y}$. An estimate of a vector parameter is *consistent* if it converges in probability to the true parameter value as the number of measurements used to form the estimate increases. This is expressed as

$$\lim_{N \to \infty} P\left(|\mathbf{Y} - \hat{\mathbf{Y}}| < \epsilon\right) = 1$$

for all ϵ.

The relative efficiency of two estimates, 1 and 2, if both estimators are based on the same sample size, can be compared by (Fukunaga, 1972, p. 124)

$$\eta = E\left(\frac{|\hat{\mathbf{Y}}_1 - \mathbf{Y}|^2}{|\hat{\mathbf{Y}}_2 - \mathbf{Y}|^2}\right)$$

If $\eta \leq 1$ for all $\hat{\mathbf{Y}}_2$, then $\hat{\mathbf{Y}}_1$ is an efficient estimate.

Estimate of Mean and Covariance Matrix

If N samples of real vector $\mathbf{X}$ are obtained, their mean value, assuming all values of $\mathbf{X}$ equally probable, is

$$\hat{\mathbf{m}} = \frac{1}{N}\sum_1^N \mathbf{x}_n$$

The expectation of this parameter is

$$E(\hat{\mathbf{m}}) = \frac{1}{N}\sum_1^N E(\mathbf{x}_n)$$

But the expectations of all $\mathbf{x}_n$ are the same, $\mathbf{m}$. Therefore $E(\hat{\mathbf{m}}) = \mathbf{m}$. The estimate of the mean of $\mathbf{x}$ is unbiased.

A tentative estimate of the covariance matrix is

$$\hat{\mathbf{C}} = \frac{1}{N}\sum_{1}^{N}(\mathbf{x}_n - \hat{\mathbf{m}})(\mathbf{x}_n - \hat{\mathbf{m}})^T$$

The estimate of the mean is used in the equation, since the true mean is unknown. This estimate can be rearranged in a straightforward manner and the expectation shown to be

$$E(\hat{\mathbf{C}}) = \frac{N-1}{N}\mathbf{C}$$

The estimate is biased. The bias can be removed by using as an estimate,

$$\hat{\mathbf{C}} = \frac{1}{N-1}\sum_{1}^{N}(\mathbf{x}_n - \hat{\mathbf{m}})(\mathbf{x}_n - \hat{\mathbf{m}})^T$$

Maximum Likelihood Estimation

One method of estimating a parameter from sample vector $\mathbf{Z}$ is to choose the parameter value which maximizes the probability that $\mathbf{Z}$ will be the result of measurements. We can do this by maximizing the probability density $p_{\mathbf{Z}|\mathbf{Y}}(\mathbf{z}|\mathbf{y})$. Given $\mathbf{z}$, we find the value of $\mathbf{y}$ most likely to give the known $\mathbf{z}$. Considered as a function of the parameter $\mathbf{y}$ rather than the data $\mathbf{z}$, this probability density is the *likelihood function*. It can be maximized by setting its gradient with respect to the elements of $\mathbf{y}$ to zero,

$$\nabla_{\mathbf{y}}\left[p_{\mathbf{Z}|\mathbf{Y}}(\mathbf{z}|\mathbf{y})\right] = \mathbf{0}$$

This is the *likelihood equation*. Since the logarithm is a monotonically increasing function, the *log likelihood equation*,

$$\nabla_{\mathbf{y}}\left\{\ln[p_{\mathbf{Z}|\mathbf{Y}}(\mathbf{z}|\mathbf{y})]\right\} = \mathbf{0}$$

can be solved instead.

The solution to one of these equations is the maximum-likelihood estimate of the parameter $\mathbf{y}$. The maximum-likelihood estimate is consistent and asymptotically the most efficient (Therrien, 1989, p. 113).

Bayes Estimation

If a parameter vector is estimated incorrectly, a cost is incurred. Cost functions are problem-dependent and cannot be specified in general, but two useful ones are:

$$C[\mathbf{Y}, \hat{\mathbf{Y}}(\mathbf{Z})] = |\mathbf{Y} - \hat{\mathbf{Y}}(\mathbf{Z})|^2 \qquad \text{squared-error cost}$$

$$C[\mathbf{Y}, \hat{\mathbf{Y}}(\mathbf{Z})] = \begin{cases} 1 & |\mathbf{Y} - \hat{\mathbf{Y}}(\mathbf{Z})| > \delta > 0 \\ 0 & \text{otherwise} \end{cases} \qquad \text{uniform cost}$$

where $\mathbf{Y}$ is the true parameter vector and $\hat{\mathbf{Y}}$ its estimate. δ is an arbitrarily selected small constant.

The average cost of choosing the parameter vector incorrectly is

$$\overline{C[\mathbf{Y}, \hat{\mathbf{Y}}(\mathbf{Z})]} = \int_{-\infty}^{\infty} \int_{-\infty}^{\infty} C[\mathbf{y}, \hat{\mathbf{y}}(\mathbf{z})] p_{\mathbf{Y},\mathbf{Z}}(\mathbf{y}, \mathbf{z})\, d\mathbf{y}\, d\mathbf{z} = \int_{-\infty}^{\infty} I(\hat{\mathbf{y}}) p_{\mathbf{Z}}(\mathbf{z})\, d\mathbf{z}$$

where

$$I(\hat{\mathbf{y}}) = \int_{-\infty}^{\infty} C[\mathbf{y}, \hat{\mathbf{y}}(\mathbf{z})] p_{\mathbf{Y}|\mathbf{Z}}(\mathbf{y}|\mathbf{z})\, d\mathbf{y}$$

This integral is the *conditional risk*. If the cost function is nonnegative, the conditional-risk integral is nonnegative. The average cost can therefore be minimized by minimizing the conditional risk for each $\mathbf{z}$.

For the squared-error cost function, the conditional-risk integral is

$$I(\hat{\mathbf{y}}) = \int_{-\infty}^{\infty} |\mathbf{y} - \hat{\mathbf{y}}|^2 p_{\mathbf{Y}|\mathbf{Z}}(\mathbf{y}|\mathbf{z})\, d\mathbf{y}$$

Setting the gradient of this integral with respect to the components of $\hat{\mathbf{y}}$ to zero leads to

$$\hat{\mathbf{y}}(\mathbf{z}) = \int_{-\infty}^{\infty} \mathbf{y}\, p_{\mathbf{Y}|\mathbf{Z}}(\mathbf{y}|\mathbf{z})\, d\mathbf{y}$$

A second differentiation of the conditional risk integral shows that this estimate gives a minimum of the integral and not a maximum. This parameter estimate is called the *mean-square estimate* (Therrien, 1989, p. 117).

The conditional risk integral for the uniform cost function is

$$I(\hat{\mathbf{y}}) = 1 - \int_{R_\delta} p_{\mathbf{Y}|\mathbf{Z}}(\mathbf{y}|\mathbf{z})\, d\mathbf{y}$$

The integration is to be carried out over that region of parameter space where $|\mathbf{y} - \hat{\mathbf{y}}| \leq \delta$. δ is small and the region of space R_δ is small. A good approximation to the conditional risk integral is, therefore,

$$\int_{R_\delta} p_{\mathbf{Y}|\mathbf{Z}}(\mathbf{y}|\mathbf{z})\, d\mathbf{y} \approx p_{\mathbf{Y}|\mathbf{Z}}(\hat{\mathbf{y}}|\mathbf{z}) R_\delta$$

To minimize the conditional risk, this integral should be a maximum. Then the rule to follow for uniform cost is to maximize the posterior density, or

$$p_{\mathbf{Y}|\mathbf{Z}}(\mathbf{y}|\mathbf{z})|_{\mathbf{y}=\hat{\mathbf{y}}(\mathbf{z})} = \text{maximum}$$

This is the *maximum a posteriori (MAP) estimate*.

9.6. NONPARAMETRIC CLASSIFICATION

A nonparametric classifier does not assume a form for the class-conditional density functions of the feature vectors. Two such classification methods are considered here. In the first, a validated vector set is used to estimate the probability density for each class, and the class assignment is based on the densities. The classifier becomes the Bayes classifier if the density estimates converge to the true densities for an infinite number of samples, and the error is the Bayes error (Fukunaga, 1990, p. 300). In the second, the nearest neighbors of an unknown vector are found, and the vector is assigned to the class whose validated vectors are the nearest neighbors of the unknown.

Histograms

The probability density for a class can be approximated by constructing a histogram using vectors known to belong to the class. To form a histogram in M variables, the M-dimensional classification space is partitioned into equal hypervolume cells and data from a verified set are used to determine and store the number of data points in each cell. If each vector element is divided into P intervals, there will be P^M cells in the histogram for each class. The number of vectors in the validated set for a class is apt to be far fewer than this number, and most of the histogram cells will be empty. An approach that avoids this difficulty is to let the validated-set data determine the cell locations and sizes, but that method is not discussed here.

Parzen Methods

Let scalar variable X belong to class ω_j and be represented by measured values $x_1, x_2, \ldots, x_N$. To simplify the notation, make the class membership implicit and write the probability distribution of $\mathbf{X}$ as

$$F_X(x) = F_{X|\omega_j}(x|\omega_j)$$

An estimator for $F(x)$ is the number of sample points whose measured values are less than x, symbolized by $n(x)$, divided by the total number of sample points N,

$$\hat{F}_X(x) = \frac{n(x)}{N}$$

An approximation to the probability density is

$$\hat{p}_X(x) = \frac{\hat{F}_X(x+h) - \hat{F}_X(x-h)}{2h} = \frac{n(x+h)/N - n(x-h)/N}{2h}$$

where $2h$ is a small interval along the line on which the sample points are plotted.

Define a function

$$k(u) = \begin{cases} 1/2h & |u| \le h \\ 0 & \text{otherwise} \end{cases}$$

With this definition, the density estimate can be written as

$$\hat{p}_X(x) = \frac{1}{N}\sum_{1}^{N} k(x - x_i)$$

The function $k(x - x_i)$ is called a *kernel* or *Parzen window*, and the given form is a rectangular kernel. With the use of a rectangular kernel, all points in the interval $(x - h)$ to $(x + h)$ contribute equally to the density and points outside the interval contribute nothing. The resulting pdf is discontinuous. To achieve a smooth pdf, a smooth kernel instead of the rectangular one can be used. Two common smooth kernels are

$$k(x - x_i) = \frac{1}{\sqrt{2\pi}h}\exp\left[-\frac{(x - x_i)^2}{2h^2}\right] \qquad \text{Gaussian}$$

$$k(x - x_i) = \frac{1}{2h}\exp\left[-\frac{|x - x_i|}{h}\right] \qquad \text{two-sided exponential}$$

In order that $\hat{p}_X(x)$ be a probability density function, the kernel must be a probability density, the choice of which is a matter of judgment. A rectangular kernel leads to the easiest computations but gives a pdf with discontinuities and points of zero density. A normal kernel shape removes the discontinuities, and since all sample points contribute to the density estimate everywhere, no zero density regions occur. The necessary calculations can be long.

The value of h associated with the rectangular kernel is one half the kernel width. For the other kernels, h is the *spread* or *smoothing factor* (Hand, 1981, p. 26). A factor in the choice of h is prior knowledge of the density. A too-small value will result in a density with many peaks, while a too-large value will cause detail of the pdf to be lost.

For an M-dimensional vector $\mathbf{x}$, it is appropriate to replace the length interval of the one-variable case by a hypercube with center at $\mathbf{x}$ and volume $V = (2h)^M$. An approximation to the density is

$$p_{\mathbf{X}}(\mathbf{x}) = \frac{\text{number of sample points in volume } V}{NV}$$

where N is the number of independent samples.

Proceeding as in the one-variable case, we define a window function as

$$k(\mathbf{u}) = \begin{cases} 1/V & |u_i| \le h \quad i = 1, 2, \ldots, M \\ 0 & \text{otherwise} \end{cases}$$

The probability density can then be written as

$$p_{\mathbf{X}}(\mathbf{x}) = \frac{1}{N} \sum_{1}^{N} k(\mathbf{x} - \mathbf{x}_i)$$

where the subscript refers to the ith sample vector.

Other kernels can be used. An M-dimensional Gaussian kernel leads to a probability density, if the same variance is assumed for all components of the vector (Hand, 1981, p. 26),

$$p_{\mathbf{X}}(\mathbf{x}) = \frac{1}{N} \sum_{1}^{N} k(\mathbf{x} - \mathbf{x}_i) = \frac{1}{(2\pi)^{M/2} \sigma^M N} \sum_{i=1}^{N} \prod_{j=1}^{M} \exp\left[\frac{(x_j - x_{ij})^2}{2\sigma^2} \right]$$

A probability density estimate that uses different h values along the vector space axes might be justified if the sample point spread along some axes is significantly different from the spread along other axes. The more general form using a normal kernel is

$$p_{\mathbf{X}}(\mathbf{x}) = \frac{1}{(2\pi)^{M/2} N [\det(\mathbf{C})]^{1/2}} \sum_{1}^{N} \exp\left[-\frac{1}{2} (\mathbf{x} - \mathbf{x}_i)^T \mathbf{C}^{-1} (\mathbf{x} - \mathbf{x}_i) \right]$$

where $\mathbf{C}$ is a chosen covariance matrix, M the number of components of each vector $\mathbf{x}$, and N the number of sample vectors in the class being considered (Hand, 1981, p. 26).

The k-Nearest-Neighbors Algorithm

Targets can be classified by comparing their classification vectors directly to those of validated vector classes, without concern for the class densities. Assume that a verified set of vectors belongs to ω_i, one of Q classes with N_i vectors in the ith class. If vector $\mathbf{x}$, to be classified, belongs to class ω_i, it is likely that a member of the verified set of vectors for class ω_i will be the nearest neighbor of $\mathbf{x}$. If both the nearest neighbor and the next nearest neighbor belong to ω_i, the assignment of $\mathbf{x}$ to ω_i is on even firmer ground.

To implement the algorithm, determine radius r of a hypersphere containing k_i neighbors from class ω_i of the unclassifed vector $\mathbf{x}$, with k_i a chosen number and $k_i \le N$. This radius determines volume V_i. The probability that a sample point of the verified set for class ω_i falls in a neighborhood of $\mathbf{x}$ is the integral over the neighborhood space,

$$P = \int_{V_i} p_{\mathbf{X}|\omega_i}(\mathbf{x}|\omega_i)\, d\mathbf{x} \approx p_{\mathbf{X}|\omega_i}(\mathbf{x}|\omega_i) V_i$$

An estimate of this probability is the ratio of sample points k_i in the region divided by the number of sample points N_i in the class. An estimate of the class-conditional probability density at $\mathbf{x}$ is, therefore,

$$\hat{p}_{\mathbf{X}|\omega_i}(\mathbf{x}|\omega_i) = \frac{k_i/N_i}{V_i} = \frac{k_i}{N_i V_i} \tag{9.10}$$

This is the *k-nearest neighbor* or *k-NN* estimate of the true pdf at $\mathbf{x}$. The terms used in this test are the k-NN estimates of the probability density and the prior probabilities that the object to be classified belongs to a particular class. The prior probability for class ω_i is N_i/N, where N is the number of objects in all classes. If the density estimate of (9.10) and this prior probability are substituted into Bayes rule, (9.2), the minimum-error-probability rule is seen to be: Assign $\mathbf{x}$ to class ω_i if

$$\frac{k_i}{V_i} > \frac{k_j}{V_j} \qquad i \neq j$$

where k_i is the number of samples in volume V_i and k_j is the number in ω_j.

This rule can be applied in various ways. For one approach, choose equal volumes for all classes and if $k_i > k_j$ for all j, choose class ω_i. If the chosen volume is small enough to be appropriate for regions of high density, it may be too small to allow a satisfactory estimate of the density in sparse regions. For that reason, it is more common to choose equal values for all k_n and to choose ω_i if $V_i < V_j$ for all j. This form of the k-NN decision rule is called the *grouped form*.

A common form of the k-nearest neighbor decision rule is the *pooled form*. In its use, a volume is chosen to include k total samples in all classes. Then the decision rule becomes: If $k_i > k_j$ for all j in the volume, assign the unknown object to class ω_i.

The estimate of the probability density, (9.10), will have less variation if k_i is chosen to be large. On the other hand, V_i should be small to reduce the errors introduced by averaging. It has been noted that two conditions, that the limit of k_i approach infinity and the limit of k_i/N_i approach zero as N_i approaches infinity, are necessary and sufficient for the estimate to converge in probability to the true density at all points of continuity of the pdf (Hand, 1981, p. 32). These two conditions are met for the choice $k_i = N_i^{1/2}$. In a form of the k-NN algorithm for which the same volume is used for all classes, volume V cannot be chosen so that k_i meets these criteria for all classes, and the volume chosen must be a compromise.

The Nearest-Neighbor Algorithm

For ease of computation, it is common to use $k = 1$ in the k-NN algorithm. An unknown object represented by $\mathbf{x}$ is assigned to the class which has one verified vector closest to $\mathbf{x}$. The method is called the 1-NN algorithm or simply the nearest-neighbor algorithm.

The Reject Option

The need for a reject option arises if a classification rule does not yield an unequivocal decision. It may be possible to defer a decision until more information is available, or it may be desirable to discard the doubtful measurement. It is particularly easy to use the reject option for k-NN classification. In the pooled form we assign an object to class ω_i if, in a chosen volume, $k_i > k_j$ for all j. A reject class is established by requiring that k_i be greater than some chosen value if the object is not to be rejected.

Reducing the Verified Set

Shown in Fig. 9.1 are verified points belonging to two overlapping classes and the boundary determined by the Bayes classification procedure. A sample point that belongs to class ω_1 is represented by an o and one belonging to ω_2 by a +. Some class-1 samples appear on the right side of the boundary and some class-2 samples on the left. An unknown object vector that lies near one of these points may be incorrectly classified by a nearest-neighbors algorithm. It is therefore desirable to remove the validated samples that do not lie in their own Bayes acceptance region before the classification of unknown objects is carried out. Short of determining the class boundary by the Bayes process, there is no perfect way of removing the "misplaced" verified set samples. An algorithm that preprocesses the verified set data and helps to remove the validated samples not lying in their Bayes acceptance region is described by Devijver and Kittler (1982, p. 115). It removes some or all of the misplaced points and produces classes with less overlap.

A second way to remove outliers from each verified class is to find the distance from every sample point in class i to every other point in all classes. If the closest sample is in class i, do nothing. If the closest sample is in class j, tag the sample being examined, but do not remove it. Repeat for all samples in all classes. When finished, remove the tagged samples. The number of distances to be found in this editing process increases as the square of the number of samples in each verified class and can require considerable computation time.

Another reduction in the size of the verified vector set is feasible without significantly affecting the performance of the k-NN algorithm. An unknown vector that lies deep in the interior of a cluster of samples for a class is easy to classify. The difficult decisions are for unknowns that lie near the boundaries between classes, and for those the interior points of a verified set are never the nearest neighbors. Removal of the interior points from the verified set makes implementation of the nearest-neighbors classification less time-consuming (Devijver and Kittler, 1982, p. 121).

Errors for the 1-NN Algorithm

Let $\mathbf{x}$ represent the object to be classified and assume that it belongs to class ω_i. Let $\hat{\mathbf{x}}$ represent the nearest verified neighbor and assume that it belongs to class ω_j. Then $\mathbf{x}$ will be assigned to ω_j, and an error will occur if i and j are not equal. For the two-class problem, let

$$H_i \text{ be the hypothesis that } \mathbf{x} \text{ belongs to } \omega_i \qquad i = 1, 2$$

$$\hat{H}_i \text{ be the hypothesis that } \hat{\mathbf{x}} \text{ belongs to } \omega_i \qquad i = 1, 2$$

The probabilities that these hypotheses are true are

$$P(H_i) = P(\omega_i|\mathbf{x}) \qquad i = 1, 2$$
$$P(\hat{H}_i) = P(\omega_i|\hat{\mathbf{x}}) \qquad i = 1, 2$$

If the probabilities are independent, an error is made if hypotheses H_i and $\hat{H}_j$, with i not equal to j, are simultaneously true, because the decision has been made to assign both vectors to the same class. The probability of error is

$$P(H_1, \hat{H}_2) + P(H_2, \hat{H}_1) = P(H_1)P(\hat{H}_2) + P(H_2)P(\hat{H}_1)$$

Note that

$$P(H_1) + P(H_2) = P(\hat{H}_1) + P(\hat{H}_2) = 1$$

Then the two-class error probability is

$$\begin{aligned}\epsilon_{NN}(\mathbf{x}) &= P(H_1)\left[1 - P(\hat{H}_1)\right] + P(H_2)\left[1 - P(\hat{H}_2)\right] \\ &= P(\omega_1|\mathbf{x})\left[1 - P(\omega_1|\hat{\mathbf{x}})\right] + P(\omega_2|\mathbf{x})\left[1 - P(\omega_2|\hat{\mathbf{x}})\right]\end{aligned}$$

If the number of sample points in the validated set is large, we expect $\mathbf{x}$ and $\hat{\mathbf{x}}$ to be close to each other, and the probability that H_i is true given $\mathbf{x}$ will closely approximate the probability that $\hat{H}_i$ is true given $\hat{\mathbf{x}}$, or $P(\omega_i|\mathbf{x}) \approx P(\omega_i|\hat{\mathbf{x}})$. With this approximation, the two-class error probability becomes

$$\epsilon_{NN}(\mathbf{x}) = \sum_1^2 P(\omega_i|\mathbf{x})\left[1 - P(\omega_i|\mathbf{x})\right]$$

It is straightforward to extend this to Q classes, for which the probability of error is

$$\epsilon_{NN}(\mathbf{x}) = \sum_{1}^{Q} P(\omega_i|\mathbf{x})\,[1 - P(\omega_i|\mathbf{x})] = \sum_{1}^{Q} P(\omega_i|\mathbf{x}) - \sum_{1}^{Q} [P(\omega_i|\mathbf{x})]^2$$

$$= 1 - \sum_{1}^{Q} [P(\omega_i|\mathbf{x})]^2 = 1 - [P(\omega_m|\mathbf{x})]^2 - \sum_{i \neq m}^{Q} [P(\omega_i|\mathbf{x})]^2 \qquad (9.11)$$

where ω_m is the class having the greatest probability of correct assignment if $\mathbf{x}$ is assigned to it. If $\mathbf{x}$ is assigned to class ω_m, the Bayes error is

$$\epsilon_B = 1 - P(\omega_m|\mathbf{x}) = \sum_{i \neq m}^{Q} P(\omega_i|\mathbf{x}) \qquad (9.12)$$

Consider

$$F = \sum_{1}^{N} u_i^2$$

If the sum of the u_i is a constant C, the minimum value of F occurs if all u_i are equal and given by C/N. If this result is used in (9.11), in conjunction with (9.12), it is straightforward to show that the upper bound on the expected value of $\epsilon_{NN}(\mathbf{x})$ is

$$\epsilon_{NN} \leq 2\epsilon_B - \frac{Q}{Q-1}\epsilon_B^2$$

If the Bayes error is small, the last term in the upper error bound can be neglected, and the 1-NN error bounds are the Bayes error and twice the Bayes error.

Bounds for the *k*-NN Algorithm

The lower error bound for the k-NN algorithm is the Bayes error. The upper error bound has been found for the two-class problem for various values of k (Therrien, 1989, p. 151). As expected, values of k greater than 1 decrease the error probability from the 1-NN error value.

B. CLASSIFICATION BY MATRIX DECOMPOSITION

A useful set of feature vectors for a target can be found by decomposition of the target's scattering matrix. "Scattering matrix" refers to any of the matrices that give either the scattered electric field or received power. The discussion is restricted to backscattering.

9.7. COHERENT DECOMPOSITION

The Sinclair matrix of a target can be written for backscattering as

$$\mathbf{S} = \begin{bmatrix} a+b & c \\ c & a-b \end{bmatrix} \tag{9.13}$$

S can be considered the sum of matrices representing canonical targets. Krogager (1993, p. 91) suggested a breakup into three Sinclair matrices, each representing a physical target. Write **S** as

$$\mathbf{S} = a\begin{bmatrix} 1 & 0 \\ 0 & 1 \end{bmatrix} + b\left\{\begin{bmatrix} 1 & 0 \\ 0 & -1 \end{bmatrix} + \frac{c}{b}\begin{bmatrix} 0 & 1 \\ 1 & 0 \end{bmatrix}\right\} \qquad b \neq 0$$

Let

$$d = \frac{c}{b} = d' + jd''$$

Then,

$$\mathbf{S} = a\begin{bmatrix} 1 & 0 \\ 0 & 1 \end{bmatrix} + b\left\{\begin{bmatrix} 1 & 0 \\ 0 & -1 \end{bmatrix} + d'\begin{bmatrix} 0 & 1 \\ 1 & 0 \end{bmatrix} + d''\begin{bmatrix} 0 & j \\ j & 0 \end{bmatrix}\right\}$$

If d'' is positive, rewrite this as

$$\mathbf{S} = a\begin{bmatrix} 1 & 0 \\ 0 & 1 \end{bmatrix} + b\begin{bmatrix} 1-d'' & d' \\ d' & -(1-d'') \end{bmatrix} + bd''\begin{bmatrix} 1 & j \\ j & -1 \end{bmatrix} \tag{9.14}$$

and if it is negative, write the Sinclair matrix as

$$\mathbf{S} = a\begin{bmatrix} 1 & 0 \\ 0 & 1 \end{bmatrix} + b\begin{bmatrix} 1+d'' & d' \\ d' & -(1+d'') \end{bmatrix} - bd''\begin{bmatrix} 1 & -j \\ -j & -1 \end{bmatrix} \tag{9.15}$$

The first matrices in (9.14) and (9.15) represent a sphere. If the second matrices are combined as

$$\begin{bmatrix} 1-|d''| & d' \\ d' & -(1-|d''|) \end{bmatrix}$$

and compared to the Sinclair matrix of a dihedral corner reflector (Mott, 1992, p. 350),

$$\mathbf{S} = \begin{bmatrix} -\cos 2\theta & \sin 2\theta \\ \sin 2\theta & \cos 2\theta \end{bmatrix}$$

it will be seen that the second matrices represent tilted dihedral corner reflectors with the angle of rotation of the dihedral fold line given by

$$\tan 2\theta = -\frac{d'}{1-|d''|}$$

The third matrix of (9.14) and (9.15) represents a helix-type target, right handed for the positive sign used in (9.14) and left handed for the negative sign of (9.15). This idealized helical scatterer should not be associated with a wire helix. Such a structure can act as an antenna radiating a wave that is approximately right- or left-circular, but the same structure will not in general scatter a wave that has only a right- or left-circular component. We therefore think of the third matrix as representing an idealized target whose scattered wave has only a circularly polarized component.

The Euclidean norms of the three component Sinclair matrices of (9.14) or (9.15), when normalized by the Euclidean norm of **S**, show the relative strengths of the component scatterers. Classification vector **X** can be formed with sphere, diplane, and helix components:

Sphere:

$$X_1 = \frac{|a|}{(|a|^2 + |b|^2 + |c|^2)^{1/2}}$$

Dihedral corner:

$$X_2 = \frac{|b|\left[(1 - |d''|)^2 + (d')^2\right]^{1/2}}{(|a|^2 + |b|^2 + |c|^2)^{1/2}}$$

Helix:

$$X_3 = \frac{\sqrt{2}|b||d''|}{(|a|^2 + |b|^2 + |c|^2)^{1/2}}$$

These vector elements are formed in the same manner, represent different but intuitively-comparable properties of the target matrix, and are numerically comparable. **X** is therefore a useful classification vector. Note that all elements of **X** are real.

Decomposition in Circular Components

A scattered wave in left- and right-circular components is, from Section 3.7,

$$\mathbf{E}^s = \begin{bmatrix} E_L^s \\ E_R^s \end{bmatrix} = \frac{1}{\sqrt{4\pi}\, r}\mathbf{A}\mathbf{E}^i = \frac{1}{\sqrt{4\pi}\, r}\begin{bmatrix} A_{LL} & A_{LR} \\ A_{LR} & A_{RR} \end{bmatrix}\begin{bmatrix} E_L^i \\ E_R^i \end{bmatrix}$$

where the symmetry of **A** for backscattering is indicated. The circular components of **A** are based on coordinate systems 1 and 2 of Fig. 7.1.

We could decompose the target matrix into the sphere, dihedral corner, and helix we used in rectangular coordinates, but a different and interesting breakup is possible. Write **A** as

$$\mathbf{A} = \begin{bmatrix} A_{LL} & A_{LR} \\ A_{LR} & A_{RR} \end{bmatrix} = A_{LR}\begin{bmatrix} 0 & 1 \\ 1 & 0 \end{bmatrix} + A_{LL}\begin{bmatrix} 1 & 0 \\ 0 & 0 \end{bmatrix} + A_{RR}\begin{bmatrix} 0 & 0 \\ 0 & 1 \end{bmatrix} \tag{9.16}$$

The component matrices of (9.16) represent a sphere, a left-, and a right-helix. Their relative strengths, characterized by their Euclidean norms compared to the Euclidean norm of **A**, can be used to form classification vector **Y** with elements:

Sphere:

$$Y_1 = \frac{\sqrt{2}|A_{LR}|}{\left(|A_{LL}|^2 + 2|A_{LR}|^2 + |A_{RR}|^2\right)^{1/2}}$$

Left helix:

$$Y_2 = \frac{|A_{LL}|}{\left(|A_{LL}|^2 + 2|A_{LR}|^2 + |A_{RR}|^2\right)^{1/2}}$$

Right helix:

$$Y_3 = \frac{|A_{RR}|}{\left(|A_{LL}|^2 + 2|A_{LR}|^2 + |A_{RR}|^2\right)^{1/2}}$$

Comparison of classification vectors **X** and **Y** and the canonical targets into which **S** was decomposed emphasizes the desirability of not describing a target by one characteristic alone. A complete classification vector, **X** or **Y**, is necessary for classification in the general case.

9.8. DECOMPOSITION OF POWER-TYPE MATRICES

A notable decomposition of a 4 × 4 power-type matrix was that of Huynen (1970, p. 157). Since Huynen's first publication, his assumptions and methods have been extended and commented on by Huynen himself and by others (Cloude and Pottier, 1996; Krogager, 1993).

Huynen separated the Kennaugh matrix into two matrices, one representing a coherently scattering target and the other a "nonsymmetric noise target". In Section 9.9, a matrix for a depolarizing target is separated, for a specified incident wave polarization, into a matrix representing a coherently scattering target and one representing a completely depolarizing target. Huynen's decomposition differs from that of Section 9.9 in two important respects: It is independent of the incident wave polarization, and the "noise target" is not completely depolarizing.

With a minor difference noted below, Huynen used the Kennaugh matrix as

$$\mathbf{K}_H = \begin{bmatrix} A_0 + B_0 & F & C & H \\ F & -A_0 + B_0 & G & D \\ C & G & A_0 + B & E \\ H & D & E & A_0 - B \end{bmatrix} \tag{9.17}$$

In his 1970 study, Huynen used $-E$ rather than E in the matrix. In his later work, he used the form given here. The matrix is symmetric and restricted to backscattering. Coordinates $x_1 y_1 z_1$ of Fig. 7.1 are used for both incident and scattered waves.

Huynen used the Stokes vector order,

$$\mathbf{G}_H = \left\langle \begin{bmatrix} |E_x|^2 + |E_y|^2 \\ 2\mathrm{Im}(E_x^* E_y) \\ |E_x|^2 - |E_y|^2 \\ 2\mathrm{Re}(E_x^* E_y) \end{bmatrix} \right\rangle$$

with a similar form for the Stokes vector of an antenna. If the antenna Stokes vector given by (7.15) is used instead of Huynen's form, the matrix of (9.17) must be rearranged. It becomes

$$\mathbf{K} = \begin{bmatrix} A_0 + B_0 & C & H & F \\ C & A_0 + B & E & G \\ H & E & A_0 - B & D \\ F & G & D & -A_0 + B_0 \end{bmatrix} \tag{9.18}$$

Huynen interpreted the elements of his form of the Kennaugh matrix to be descriptive of target properties (Huynen, 1970; Krogager, 1993, p. 82), but the interpretation is unnecessary in the classification procedures outlined here.

If the Kennaugh matrix, (9.18), describes a coherently scattering target, only five of the parameters are independent, and relationships among the elements must exist. Huynen (1970, p. 138) derived four equations that relate the elements for a coherently scattering target,

$$2A_0(B_0 + B) = C^2 + D^2 \tag{9.19}$$

$$2A_0 E = CH - DG \tag{9.20}$$

$$2A_0(B_0 - B) = G^2 + H^2 \tag{9.21}$$

$$2A_0 F = CG + DH \tag{9.22}$$

Note: Huynen (1970) wrote the right side of (9.20) as $DG - CH$. Equation (9.20) agrees with Huynen's later work.

The relationships of (9.19)–(9.22) can be verified by forming the Kennaugh matrix of a coherently scattering target from the Sinclair matrix of (9.13) and finding the parameters of (9.18) by comparison.

In the Huynen decomposition, the Kennaugh matrix is divided as $\mathbf{K} = \mathbf{K}_T + \mathbf{K}_N$. Matrix $\mathbf{K}_T$ is required to represent a single coherently scattering target. The form of $\mathbf{K}_N$ is assumed to be that which appears in

$$\mathbf{K} = \mathbf{K}_T + \mathbf{K}_N = [K_{ij}] \qquad i, j = 1, 2, 3, 4$$

$$= \begin{bmatrix} A_0 + B_0^T & C^T & H^T & F^T \\ C^T & A_0 + B^T & E^T & G^T \\ H^T & E^T & A_0 - B^T & D^T \\ F^T & G^T & D^T & -A_0 + B_0^T \end{bmatrix}$$

$$+ \begin{bmatrix} B_0^N & 0 & 0 & F^N \\ 0 & B^N & E^N & 0 \\ 0 & E^N & -B^N & 0 \\ F^N & 0 & 0 & B_0^N \end{bmatrix} \tag{9.23}$$

Huynen chose these forms for the constituent matrices and has discussed his reasons for doing so. Krogager (1993) provides additional insight and gives additional references. Clearly, other matrix separations are feasible. We find immediately from the separated matrices that

$$C^T = K_{12}$$
$$D^T = K_{34}$$
$$G^T = K_{24}$$
$$H^T = K_{13}$$

Also,

$$A_0 + B_0^T + B_0^N = K_{11} \tag{9.24}$$
$$-A_0 + B_0^T + B_0^N = K_{44} \tag{9.25}$$
$$A_0 + B^T + B^N = K_{22} \tag{9.26}$$
$$A_0 - B^T - B^N = K_{33} \tag{9.27}$$
$$F^T + F^N = K_{14} \tag{9.28}$$
$$E^T + E^N = K_{23} \tag{9.29}$$

We require that the elements of $\mathbf{K}_T$ satisfy (9.19)–(9.22) with notational changes,

$$2A_0(B_0^T + B^T) = (C^T)^2 + (D^T)^2 = K_{12}^2 + K_{34}^2 \tag{9.30}$$
$$2A_0(B_0^T - B^T) = (G^T)^2 + (H^T)^2 = K_{24}^2 + K_{13}^2 \tag{9.31}$$
$$2A_0E^T = C^T H^T - D^T G^T = K_{12}K_{13} - K_{34}K_{24} \tag{9.32}$$
$$2A_0F^T = C^T G^T + D^T H^T = K_{12}K_{24} + K_{34}K_{13} \tag{9.33}$$

of the scattered wave be one; they may also be inferred from

$$\mathbf{D}_c = \mathbf{D}^{(2)} = \kappa_{Tb} = \begin{bmatrix} |T_{xx}|^2 & T_{xx}T_{xy}^* & T_{xx}^*T_{xy} & |T_{xy}|^2 \\ -T_{xx}T_{xy}^* & T_{xx}T_{yy}^* & -|T_{xy}|^2 & T_{xy}T_{yy}^* \\ -T_{xx}^*T_{xy} & -|T_{xy}|^2 & T_{xx}^*T_{yy} & T_{xy}^*T_{yy} \\ |T_{xy}|^2 & -T_{xy}T_{yy}^* & -T_{xy}^*T_{yy} & |T_{yy}|^2 \end{bmatrix} \tag{9.34}$$

The superscript on the second use of **D** anticipates a matrix separation that follows.

The coherency vector of the wave scattered from the target, neglecting a scalar multiplier, is

$$\mathbf{J} = \mathbf{D}(\mathbf{E}^i \otimes \mathbf{E}^{i*}) \tag{9.35}$$

where $\mathbf{E}^i$ is the incident field vector.

We noted in Section 7.5 that the coherency vector of a wave can be separated into a coherency vector for an unpolarized wave and one for a completely polarized wave. If we separate the scattered wave from a target in that manner, we can associate the unpolarized and completely polarized parts with separate targets whose matrices add to form target matrix **D**. The separation of the coherency vector and the corresponding decomposition of the target matrix are valid only for the specified incident wave polarization.

Linear-Horizontal Incident Wave

A horizontal (x-directed) incident wave, with coherency vector $[\,1\ \ 0\ \ 0\ \ 0\,]^T$, leads to a scattered-wave coherency vector,

$$\mathbf{J}_H = \mathbf{D}\mathbf{J}^i = \begin{bmatrix} J_{H11} \\ J_{H12} \\ J_{H21} \\ J_{H22} \end{bmatrix} = \begin{bmatrix} D_{11} \\ D_{21} \\ D_{31} \\ D_{41} \end{bmatrix}$$

The coherency vector can be separated into two parts, the first representing an unpolarized wave and the second a completely polarized wave,

$$\mathbf{J}_H = \mathbf{J}_H^{(1)} + \mathbf{J}_H^{(2)}$$

with component matrices given by

$$\mathbf{J}_H^{(1)} = \begin{bmatrix} A_H \\ 0 \\ 0 \\ A_H \end{bmatrix} \quad \text{unpolarized wave} \qquad \mathbf{J}_H^{(2)} = \begin{bmatrix} B_H \\ F_H \\ F_H^* \\ C_H \end{bmatrix} \quad \text{polarized wave}$$

The constants are found from (Mott, 1992, p. 379)

$$A_H = \frac{1}{2}(J_{H11} + J_{H22}) - \frac{1}{2}\left[(J_{H11} + J_{H22})^2 - 4(J_{H11}J_{H22} - J_{H12}J_{H21})\right]^{1/2}$$

$$= \frac{1}{2}(D_{11} + D_{14}) - \frac{1}{2}\left[(D_{11} + D_{14})^2 - 4(D_{11}D_{14} - D_{12}D_{12}^*)\right]^{1/2} \quad (9.36)$$

$$B_H = J_{H11} - A_H = D_{11} - A_H \quad (9.37)$$

$$C_H = J_{H22} - A_H = D_{14} - A_H \quad (9.38)$$

$$F_H = J_{H12} = J_{H21}^* = -D_{12} \quad (9.39)$$

D is separable,

$$\mathbf{D} = \mathbf{D}^{(1)} + \mathbf{D}^{(2)}$$

Each matrix can be associated with a coherency vector,

$$\mathbf{J} = \mathbf{J}^{(1)} + \mathbf{J}^{(2)} = \mathbf{D}(\mathbf{E}^i \otimes \mathbf{E}^{i*})$$

where

$$\mathbf{J}^{(1)} = \mathbf{D}^{(1)}(\mathbf{E}^i \otimes \mathbf{E}^{i*})$$
$$\mathbf{J}^{(2)} = \mathbf{D}^{(2)}(\mathbf{E}^i \otimes \mathbf{E}^{i*})$$

From these equations, we recognize that the first columns of the component matrices are given by

$$\mathbf{J}_H^{(1)} = \begin{bmatrix} D_{11}^{(1)} \\ D_{21}^{(1)} \\ D_{31}^{(1)} \\ D_{41}^{(1)} \end{bmatrix} = \begin{bmatrix} A_H \\ 0 \\ 0 \\ A_H \end{bmatrix} \qquad \mathbf{J}_H^{(2)} = \begin{bmatrix} D_{11}^{(2)} \\ D_{21}^{(2)} \\ D_{31}^{(2)} \\ D_{41}^{(2)} \end{bmatrix} = \begin{bmatrix} B_H \\ F_H \\ F_H^* \\ C_H \end{bmatrix}$$

We may now write

$$\mathbf{D} = \begin{bmatrix} A_H & \cdots & \cdots & \cdots \\ 0 & \cdots & \cdots & \cdots \\ 0 & \cdots & \cdots & \cdots \\ A_H & \cdots & \cdots & \cdots \end{bmatrix} + \begin{bmatrix} B_H & \cdots & \cdots & \cdots \\ F_H & \cdots & \cdots & \cdots \\ F_H^* & \cdots & \cdots & \cdots \\ C_H & \cdots & \cdots & \cdots \end{bmatrix}$$

If we apply the element relationships of (9.34) for a coherently scattering target to the second matrix, our knowledge of **D** allows us to write

$$\mathbf{D} = \begin{bmatrix} A_H & D_{12} + F_H & D_{12}^* + F_H^* & A_H \\ 0 & \cdots & D_{23} + C_H & \cdots \\ 0 & D_{23} + C_H & \cdots & \cdots \\ A_H & \cdots & \cdots & \cdots \end{bmatrix} + \begin{bmatrix} B_H & -F_H & -F_H^* & C_H \\ F_H & \cdots & -C_H & \cdots \\ F_H^* & -C_H & \cdots & \cdots \\ C_H & \cdots & \cdots & \cdots \end{bmatrix}$$

These matrices cannot be determined uniquely without further constraints. One constraint is that the (44) element of the second matrix must not be negative. If $D_{44} \geq A_H$, it is convenient to use A_H as the real (44) element of the first matrix and $D_{44} - A_H$ as the corresponding real element of the second matrix. With this arbitrary choice, the remaining elements of both matrices can be found.

With the use of (9.34), it can be seen that a relationship between the (24), (14), and (44) elements leads to the magnitude of $D_{24}^{(2)}$. If we choose its phase angle as that of D_{24}, we can write

$$D_{24}^{(2)} = \sqrt{C_H(D_{44} - A_H)}\, e^{j\Phi_{24}}$$

In a similar manner we can obtain

$$D_{22}^{(2)} = \sqrt{B_H(D_{44} - A_H)}\, e^{j\Phi_{22}}$$

The elements of the component matrices of **D** are now completely determined; it can be written as

$$\mathbf{D} = \begin{bmatrix} A_H & D_{12} + F_H & D_{12}^* + F_H^* & A_H \\ 0 & D_{22} - D_{22}^{(2)} & D_{23} + C_H & D_{24} - D_{24}^{(2)} \\ 0 & D_{23} + C_H & D_{22}^* - D_{22}^{(2)*} & D_{24}^* - D_{24}^{(2)*} \\ A_H & -D_{24} + D_{24}^{(2)} & -D_{24}^* + D_{24}^{(2)*} & A_H \end{bmatrix} + \begin{bmatrix} B_H & -F_H & -F_H^* & C_H \\ F_H & D_{22}^{(2)} & -C_H & D_{24}^{(2)} \\ F_H^* & -C_H & D_{22}^{(2)*} & D_{24}^{(2)*} \\ C_H & -D_{24}^{(2)} & -D_{24}^{(2)*} & D_{44} - A_H \end{bmatrix} \quad (9.40)$$

Additional constraints on **D** are that B_H and C_H be real and positive if the second matrix represents a coherently-scattering target. It is therefore required that A_H be smaller than D_{11}, D_{14}, and D_{44}. It should be apparent from the manner in which this decomposition was developed that a small value of A_H corresponds to a high degree of polarization of the scattered wave, and it is left as an exercise to derive this relationship. If the constraints on A_H are not met, decomposition with a linear-vertical wave may be successful. If not, **D** can be transformed to a different polarization basis and the decomposition carried out as suggested by the final subsection of this section.

Linear-Vertical Incident Wave

If a similar decomposition is carried out with a y-directed wave, the component matrices are

$$\mathbf{D} = \begin{bmatrix} A_V & D_{12} - \overline{D}_{12}^{(2)} & D_{12}^* - \overline{D}_{12}^{(2)*} & A_V \\ -D_{12} + \overline{D}_{12}^{(2)} & D_{22} - \overline{D}_{22}^{(2)} & D_{23} + B_V & 0 \\ -D_{12}^* + \overline{D}_{12}^{(2)*} & D_{23} + B_V & D_{22}^* - \overline{D}_{22}^{(2)*} & 0 \\ A_V & -D_{24} + F_V & -D_{24}^* + F_V^* & A_V \end{bmatrix}$$
$$+ \begin{bmatrix} D_{11} - A_V & \overline{D}_{12}^{(2)} & \overline{D}_{12}^{(2)*} & B_V \\ -\overline{D}_{12}^{(2)} & \overline{D}_{22}^{(2)} & -B_V & F_V \\ -\overline{D}_{12}^{(2)*} & -B_V & \overline{D}_{22}^{(2)*} & F_V^* \\ B_V & -F_V & -F_V^* & C_V \end{bmatrix} \tag{9.41}$$

where

$$A_V = \frac{1}{2}(D_{14} + D_{44}) - \frac{1}{2}\left[(D_{14} + D_{44})^2 - 4(D_{14}D_{44} - D_{24}D_{24}^*)\right]^{1/2} \tag{9.42}$$

$$B_V = D_{14} - A_V \tag{9.43}$$

$$C_V = D_{44} - A_V \tag{9.44}$$

$$F_V = D_{24} \tag{9.45}$$

and

$$\overline{D}_{12}^{(2)} = \sqrt{B_V(D_{11} - A_V)}e^{j\Phi_{12}}$$
$$\overline{D}_{22}^{(2)} = \sqrt{C_V(D_{11} - A_V)}e^{j\Phi_{22}}$$

Decomposition with a linear-vertical wave requires that A_V be smaller than D_{11}, D_{14}, and D_{44}. If it is not, decomposition with a linear-horizontal wave or with a transformed **D** matrix, as outlined in the following subsection, may be effective.

The inequality of the component matrices of **D** for the two illuminations shows what was noted previously: The decomposition is not independent of the polarization of the incident wave. We noted also that, for a chosen polarization, an arbitrary value was assigned to one matrix element. Since we are concerned with target classification, we need not attempt a decomposition of **D** that is valid for all incident waves. Classification of a target with a specific incident wave polarization is sufficient for many purposes.

Consider a target that is a coherently-scattering target combined with an independent target whose scattered wave is unpolarized for all incident wave polarizations. For such a target, **D** is

$$\mathbf{D} = \begin{bmatrix} A & 0 & 0 & A \\ 0 & 0 & 0 & 0 \\ 0 & 0 & 0 & 0 \\ A & 0 & 0 & A \end{bmatrix} + \begin{bmatrix} D_{11} - A & D_{12} & D_{12}^* & D_{14} - A \\ -D_{12} & D_{22} & A - D_{14} & D_{24} \\ -D_{12}^* & A - D_{14} & D_{22}^* & D_{24}^* \\ D_{14} - A & -D_{24} & -D_{24}^* & D_{44} - A \end{bmatrix} \qquad (9.46)$$

where $A = A_H = A_V$ can be found from (9.36) or (9.42). This separation into matrices representing a completely depolarizing target and a coherently scattering target is valid for all polarizations of the incident wave. If the target does not meet the requirement given at the beginning of this paragraph, A_H and A_V are not equal, and the second matrix of (9.46) does not represent a coherently scattering target for all incident waves.

A Sinclair matrix can be found from $\mathbf{D}^{(2)}$, with the restrictions noted above. The degree of polarization of the scattered wave can also be found. These quantities can then be used for classification.

Decomposition in a General Polarization Basis

Decomposition of the **D** matrix has been discussed for convenience with linearly polarized incident waves, either horizontal or vertical, and with the matrix expressed in rectangular components, but it does not need to be restricted in this manner. The electric field can be transformed from a rectangular polarization basis to a desired orthonormal basis with the unitary matrix **U** of (6.28). A coherency vector in the new basis is related to that in the old by (6.29). Received power in the new basis can be equated to that in the old, and, from (6.29) and (7.21), the power in the two bases is

$$W = \frac{Z_0^2 I^2}{128\pi R_a \lambda^2 r_1^2 r_2^2} \hat{\mathbf{J}}_{Ar}^T \mathbf{diag}(1, -1, -1, 1) \hat{\mathbf{D}}_T \hat{\mathbf{J}}_{At}$$
$$= \frac{Z_0^2 I^2}{128\pi R_a \lambda^2 r_1^2 r_2^2} \mathbf{J}_{Ar}^T \mathbf{diag}(1, -1, -1, 1) \mathbf{D}_T \mathbf{J}_{At}$$

where

$$\mathbf{D}_T = \mathbf{diag}(1, -1, -1, 1)(\mathbf{U} \otimes \mathbf{U}^*)^T \mathbf{diag}(1, -1, -1, 1) \hat{\mathbf{D}}_T (\mathbf{U} \otimes \mathbf{U}^*)$$

The coherency vector of a scattered wave is given in the new basis by an equation like that of (7.31) and is

$$\hat{\mathbf{J}}^s = \frac{1}{4\pi r_2^2} \hat{\mathbf{D}}_T \hat{\mathbf{J}}^i \qquad (9.47)$$

where the superscripts refer to scattered and incident waves and r_2 is the distance from target to receiver.

The elements of $\hat{\mathbf{D}}_T$ are not all independent. Relationships among the elements can be determined from the requirement that received power be real and non-negative. Additional relationships can be found from the requirement that $\mathbf{diag}(1, -1, -1, 1)\hat{\mathbf{D}}_T$ be symmetric for backscattering. A final group of element values can be found for a coherently-scattering target from the requirement that the degree of polarization of the scattered wave be one. The coherency vector elements in the old and new bases have the same relationships, as noted in Section 6.2. If these factors are taken in account, it is clear that the element relationships in $\hat{\mathbf{D}}_T$ are the same as those in $\mathbf{D}_T$ and that the two matrices can be decomposed by the same procedure. The same equations can be used in the process, with one difference. Instead of linear horizontal and vertical incident waves, the incident waves are chosen for the decomposition of $\hat{\mathbf{D}}_T$ to make all terms, except the first (or last), element of $\hat{\mathbf{J}}^i$ zero.

9.10. POLARIZED CLUTTER

The procedure of the previous section removes the unpolarized components of the scattered wave. It does not follow that "clutter" is being removed. Clutter is "a return or group of returns that is unwanted in the radar situation being considered" (Currie, 1987, p. 281). The definition does not specify that the clutter return be unpolarized. Clutter returns may add coherently to the return from a target of interest and to the target's Sinclair matrix. If so, clutter will not be removed by the decomposition, and the Sinclair matrix of the target cannot be recovered. If the target and clutter characteristics are known or can be estimated, the clutter can be suppressed by choosing antenna polarizations appropriately, but classification of the target by means of its Sinclair matrix is not feasible if coherent scattering from clutter is significant.

9.11. A SIMILAR DECOMPOSITION

Cloude and Pottier (1996) discussed a decomposition that has points of similarity to the method discussed in Section 9.9. They noted that a 4×4 target coherency matrix for noise has the form of the identity matrix and that noise can be removed from the coherency matrix of a target by subtraction of a multiple of the identity matrix from the eigenvalues of the measured matrix. This procedure does not yield the degrees of polarization for specified antenna polarizations, although the polarimetric entropy developed by Cloude and Pottier bears an approximate relationship to an averaged degree of polarization.

9.12. POLARIMETRIC SIMILARITY CLASSIFICATION

An efficient algorithm that iteratively maximizes the power contrast ratio,

$$R_{ab} = \frac{\mathbf{Y}^T \mathbf{A} \mathbf{X}}{\mathbf{Y}^T \mathbf{B} \mathbf{X}} \tag{9.48}$$

was discussed in Section 8.5. **A** and **B** are Kennaugh matrices, and **X** and **Y** are the Stokes vectors of the transmitting and receiving antennas, respectively.

The algorithm that maximizes (9.48) can be used to classify targets represented by their Kennaugh matrices. Let **A** represent the unknown target and **B** a known target such as a sphere. The matrices are normalized. If target **A** closely resembles a sphere, the power contrast ratio will be small, with our assumption that **B** is a sphere.

A straightforward classification procedure is to determine R_{ab} successively, with **B** representing a sphere, a left-helix, and a right-helix. The ratios provide three classification parameters. An alternative set of constituent targets, which included a dihedral corner, was given in Section 9.7, but the dihedral corner includes a tilt angle that cannot be incorporated into this classification procedure easily.

REFERENCES

R. M. Barnes, "Roll-Invariant Decompositions for the Polarization Covariance Matrix", *Proceedings of the Polarimetric Technology Workshop*, Redstone Arsenal, Alabama, August 1988.

S. R. Cloude, "Uniqueness of Target Decomposition Theorems in Radar Polarimetry", in *Direct and Inverse methods in Radar Polarimetry*, W-M. Boerner et al., eds., Kluwer Academic Publishers, Dordrecht, The Netherlands, 1992.

S. R. Cloude, and E. Pottier, "A Review of Target Decomposition Theorems in Radar Polarimetry", *IEEE Trans. GRS*, **34**(2), 498–518 (March 1996).

N. C. Currie, "Clutter Characteristics and Effects", Chapt. 10, in *Principles of Modern Radar*, J. L. Eaves and E. K. Reedy, eds., Van Nostrand Reinhold, New York, 1987.

P. A. Devijver and J. Kittler, *Pattern Recognition, a Statistical Approach*, Prentice-Hall, Englewood Cliffs, NJ, 1982.

K. Fukunaga, *Introduction to Statistical Pattern Recognition*, Academic, New York, 1972.

K. Fukunaga, *Statistical Pattern Recognition*, Second edition, Academic, New York, 1990.

D. J. Hand, *Discrimination and Classification*, Wiley, Chichester, 1981.

J. R. Huynen, *Phenomenological Theory of Radar Targets*, Drukkerij Bronder-Offset, Rotterdam, 1970.

E. Krogager, *Aspects of Polarimetric Radar Imaging*, Dissertation, Technical University of Denmark, 1993.

H. Mott, *Antennas for Radar and Communications: A Polarimetric Approach*, Wiley-Interscience, New York, 1992.

M. I. Skolnik, *Introduction to Radar Systems*, McGraw-Hill, New York, 1962.

C. W. Therrien, *Decision Estimation and Classification*, Wiley, New York, 1989.

H. L. van Trees, *Detection, Estimation, and Modulation Theory*, Wiley-Interscience, New York, 1968.

PROBLEMS

Note: A general-purpose mathematics program is desirable for solving some of the following problems.

9.1. Objects represented by scalar variable X are to be assigned to classes ω_1, ω_2, and ω_3 whose densities are Gaussian with means 1.2, 1.4, and 1.6, all with variances 0.04. The prior probabilities that an object belongs to one of the classes are 0.3, 0.4, and 0.3, respectively. The measured value x for an object is 1.55. To which class should it be assigned?

9.2. The Sinclair matrix of a target is

$$\begin{bmatrix} 2 & (1+j)/\sqrt{2} \\ (1+j)/\sqrt{2} & 4 \end{bmatrix}$$

Coherently decompose this matrix and find classification vector **X** according to the procedure of Section 9.7. Find classification vector **Y** which compares this target to a spherical target and left and right helices.

9.3. Separate the Kennaugh matrix $\mathbf{K}_a$ of Problem 8.1, using Huynen's decomposition method. Find the Sinclair matrix corresponding to $\mathbf{K}_T$ for a coherently scattering target.

9.4. Find the relationship between A_H of Section 9.9 and the degree of polarization of the scattered wave.

9.5. Decompose the matrix $\mathbf{D}_T$ of Problem 7.7 into a matrix representing a coherently scattering target and one representing a completely depolarizing target. Examine $\mathbf{D}_T$ before carrying out the decomposition and choose the incident wave polarization (linear x or linear y) most likely to lead to a successful decomposition. Find the Jones matrix corresponding to the coherently scattering target.

9.6. Simulate a nearest-neighbor classification using classification vector **X** of Section 9.7. Generate two "validated" target classes, with 100 vectors each, with Gaussian elements having means m and standard deviations σ:

	Class 1	Class 2
X_1	$m = 0.762$, $\sigma = 0.012$	$m = 0.749$, $\sigma = 0.012$
X_2	$m = 0.533$, $\sigma = 0.016$	$m = 0.550$, $\sigma = 0.016$
X_3	$m = 0.377$, $\sigma = 0.011$	$m = 0.388$, $\sigma = 0.011$

Generate a class of 100 targets with descriptive vectors

X_1	$m = 0.764$, $\sigma = 0.019$
X_2	$m = 0.512$, $\sigma = 0.030$
X_3	$m = 0.432$, $\sigma = 0.018$

Assign these targets to one of the validated classes. Count the number assigned to each class.

9.7. Two-element random vector **X** has mean $[\,1 \;\; 1\,]^T$ and covariance matrix

$$\begin{bmatrix} 2 & -1 \\ -1 & 1 \end{bmatrix}$$

Find the Euclidean distance between two sample points with vectors $[\,1 \;\; -1\,]^T$ and $[\,2 \;\; 1\,]^T$. Find the Mahalonobis distance between the sample points. Show that the Mahalonobis distance is unaffected by a change of scale of the vectors; for example, if **X** is a length, show that measuring it in centimeters rather than meters does not affect the Mahalonobis distance.

APPENDIX A

FADING AND SPECKLE

It was noted in Section 5.3 that scattering from terrain exhibits the phenomena of fading and speckle. These properties can be illustrated if we assume that a terrain cell has many scattering centers randomly distributed throughout it and that no scatterer has a cross-section significantly greater than the average of all scatterers. This limiting case does not represent all real terrain.

The real and imaginary parts of the received voltage from the nth scatterer in a terrain cell with N scattering centers are

$$v_r = \sum_{1}^{N} a_n \cos\phi_n \tag{A.1}$$

$$v_i = \sum_{1}^{N} a_n \sin\phi_n \tag{A.2}$$

If another terrain cell is examined, or if the same cell is examined at another aspect angle, a_n and ϕ_n will be different. To account for this, we define random variables A_n and Φ_n, for which a_n and ϕ_n are sample values. We also define random variables V_r and V_i, for which v_r and v_i are sample values. They are related by

$$V_r = \sum_{1}^{N} A_n \cos\Phi_n$$

Remote Sensing with Polarimetric Radar, by Harold Mott

$$V_i = \sum_1^N A_n \sin \Phi_n$$

A_n is nonnegative and Φ_n is distributed uniformly over a 2π range. Complex random variable V is defined as

$$V = V_r + jV_i = Ae^{j\Phi} \tag{A.3}$$

with the random variables taking on values v, a, and ϕ. The pdf of V is the joint pdf of V_r and V_i if $E(V_r V_i) = 0$, which is clearly true.

The means of V_r and V_i are zero. If no scatterer has a cross-section significantly greater than the average for all scatterers, it is to be expected from the central-limit theorem that V_r and V_i are approximately Gaussian. Schwarz (1959) gives the density of V_i as

$$p_{V_i}(v_i) = \frac{1}{\sqrt{2\pi}\,\sigma} e^{-v_i^2/2\sigma^2}$$

where

$$\sigma^2 = \frac{1}{2} \sum_1^N a_n^2$$

The density of V_r is given by a similar equation, and the variance is the same. The Gaussian pdfs are valid only for large N and for V_r and V_i near zero. They are clearly incorrect for V_r and V_i greater than the sum of the A_n. Given a value for V_r, a corresponding value for V_i cannot be found from (A.2). For N finite, $v_r^2 + v_i^2$ has a maximum value, which shows a lack of independence between V_r and V_i. In the region of interest, this relationship is irrelevant, and it is assumed here that V_r and V_i are independent. Their joint pdf can therefore be written as

$$p_{V_r,V_i}(v_r, v_i) = p_{V_r}(v_r) p_{V_i}(v_i) = \frac{1}{2\pi\sigma^2} e^{-(v_r^2+v_i^2)/2\sigma^2}$$

From (A.3) and the statement following it,

$$p_V(v) = p_V(ae^{j\phi}) = p_{V_r,V_i}(v_r, v_i)$$

The joint density of A and Φ can be found by the transformation,

$$p_{A,\Phi}(a, \phi) = p_{V_r,V_i}(v_r, v_i) \left| \det \begin{bmatrix} \dfrac{\partial v_r}{\partial a} & \dfrac{\partial v_r}{\partial \phi} \\ \dfrac{\partial v_i}{\partial a} & \dfrac{\partial v_i}{\partial \phi} \end{bmatrix} \right|$$

$$= p_{V_r,V_i}(v_r, v_i) \left| \det \begin{bmatrix} \cos\phi & -a\sin\phi \\ \sin\phi & a\cos\phi \end{bmatrix} \right| = a p_{V_r,V_i}(v_r, v_i)$$

where the vertical bars signify the magnitude. With the recognition that $a^2 = v_r^2 + v_i^2$, the joint density becomes

$$p_{A,\Phi}(a, \phi) = \frac{a}{2\pi\sigma^2} e^{-a^2/2\sigma^2}$$

The marginal densities are

$$p_A(a) = \int_0^{2\pi} p_{A,\Phi}(a, \phi)\, d\phi = \frac{a}{\sigma^2} e^{-a^2/2\sigma^2}$$

$$p_\Phi(\phi) = \int_0^{\infty} p_{A,\Phi}(a, \phi)\, da = \frac{1}{2\pi}$$

The probability density of the envelope is Rayleigh and that of the phase uniform.

The power developed in a one-ohm resistor is $w = a^2$. The transformation of $p_A(a)$ to $p_W(w)$ is carried out by

$$p_W(w) = p_A(a) \left| \frac{da}{dw} \right| = \frac{1}{2\sigma^2} e^{-w/2\sigma^2} \tag{A.4}$$

The signal power has exponential density with mean value,

$$m_W = \int_0^{\infty} w p_W(w)\, dw = 2\sigma^2$$

Using this, (A.4) can be written as

$$p_W(w) = \frac{1}{m_W} e^{-w/m_W}$$

The probability that the power received lies between ± 3 dB of the mean is 0.47. The probability that the measured power is less than the mean is 0.63.

If a terrain map is produced with the brightness of map pixels corresponding to received power from terrain cells, pixel brightness may differ significantly from the mean. This variation in brightness is called speckle. Speckle may be reduced by averaging the returned powers from adjacent terrain cells.

REFERENCE

M. Schwarz, *Information Transmission, Modulation, and Noise*, McGraw-Hill, New York, 1959.

APPENDIX B

PROBABILITY AND RANDOM PROCESSES

Waves scattered to a radar receiver generally cover a band of frequencies and are partially polarized. Electromagnetic noise also reaches the receiver or is generated therein. If these phenomena are to be understood, the received signals must be considered as random processes. This appendix presents the fundamental concepts necessary for understanding partially polarized waves and the targets that scatter such waves.

Classification of radar targets also requires the application of the principles of probability theory and random variables, and this appendix gives a basis for target classification.

B.1. PROBABILITY

An *experiment* is a measurement procedure under predetermined conditions. A *trial* of the experiment is the making of a measurement. Each trial has an *outcome*. The transmission of a radar pulse and measurement of the received energy by a radar receiver is a trial.

Experiments with Random Outcomes

If trials of an experiment are made, the outcomes are random; that is, they differ unpredictably from one outcome to the next. The outcomes form a set, a collection of objects that need not be material and may be a number. The outcomes can be grouped into classes which meet criteria of our choosing. Such a class is called a *result*. A result of trials of the experiment mentioned above might be the set of outcomes for which measured energy lies between values a and b. The outcomes grouped into a class are a subset of the set of all outcomes.

If a number of trials, N, of an experiment are made and the outcomes assigned to results $A, B, \ldots$ according to specific rules, the number of outcomes assigned to result A is N_A. The ratio of the number of outcomes assigned to result A to the total number of outcomes is the *relative frequency* of result A,

$$f_A = \frac{N_A}{N} \tag{B.1}$$

If several groups of trials are made, the average value of the outcomes may vary greatly from one group to the next if the number of trials in each group is small, but as the number of trials in each group becomes larger it tends to stabilize near some constant. This illustrates the statistical regularity that appears to be a fundamental attribute of physical processes (Wozencraft and Jacobs, 1965, p. 13). It is this regularity that allows us to acquire useful information in a world in which many measured quantities are random.

A Model of Experiments Having Random Outcomes

Probability theory was developed for certain experiments having random outcomes and is well suited for studying the transmission of a radar pulse and the measurement of received energy. The probability theory model has the following properties:

There is a *sample space*, which is a collection of objects. The objects in the sample space form a set. We denote this space by Ω and an object, called a *sample point*, by ω and write $\Omega = \{\omega\}$, where the braces denote the collection of sample points.

The sample space is divided into *events*, each of which is a set of sample points obeying some constraint. The sample points of an event form a subset of the points of the sample space. As an example, if the sample space contains a finite number of sample points, denoted here by integers,

$$\Omega = \{0, 1, 2, 3, 4, 5, 6, 7, 8, 9\}$$

event A may contain those sample points denoted by the odd integers,

$$A = \{1, 3, 5, 7, 9\}$$

In the probability model, a *probability measure* is associated with each event. It is an assignment of real numbers to the events of the sample space. We defer the rules for making this assignment.

We considered three aspects of an experiment having random outcomes: the set of outcomes, the formation of classes of outcomes (the results), and the relative frequencies with which the results occur. If the probability model is used, the correspondences between experiment and probability model should be

The Experiment	The Probability Model
1a. The set of all outcomes	1b. The sample space
2a. The classes or results	2b. Events
3a. Relative frequencies of results	3b. Probability measure of events

The third correspondence indicates that we should use as the probability measure of an event the value around which the relative frequency of the corresponding result appears to be anchored in a large number of trials.

Some Properties of Sets

Before further discussion of the application of probability theory to remote sensing, it is desirable to consider some properties of sets.

The *complement* of a set is a set consisting of all objects, or sample points, not in the original set. The complement of set A is

$$\overline{A} = \{\omega : \omega \text{ not in } A\}$$

We use the same notation for sets that we used for events.

Sets A and B may contain points common to both or they may contain no common points, in which case they are *disjoint* sets. The *union* of A and B is the set of all points in A or in B or in both. It is denoted by

$$A \cup B = \{\omega : \omega \text{ in } A \text{ or in B}\}$$

The *intersection* of A and B is the set of all points in both A and B. it is denoted by

$$A \cap B = \{\omega : \omega \text{ in both } A \text{ and } B\}$$

The inverted cup symbol is often omitted and the intersection written as AB.

The set containing all points corresponding to an experiment has been denoted as Ω. It is sometimes called the *universal set*. Its complement is the set containing no points, called the *null set* and denoted as $\overline{\Omega} = \emptyset$

The Venn Diagram

The definitions given above can be illustrated by *Venn diagrams*, one of which is shown in Fig. B.1a The rectangle represents the universal set, with all members of the set inside its boundaries. Sets A and B are represented by plane figures of arbitrary shape and size. All members of a set are within the boundary of that set. The members of Ω lying outside the boundary of A form the set $\overline{A}$.

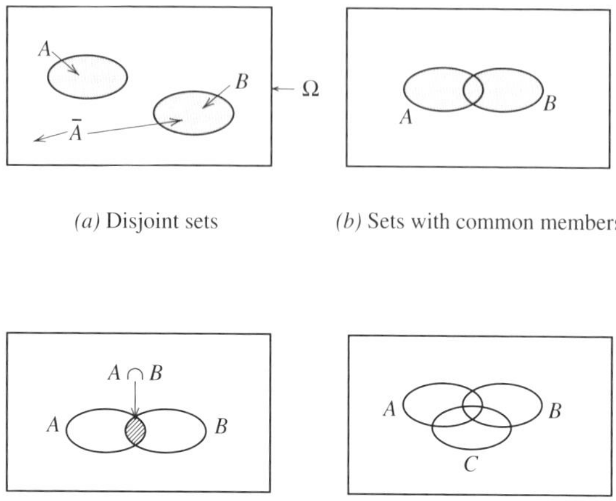

(*a*) Disjoint sets (*b*) Sets with common members

(*c*) Intersection of sets (*d*) Union and intersection

Fig. B.1. Venn diagrams.

If the set members are considered to be sample points corresponding to the outcomes of experimental trials, the geometric nature of the Venn diagram suggests, incorrectly, that the members are distributed throughout the plane with locations corresponding to the outcomes. An outcome may determine whether a sample point lies inside A, for example, but other than that its position on the Venn diagram is arbitrary.

The union of sets A and B is shown as the shaded areas of Figs. B.1*a* and *b*. In Fig. B.1*a*, sets A and B have no set members in common; they are disjoint. In Fig. B.1*b* there are sample points common to A and B. Figure B.1*c* shows the intersection of A and B as the shaded area. It is readily seen from these diagrams that

$$A \cup \overline{A} = \Omega$$

$$A\overline{A} = \emptyset$$

$$A\Omega = \Omega$$

It may be seen from the diagrams that the union and intersection operations are commutative, or

$$A \cup B = B \cup A$$

$$AB = BA$$

A closer examination of the Venn diagrams shows that

$$\overline{A \cup B} = \overline{A} \cap \overline{B}$$

$$\overline{A \cap B} = \overline{A} \cup \overline{B}$$

These two equations represent *De Morgan's laws*.

If a third set, intersecting A and B, is represented on a Venn diagram as in Fig. B.1d, it can be determined that union and intersection operations are associative and distributive, or

$$A \cup (B \cup C) = (A \cup B) \cup C$$
$$A \cap (B \cap C) = (A \cap B) \cap C$$

and

$$A \cap (B \cup C) = (A \cap B) \cup (A \cap C)$$
$$A \cup (B \cap C) = (A \cup B) \cap (A \cup C)$$

Probability: The Axioms

We now consider the third property of the probability model, the probability measure of event A. This is written as $P(A)$ and is a unique real number assigned to event A in sample space Ω. The axioms of the probability measure are taken as:

Axiom 1:

$$0 \leq P(A) \leq 1$$

Axiom 2:

$$P(\Omega) = 1$$

Note that Ω is itself the event which includes all possible events, and it is reasonable that it have the highest probability.

Axiom 3:

If N events $A_1, A_2, \ldots, A_N$ are disjoint, then

$$P(A_1 \cup A_2 \cup \ldots \cup A_N) = \sum_{1}^{N} P(A_n) \text{ if } A_m \cap A_n = \emptyset \qquad m \neq n$$

This axiom is most commonly used for two events, so that a special case of it is

$$P(A \cup B) = P(A) + P(B) \text{ if } A \cap B = \emptyset \tag{B.2}$$

Intersection

Assume that A and B are not disjoint. It can be seen from Fig. B.1c that

$$P(A) = P(A \cap B) + P(A \cap \overline{B})$$
$$P(B) = P(A \cap B) + P(\overline{A} \cap B)$$

Study of the figure will also show that

$$A \cup B = (A \cap B) \cup (A \cap \overline{B}) \cup (\overline{A} \cap B)$$

and that $A \cap B$, $A \cap \overline{B}$, and $\overline{A} \cap B$ are disjoint events. Then

$$P(A \cup B) = P(A \cap B) + P(A \cap \overline{B}) + P(\overline{A} \cap B)$$

If the term $P(A \cap B)$ is added to and subtracted from this probability and if (B.2) is used, the probability becomes

$$P(A \cup B) = P(A) + P(B) - P(A \cap B) \tag{B.3}$$

If A and B are disjoint, this form reduces to the special case of Axiom 3.

Probability: Application to Experiments

In the tabulation of correspondences of the probability model to an experiment, the probability measure of the model event corresponds to the relative frequency of the experimental result. If N trials of the experiment are made, N_A outcomes will be placed in class A, or result A, where we assume that all outcomes are grouped or classified in distinct results, $A, B, \ldots$. The relative frequency of result A is given by (B.1). Note that $0 \leq f_A \leq 1$ and that every outcome that falls in A contributes to N_A, even if it is identical to other outcomes.

Let class A correspond to event A of the probability model. We must assign probability $P(A)$ to event A, and a reasonable assignment is to say that $P(A) = f_A$. In a long sequence of independent trials, we expect the relative frequency of a result to converge to the probability of the corresponding mathematical event (Wozencraft and Jacobs, 1965, p. 17). This relationship is critical. If the outcomes of independent trials are random, the relative frequency offers the only concise information available for the experiment; if probability theory is to be applicable, the probability of an event must be the relative frequency of the corresponding result.

Consider Fig. B.1*a* with disjoint sets A and B. In N trials, N_A outcomes will fall into the class corresponding to set A (we use A to represent both the experiment's result and the probability model's event) and it is appropriate to take as corresponding values

$$P(A) = f_A = \frac{N_A}{N}$$

Similarly,

$$P(B) = f_B = \frac{N_B}{N}$$

If we consider a result that is augmented by one outcome if the outcome falls into either class, the relative frequency of the overall result is

$$f_{A \text{ or } B} = f_A + f_B = P(A) + P(B)$$

The event that corresponds to results A or B is the union of A and B, with probability $P(A \cup B)$. We set it equal to the relative frequency of the result that includes classes A or B,

$$P(A \cup B) = f_{A \text{ or } B} = P(A) + P(B) \qquad \text{if results are distinct} \tag{B.4}$$

The requirement that the results be distinct led to (B.4) rather than the more general form (B.3). This discussion has provided insight into the relative-frequency interpretation of the probability measure and has shown the applicability of the third axiom to experimental results.

It is of interest to determine if intersecting events such as those shown in Fig. B.1*b* can occur in the probability model corresponding to a physical experiment. If the class boundaries are chosen so that they do not overlap, results A and B will not contain common outcomes, and events A and B in the probability model will be disjoint; Figure B.1*b* does not apply. If we transmit a radar pulse and measure received energy W and if results A and B correspond, respectively, to $a \leq W \leq b$ and $c \leq W \leq d$, events A and B in the model will intersect if $c < b$. Sample points corresponding to outcomes placed in both results A and B lie in the common region of A and B of the Venn diagram.

In the discussions that follow, we will use primarily the language of experiments and use outcomes and classes, or results, rather than sample points and events. Because of the correspondences between experiment and the probability model, we shall freely use the concept of the probability measure.

Joint and Conditional Probabilities

Consider next a compound experiment, in contrast to the single, or simple, experiment considered to now. A compound experiment is two or more experiments whose trials correspond. If we limit the discussion to two experiments for simplicity, the nth trial of experiment 1 corresponds to the nth trial of experiment 2. As an example, consider the transmission of a radar pulse and the measurement of the scattered energy by a receiver. Experiment 1 is radiation with reception by receiver 1. Experiment 2 is radiation by the same transmitter of the same pulse and reception by receiver 2, at a different location. Let result A of experiment 1 be energy to receiver 1 between a and b inclusive and result $\overline{A}$ be received energy outside this range. Result $\overline{A}$ may be separated into as many classes as desired, but we need not consider that here. Let result B of experiment 2 be energy to receiver 2 between c and d inclusive and result $\overline{B}$ be received energy outside this range. Four results are possible for this compound experiment:

Energy to receiver 1 is in result A.
    Energy to receiver 2 is in result B.
    Energy to receiver 2 is in result $\overline{B}$.

Energy to receiver 1 is in result $\overline{A}$.
Energy to receiver 2 is in result B.
Energy to receiver 2 is in result $\overline{B}$.

In N trials of the compound experiment, let the number of trials for which receiver 1 measures energy in class A and receiver 2 measures energy in class B be $N(A, B)$. Then,

$$f_N(A, B) = \frac{N(A, B)}{N}$$

is the *joint relative frequency* of results A and B.

The relative frequency with which result B occurs, after it is known that result A occurred, is of interest. It is called the *conditional relative frequency* and symbolized by $f_N(B|A)$. The conditional relative frequency is the number of outcomes contributing to $N(A, B)$ divided by the number of outcomes contributing to result A. Then,

$$f_N(B|A) = \frac{N(A, B)}{N(A)} = \frac{N(A, B)/N}{N(A)/N} = \frac{f_N(A, B)}{f_N(A)} \tag{B.5}$$

If the occurrence of result B is independent of the occurrence of result A, then $f_N(B|A) = f_N(B)$ and $f_N(A, B) = f_N(A)f_N(B)$.

In relating the single experiment to the probability model, we set $P(A)$ equal to the relative frequency N_A/N. We do the same for the compound experiment and define the *joint probability* of A and B as

$$P(A, B) = P(A \cap B) = f_N(A, B)$$

where A and B are events whose intersection need not be the null set.

A Venn diagram can be used to illustrate the events of a compound experiment. The intersection of two events for a single experiment implies an overlap of class boundaries. Intersection of events for a compound experiment does not imply an overlap of class boundaries if one event encompasses the outcomes for one single experiment of the compound experiment and the other event includes the outcomes of another single experiment.

Following the form of the definition of the conditional relative frequency, we define the *conditional probability* of event B by

$$P(B|A) = \frac{P(A \cap B)}{P(A)} \tag{B.6}$$

Note the correspondence between this probability relationship and the relative frequency equation, B.5.

If the result of a trial of one of these simple experiments does not affect the result of another experiment, then the relative frequency $f_N(B|A)$ can be expected to be very close to $f_N(B)$. In this case, we assume $P(B|A) = P(B)$

and $P(A \cap B) = P(A)P(B)$. The last equation can be used to define statistically independent events. A simple example of such events is the tossing of two coins in a compound experiment, in which the result of one coin's toss does not affect that of the other.

Bayes Rule

If the event symbols are interchanged in the definition of conditional probability, there results

$$P(A|B) = \frac{P(A \cap B)}{P(B)}$$

where we use the commutativity of $A \cap B$.

Combining the equations in the definition of conditional probability gives

$$P(A|B)P(B) = P(B|A)P(A) \tag{B.7}$$

This is a special case of Bayes rule.

Additional Theorems

Let A be a given event and let B_m be a group of disjoint events with $P(B_m B_n) = 0$ if $m \neq n$. We make the reasonable assumption that the third axiom of probability theory is valid for conditional probabilities,

$$P(B_1 \cup B_2 \cup \ldots \cup B_M|A) = \sum_1^M P(B_m|A)$$

If the B_m are exhaustive; that is, if

$$\bigcup_1^M B_m = \Omega$$

where the cup symbol signifies the union of all B_m, then

$$\sum_1^M P(B_m|A) = 1$$

Another theorem can be developed readily. Let A_i be disjoint and exhaustive events,

$$\bigcup_1^\infty A_i = \Omega$$

$$P(A_i \cap A_j) = 0 \quad i \neq j$$

Let event B be written as

$$B = B \cap \Omega = B \cap \bigcup_{1}^{\infty} A_i = \bigcup_{1}^{\infty} A_i \cap B$$

Applying axiom 3,

$$P(B) = \sum_{1}^{\infty} P(A_i \cap B) = \sum_{1}^{\infty} P(A_i)P(B|A_i)$$

This is called the *theorem of total probability*.

Finally, we look at Bayes rule once more. We may write (B.7) as

$$P(A_i|B) = \frac{P(A_i)P(B|A_i)}{P(B)}$$

where A_i is one of a group of events.

Using the theorem of total probability, with the assumption that the A_i are disjoint and exhaustive, allows this to be written as

$$P(A_i|B) = \frac{P(A_i)P(B|A_i)}{\sum_1^{\infty} P(A_k)P(B|A_k)} = \frac{P(A_i)P(B|A_i)}{\sum_1^{\infty} P(A_k \cap B)}$$

This relationship expresses *Bayes theorem*.

Hypotheses

Before leaving this discussion of probability theory, we introduce another point of view of probability that will be useful in later discussions. We have used phrases such as "the occurrence of event A". That is a short way of saying that the outcome of a given trial of an experiment is classified in result A. We used $P(A)$ as the probability of the occurrence of event A. This again should be interpreted as the probability that a given trial of the experiment yields an outcome that is grouped in result A. Sometimes thought can be clarified by defining hypotheses corresponding to events. For example, H_A is the hypothesis that event A occurs, in the sense noted above. Then $P(H_A)$ is the probability that hypothesis H_A is true. We do not think of an occurrence but think instead of truth or falsity. We can use $\overline{H}_A$ to be the hypothesis that event A does not occur, or that H_A is false. Then,

$$P(H_A) = 1 - P(\overline{H}_A)$$

B.2. RANDOM VARIABLES

In a formal sense, a random variable is a mapping over the sample space into the real line (Urkowitz, 1983, p. 199) or a real function of the elements of the sample space (Peebles, 1980, p. 33). A useful interpretation is that it is a set of values associated with the outcomes of a physical process (Urkowitz, 1983, p. 199).

It was noted previously that a real number can be associated with each sample point. For the example that we cited, energy received from a radar pulse, it is natural to associate with each sample point the energy received. Another natural association would be the number of spots on the upper face of a die. If the trial outcome is not a number, as would be the outcome of tossing coins, a number may be assigned arbitrarily, one for the outcome of a single toss a head and another for the outcome a tail.

We take the number associated with each sample point, which arises from the outcome of the real-world trial, as the value to be mapped to the real line to create the real random variable. The resulting random variable may be continuous, as it would be if the measured pulse energy of our example were used to form the variable. It may be a discrete random variable if the values associated with the sample points and mapped to the real line are discrete, as in the rolling of a die. Continuous random variables are of greater interest in remote sensing than discrete ones. The random variable may be real or complex, but unless it is stated that a variable is complex, it is used here as real. We use a capital letter, for example X, for a random variable and a corresponding lower case letter, for example x, for the value that the random variable takes.

If the function mapped from the sample space to the real line is denoted by $X(\omega)$, it must meet certain conditions to be a random variable (Peebles, 1980, p. 35). Every point in Ω must correspond to only one value of X. Further, the set $\{X \leq x\}$ must be an event for real x. Finally, $P\{X = -\infty\} = P\{X = \infty\} = 0$. These conditions are met in the problems of concern to remote sensing.

We define a *probability distribution function* or *cumulative distribution function (cdf)* as the probability that random variable X does not exceed x, or

$$F_X(x) = P(X \leq x)$$

The subscript X denotes the random variable being considered. If there is no possibility of mistaking the random variable, the subscript may be omitted. Since the cdf is a probability, it must lie between 0 and 1, inclusive. For x approaching negative infinity, the cdf must be 0. For x approaching positive infinity it approaches 1. Then,

$$0 \leq F_X(x) \leq 1$$

$$F_X(-\infty) = 0$$

$$F_X(\infty) = 1$$

Consider random variable X lying in the disjoint intervals defined by $X \leq x_1$ and $x_1 < X \leq x_2$. Then,

$$P(X \leq x_2) = P(X \leq x_1) + P(x_1 < X \leq x_2)$$

or

$$P(x_1 < X \leq x_2) = F_X(x_2) - F_X(x_1)$$

Since probabilities are nonnegative, $F_X(x_2) > F_X(x_1)$ if $x_2 > x_1$. It may be shown that $F_X(x)$ is continuous from the right; that is, for $\delta > 0$,

$$\lim_{\delta \to 0} F_X(x + \delta) = F_X(x)$$

Probability Density Functions

The *probability density function (pdf)* of a continuous random variable is the derivative of the cdf,

$$p_X(x) = \frac{dF_X(x)}{dx}$$

Since F_X is a nondecreasing function, $p_x(x) \geq 0$.

If the probability density is integrated from $-\infty$ to x, the result is

$$\int_{-\infty}^{x} p_X(u)\, du = F_X(x) - F_X(-\infty) = F_X(x)$$

It follows that

$$P(x_1 < X \leq x_2) = F_X(x_2) - F_X(x_1) = \int_{x_1}^{x_2} p_X(x)\, dx$$

It is of interest to ascertain the probability that a continuous random variable X lies near a specified value, x_0. This requires a specification of what is meant by "near", and we specify it by an interval Δx. If we let $x_1 = x_0$ and $x_2 = x_0 + \Delta x$ in this equation, we obtain

$$P(x_0 < X \leq x_0 + \Delta x) = \int_{x_0}^{x_0+\Delta x} p_X(x)\, dx \approx p_X(x_0)\Delta x$$

This leads to an interpretation of the probability density function of a random variable at x_0 as the probability that the value of the variable lies within a small range at x_0, divided by the range.

When a trial is made of an experiment having random outcomes, the outcome cannot be predicted successfully. The information that we have about a random variable representing the output is contained in the cumulative distribution function or, for a continuous random variable, in the probability density function.

Average of a Discrete Random Variable

Let N samples, with N large, of random variable X take on the values $x_1, x_2, \ldots, x_M$ with corresponding probabilities $P_1, P_2, \ldots, P_M$. Then, the *statistical average* or *mean* or *expectation* of the variable is

$$E(X) = \frac{1}{N}\sum_{1}^{N} x_n = \sum_{1}^{M} x_m P_m$$

The second form for the expectation can be justified by noting that if X takes on M different values in N repetitions of the experiment, the arithmetic average is

$$\frac{1}{N}\sum_{1}^{N} x_n = \frac{1}{N}\sum_{1}^{M} N_m x_m = \sum_{1}^{M} x_m \frac{N_m}{N}$$

where N_m is the number of occurrences of the value x_m in the N trials. Since N_m/N approaches probability P_m for large N, the definition of expectation is justified.

Average of a Continuous Random Variable

Divide the interval in which X lies into a large number, M, of small intervals Δx. The probability that X lies in interval m is

$$P(x_m < X \le x_m + \Delta x) \approx p_X(x_m)\Delta x$$

Then the mean or expectation of X is

$$E(X) = \sum_{1}^{M} x_m p_X(x_m)\Delta x$$

In the limit as Δx approaches 0, this becomes

$$E(X) = \int_{-\infty}^{\infty} x p_X(x)\,dx$$

Functions of a Random Variable

A function $f(X)$ of a continuous random variable can be defined. It will have a mean value or expectation,

$$E[f(X)] = \int_{-\infty}^{\infty} f(x) p_X(x)\,dx$$

The nth *moment* of a random variable is defined by

$$\alpha_n = E(X^n) = \int_{-\infty}^{\infty} x^n p_X(x)\,dx$$

The nth *central moment* is

$$\mu_n = E[(X - E(X))^n] = \int_{-\infty}^{\infty} [x - E(X)]^n p_X(x)\,dx$$

Of particular interest is the second central moment. It is the *variance* of X and is written as

$$\mu_2 = \sigma^2 = \int_{-\infty}^{\infty} [x - E(X)]^2 p_X(x)\,dx$$

It is easy to show that

$$\sigma^2 = E(X^2) - [E(X)]^2$$

Let $Y = f(X)$ be a function of a random variable whose probability density function is known and let x and y be the values taken on by X and Y. To find the pdf of Y, note that x and y obey the same functional relationship as the variables themselves. We assume that both $f(x)$ and its inverse are single valued.

Figure B.2 illustrates a monotonically-increasing functional relationship between random variables X and Y. Values taken on by the variables are x and y, and a point on the curve is denoted by (x, y). The probability that X does not exceed x is the same as the probability that Y does not exceed y,

$$P(Y \le y) = P(X \le x) = P[X \le f^{-1}(y)]$$

These are cdfs, so $F_Y(y) = F_X[f^{-1}(y)]$. The pdf of Y is

$$p_Y(y) = \frac{dF_Y(y)}{dy} = \frac{dF_X[f^{-1}(y)]}{dx}\frac{dx}{dy} = p_X[f^{-1}(y)]\frac{dx}{dy}$$

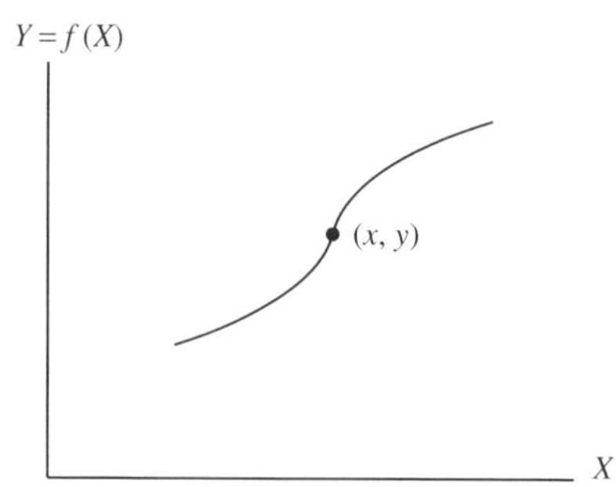

Fig. B.2. Monotonically increasing function.

Let Y now be a monotonically-decreasing function of X. For this case,

$$P(X \leq x) = P(Y \geq y) = 1 - P(Y < y) = 1 - P(Y \leq y) + P(Y = y)$$

and in terms of cdfs,

$$F_Y(y) = 1 - F_X[f^{-1}(y)] + P(Y = y)$$

Differentiating this equation gives

$$p_Y(y) = -p_X[f^{-1}(y)]\frac{dx}{dy}$$

However, the derivative dx/dy is negative, so we can write

$$p_Y(y) = p_X[f^{-1}(y)]\left|\frac{dx}{dy}\right|$$

This form holds for the monotonically-increasing function also, so it is a general relation between the pdfs of a random variable and a function of it.

Multiple Random Variables

More than one random variable is involved in many physical problems, and relationships between them exist. Let these variables be $X_1, X_2, \ldots, X_N$. Of interest is the joint probability that each of these variables is less than or equal to a specified value, and consequently we define a *joint probability distribution function (joint cdf)* as

$$F_{X_1,X_2,\ldots,X_N}(x_1, x_2, \ldots, x_N) = P(X_1 \leq x_1, X_2 \leq x_2, \ldots, X_N \leq x_N) \qquad \text{(B.8)}$$

The *joint probability density function (joint pdf)* is the derivative of this distribution function,

$$p_{X_1,X_2,\ldots,X_N}(x_1, \ldots, x_N) = \frac{\partial^N F_{X_1,X_2,\ldots,X_N}(x_1, x_2, \ldots, x_N)}{\partial x_1 \partial x_2 \ldots \partial x_N} \qquad \text{(B.9)}$$

The joint distribution function can be found by integration,

$$F_{X_1,\ldots,X_N}(x_1, \ldots, x_N) = \int_{-\infty}^{x_1} \cdots \int_{-\infty}^{x_N} p_{X_1,\ldots,X_N}(u_1, \ldots, u_N)\, du_1 \ldots du_N$$

Marginal Probability Distribution Function and Density

The *marginal probability distribution function (marginal cdf)* for random variable X_n of a multivariate case is defined like the cdf of X_n if it were the only variable, thus

$$F_{X_n}(x_n) = P(X_n \le x_n)$$

If a marginal probability density (marginal pdf) is defined as

$$p_{X_n}(x_n) = \int_{-\infty}^{\infty} \cdots \int_{-\infty}^{\infty} p_{X_1,\ldots,X_N}(u_1, \ldots, u_N)\, du_1 \ldots du_{n-1}\, du_{n+1} \ldots du_N$$

$F_{X_n}(x_n)$ can be written as

$$\begin{aligned} F_{X_n}(x_n) &= P(X_1 \le \infty, \ldots, X_n \le x_n, \ldots, X_N \le \infty) \\ &= F_{X_1,\ldots,X_N}(\infty, \ldots, x_n, \ldots, \infty) = \int_{-\infty}^{x_n} p_{X_n}(u_n)\, du_n \end{aligned}$$

Moments

Problems involving two random variables are of particular interest. Let these be X and Y with joint probability density $p_{X,Y}(x, y)$. We can find by integration of the joint density the marginal densities $p_X(x)$ and $p_Y(y)$ and the means and variances.

The *joint moments* of two random variables are defined by

$$E(X^n Y^m) = \int_{-\infty}^{\infty} \int_{-\infty}^{\infty} x^n y^m p_{X,Y}(x, y)\, dx\, dy$$

where the sum $n + m$ is the order of the moment. The second-order moment

$$R_{XY} = E(XY) = \int_{-\infty}^{\infty} \int_{-\infty}^{\infty} xy\, p_{X,Y}(x, y)\, dx\, dy$$

is the *correlation* of the two variables.

Two random variables are statistically independent if and only if

$$P(X \le x, Y \le y) = P(X \le x)P(Y \le y)$$

Then the joint distribution and density functions for statistically independent variables are

$$\begin{aligned} F_{X,Y}(x, y) &= F_X(x)F_Y(y) \\ p_{X,Y}(x, y) &= p_X(x)p_Y(y) \end{aligned}$$

The defining equation for the second moment of statistically independent variables shows that they satisfy $E(XY) = E(X)E(Y)$. Variables satisfying this relationship are uncorrelated. The statistical independence of X and Y is sufficient to show that they are uncorrelated. The converse is not necessarily true, unless X and Y are Gaussian (Peebles, 1980, p. 102).

The *joint central moments* of two random variables are defined by

$$E[(X - m_x)^n (Y - m_y)^m)] = \int_{-\infty}^{\infty} \int_{-\infty}^{\infty} (x - m_x)^n (y - m_y)^m p_{X,Y}(x, y)\, dx\, dy$$

where m_x and m_y are the means of the variables. The second-order central moment is of particular importance. It is the *covariance* of X and Y and is given by

$$C_{XY} = E[(X - m_x)(Y - m_y)] = \int_{-\infty}^{\infty} \int_{-\infty}^{\infty} (x - m_x)(y - m_y) p_{X,Y}(x, y)\, dx\, dy$$

If the product in the integral is expanded, it can be seen that

$$E[(X - m_x)(Y - m_y)] = E(XY) - m_x m_y$$

The covariance can be normalized by division with the square roots of the individual variances,

$$\sigma_x^2 = E[(X - m_x)^2]$$
$$\sigma_y^2 = E[(X - m_y)^2]$$

Doing so gives the correlation coefficient of the variables,

$$\rho = E\left[\frac{(X - m_x)}{\sigma_x} \frac{(Y - m_y)}{\sigma_y}\right]$$

It lies between 0 and 1 inclusive.

B.3. RANDOM VECTORS

Consider a problem with random variables $X_1, X_2, \ldots, X_N$. The joint probability that each of these variables is less than or equal to a specified value is given by (B.8). We simplify the notation of that equation by defining a random vector, $\mathbf{X}$, and the value it takes on, $\mathbf{x}$, with elements the scalar random variables of the equation. Then the probability distribution function (no longer referred to as joint) of $\mathbf{X}$ can be shortened to

$$F_{\mathbf{X}}(\mathbf{x}) = P(\mathbf{X} \leq \mathbf{x}) \tag{B.10}$$

The probability density function for multiple random variables can now be written as

$$p_{\mathbf{X}}(\mathbf{x}) = \frac{\partial^N}{\partial x_1 \partial x_2 \ldots \partial x_N} F_{\mathbf{X}}(\mathbf{x}) \tag{B.11}$$

The cdf may be found from the pdf by

$$F_{\mathbf{X}}(\mathbf{x}) = \int_{-\infty}^{x_1} \cdots \int_{-\infty}^{x_N} p_{\mathbf{X}}(\mathbf{u})\, du_1 \ldots du_N = \int_{-\infty}^{\mathbf{x}} p_{\mathbf{X}}(\mathbf{u})\, d\mathbf{u}$$

where the meaning of the integration with respect to $\mathbf{u}$ is clear.

As we did with a continuous scalar random variable, we can determine the probability that the continuous random vector $\mathbf{X}$ lies in the small region around point $\mathbf{x}_0$ by writing

$$P(\mathbf{x}_0 < \mathbf{X} \le \mathbf{x}_0 + \Delta\mathbf{x}) = \int_{\mathbf{x}_0}^{\mathbf{x}_0+\Delta\mathbf{x}} p_{\mathbf{X}}(\mathbf{u})\, d\mathbf{u} \approx p_{\mathbf{X}}(\mathbf{x}_0)\Delta x_1\, \Delta x_2 \ldots \Delta x_N$$

Moments of a Random Vector

The mean or expectation of a continuous random vector is defined by

$$\mathbf{m} = E(\mathbf{X}) = \int_{-\infty}^{\infty} \mathbf{x} p_{\mathbf{X}}(\mathbf{x})\, d\mathbf{x}$$

The notation signifies the operation

$$\mathbf{m} = \begin{bmatrix} \int_{-\infty}^{\infty} \cdots \int_{-\infty}^{\infty} x_1 p_{\mathbf{X}}(\mathbf{x})\, dx_1\, dx_2 \ldots dx_N \\ \cdot \\ \cdot \\ \int_{-\infty}^{\infty} \cdots \int_{-\infty}^{\infty} x_N p_{\mathbf{X}}(\mathbf{x})\, dx_1\, dx_2 \ldots dx_N \end{bmatrix}$$

If $\mathbf{Y} = \mathbf{Y}(\mathbf{X})$, the expectation of $\mathbf{Y}$ is found from

$$\mathbf{m}_{\mathbf{Y}} = \int_{-\infty}^{\infty} \mathbf{y} p_{\mathbf{Y}}(\mathbf{y})\, d\mathbf{y} = \int_{-\infty}^{\infty} \mathbf{y}(\mathbf{x}) p_{\mathbf{X}}(\mathbf{x})\, d\mathbf{x}$$

The mean of a sum of random vectors with the same dimensions is the sum of their means, or

$$\mathbf{m}\left(\sum_i \mathbf{X}_i\right) = \sum_i \mathbf{m}(\mathbf{X}_i)$$

If random vector $\mathbf{X}$ takes on only N discrete values, the mean of the vector is

$$\mathbf{m} = \sum_1^N \mathbf{x}_n P(\mathbf{X} = \mathbf{X}_n)$$

Important moments for random vectors are the mean and the correlation and covariance matrices. The *correlation matrix* for a real random vector is defined as

$$\mathbf{R} = E[\mathbf{X}\mathbf{X}^T]$$

The *covariance matrix* is defined similarly, but the vector mean is subtracted from the vector; thus for a real vector $\mathbf{X}$,

$$\mathbf{C} = E[(\mathbf{X} - \mathbf{m})(\mathbf{X} - \mathbf{m})^T] \tag{B.12}$$

The covariance and correlation matrices are related by

$$\begin{aligned}\mathbf{C} &= E[(\mathbf{X} - \mathbf{m})(\mathbf{X} - \mathbf{m})^T] = E[\mathbf{X}\mathbf{X}^T] - E(\mathbf{X})\mathbf{m}^T - \mathbf{m}E(\mathbf{X}^T) + \mathbf{m}\mathbf{m}^T \\ &= \mathbf{R} - \mathbf{m}\mathbf{m}^T\end{aligned}$$

Joint Distribution and Density Functions

The distribution function of a random vector is a joint distribution function, although it is convenient not to refer to it as one. There is a need, however, to define a joint distribution function for two vectors (which can be extended to any number of vectors). Consider random vector $\mathbf{X}$ of N elements and random vector $\mathbf{Y}$ of M elements, with vectors $\mathbf{x}$ and $\mathbf{y}$ specifying the element values taken on by $\mathbf{X}$ and $\mathbf{Y}$.

We define a joint distribution function for two vectors as

$$F_{\mathbf{X},\mathbf{Y}}(\mathbf{x}, \mathbf{y}) = P(\mathbf{X} \leq \mathbf{x} \text{ and } \mathbf{Y} \leq \mathbf{y}) \tag{B.13}$$

The joint probability density is related to the joint distribution function by

$$p_{\mathbf{X},\mathbf{Y}}(\mathbf{x}, \mathbf{y}) = \frac{\partial^{N+M}}{\partial x_1\, \partial x_2 \ldots \partial x_N\, \partial y_1\, \partial y_2 \ldots \partial y_M} F_{\mathbf{X},\mathbf{Y}}(\mathbf{x}, \mathbf{y})$$

and

$$F_{\mathbf{X},\mathbf{Y}}(\mathbf{x}, \mathbf{y}) = \int_{-\infty}^{\mathbf{x}} \int_{-\infty}^{\mathbf{y}} p_{\mathbf{X},\mathbf{Y}}(\mathbf{u}, \mathbf{v})\, d\mathbf{u}\, d\mathbf{v} \tag{B.14}$$

It follows from (B.13) that

$$F_{\mathbf{X},\mathbf{Y}}(\infty, \infty) = 1 \tag{B.15}$$

$$F_{\mathbf{X},\mathbf{Y}}(\mathbf{x}, \infty) = F_{\mathbf{X}}(\mathbf{x}) \tag{B.16}$$

$$F_{\mathbf{X},\mathbf{Y}}(\infty, \mathbf{y}) = F_{\mathbf{Y}}(\mathbf{y}) \tag{B.17}$$

From (B.14) and (B.15),

$$\int_{-\infty}^{\infty} \int_{-\infty}^{\infty} p_{\mathbf{X},\mathbf{Y}}(\mathbf{x}, \mathbf{y})\, d\mathbf{x}\, d\mathbf{y} = 1$$

Equations B.16 and B.17 can be used to find the marginal probability densities,

$$p_{\mathbf{U}}(\mathbf{u}) = \int_{-\infty}^{\infty} p_{\mathbf{X},\mathbf{Y}}(\mathbf{x}, \mathbf{y})\, d\mathbf{v} \qquad \mathbf{u}, \mathbf{v} = \mathbf{x}, \mathbf{y} \qquad \mathbf{u} \neq \mathbf{v} \tag{B.18}$$

Transformations of Random Vectors

We developed in Section B.2 the transformation between the pdfs for scalar random variable X and variable Y, a function of X. It is desirable to have a similar transformation for random vectors. For clarity, we do this first for two-element vectors. Let

$$\begin{bmatrix} X & Y \end{bmatrix}^T$$

be a random vector whose elements take on the values x and y. Let a function of this vector and its value be

$$\begin{bmatrix} U(X, Y) \\ V(X, Y) \end{bmatrix} \quad \text{and} \quad \begin{bmatrix} u(x, y) \\ v(x, y) \end{bmatrix}$$

We assume the transformation is one-to-one so that we can write x and y as functions of u and v.

The probability that the vector components fall within infinitesimal limits around chosen values is

$$P(u < U \leq u + du, v < V \leq v + dv) = P(x < X \leq x + dx, y < Y \leq y + dy)$$

which gives immediately

$$p_{U,V}(u, v)\, du\, dv = p_{X,Y}(x, y)\, dx\, dy \tag{B.19}$$

$$p_{U,V}(u, v)\, dA_{uv} = p_{X,Y}(x, y)\, dA_{xy} \tag{B.20}$$

where dA_{uv} and dA_{xy} are corresponding infinitesimal areas in the planes defined by u, v and x, y.

The infinitesimal areas are related by the magnitude of the Jacobi determinant, or *Jacobian*,

$$J(x, y/u, v) = \det \begin{bmatrix} \dfrac{\partial x}{\partial u} & \dfrac{\partial x}{\partial v} \\ \dfrac{\partial y}{\partial u} & \dfrac{\partial y}{\partial v} \end{bmatrix}$$

The area relationship is

$$dA_{xy} = |J(x, y/u, v)|\, dA_{uv} \tag{B.21}$$

where the bars denote absolute value.

It follows from (B.20) and (B.21) that the probability densities are related by

$$p_{U,V}(u, v) = |J(x, y/u, v)|\, p_{X,Y}[x(u, v), y(u, v)]$$

where the functional relationship is explicit because the right side of the equation should ultimately be expressed in terms of **u** and **v**.

The extension to greater dimensions is straightforward. Random vector **Y** is a function of **X**. They have the same number of elements and take on values **x** and **y**. We assume a one-to-one relationship between **x** and **y**.

The probability density functions for the vectors are related by

$$p_{\mathbf{Y}}(\mathbf{y}) = |J(\mathbf{x}/\mathbf{y})|\, p_{\mathbf{X}}(\mathbf{x}) \tag{B.22}$$

where the Jacobian is

$$J(\mathbf{x}/\mathbf{y}) = \det \begin{bmatrix} \dfrac{\partial x_1}{\partial y_1} & \cdots & \dfrac{\partial x_1}{\partial y_N} \\ . & \cdots & . \\ \dfrac{\partial x_N}{\partial y_1} & \cdots & \dfrac{\partial x_N}{\partial y_N} \end{bmatrix}$$

If the transformation between vectors is $\mathbf{Y} = \mathbf{AX}$, where **A** is a non-singular square matrix, then

$$J(\mathbf{x}/\mathbf{y}) = \det(\mathbf{A}^{-1}) = \frac{1}{\det(\mathbf{A})}$$

Joint Functions of Random Vectors and Events

The joint cdf of a continuous random vector **X** and the occurrence of event A is defined by

$$F_{\mathbf{X},A}(\mathbf{x}, A) = P(\mathbf{X} \le \mathbf{x} \text{ and } A)$$

The joint density is given by

$$p_{\mathbf{X},A}(\mathbf{x}, A) = \frac{\partial^N}{\partial x_1 \cdots \partial x_N} F_{\mathbf{X},A}(\mathbf{x}, A)$$

The joint cdf can also be written as

$$F_{\mathbf{X},A}(\mathbf{x}, A) = \int_{-\infty}^{\mathbf{x}} p_{\mathbf{X},A}(\mathbf{u}, A)\, d\mathbf{u}$$

The joint probability that **X** can be found in a small region around value **x** and that event A occurs is

$$P(\mathbf{x} < \mathbf{X} \le \mathbf{x} + \Delta\mathbf{x} \quad \text{and} \quad A) = p_{\mathbf{X},A}(\mathbf{x}, A)\,\Delta x_1 \ldots \Delta x_N$$

When we considered the joint probability of two vectors, we defined a marginal cdf by integrating with respect to one of the vectors. Not surprisingly, the marginal distribution function when events are involved is a summation. If $A_1, A_2, \ldots, A_M$ is a set of mutually exclusive and collectively exhaustive events, the marginal cdf for random vector **X** is

$$F_{\mathbf{X}}(\mathbf{x}) = \sum_{1}^{M} F_{\mathbf{X},A_m}(\mathbf{x}, A_m)$$

The marginal density function is

$$p_{\mathbf{X}}(\mathbf{x}) = \sum_{1}^{M} p_{\mathbf{X},A_m}(\mathbf{x}, A_m) \tag{B.23}$$

Conditional Functions of Random Vectors and Events

Consider the conditional distribution function of a random vector given that event A has occurred. For example, vector **X** might describe a measurement of variables, associated with some object, whose values depend on the class to which the object belongs. If the object is from class A, then **X** will have some distribution, and if it is from B, **X** will have a different distribution. We therefore define a conditional probability distribution function (conditional cdf),

$$F_{\mathbf{X}|A}(\mathbf{x}|A) = P[(\mathbf{X} \le \mathbf{x})|A]$$

The conditional cdf can be related to the joint cdf by (B.6). If B is the event that **X** is less than or equal to **x**, (B.6) becomes

$$P[(\mathbf{X} \le \mathbf{x})|A] = \frac{P(\mathbf{X} \le \mathbf{x} \text{ and } A)}{P(A)}$$

or

$$F_{\mathbf{X}|A}(\mathbf{x}|A) = \frac{F_{\mathbf{X},A}(\mathbf{x}, A)}{P(A)}$$

Differentiation shows that the probability density functions are related by

$$p_{\mathbf{X}|A}(\mathbf{x}|A) = \frac{p_{\mathbf{X},A}(\mathbf{x}, A)}{P(A)} \tag{B.24}$$

Interchange symbols A and B in (B.6) and let B be the event that $\mathbf{X}$ lies between $\mathbf{x}$ and $\mathbf{x} + \Delta\mathbf{x}$. Consider the conditional probability,

$$P[A|(\mathbf{x} < \mathbf{X} \leq \mathbf{x} + \Delta\mathbf{x})] = \frac{P(\mathbf{x} < \mathbf{X} \leq \mathbf{x} + \Delta\mathbf{x} \text{ and } A)}{P(\mathbf{x} < \mathbf{X} \leq \mathbf{x} + \Delta\mathbf{x})}$$

But

$$P(\mathbf{x} < \mathbf{X} \leq \mathbf{x} + \Delta\mathbf{x} \text{ and } A) \approx p_{\mathbf{X},A}(\mathbf{x}, A)\, \Delta x_1 \ldots \Delta x_N$$

$$P(\mathbf{x} < \mathbf{X} \leq \mathbf{x} + \Delta\mathbf{x}) \approx p_{\mathbf{X}}(\mathbf{x})\, \Delta x_1 \ldots \Delta x_N$$

Then,

$$P[A|(\mathbf{x} < \mathbf{X} \leq \mathbf{x} + \Delta\mathbf{x})] = \frac{p_{\mathbf{X},A}(\mathbf{x}, A)\, \Delta x_1 \ldots \Delta x_N}{p_{\mathbf{X}}(\mathbf{x})\, \Delta x_1 \ldots \Delta x_N} = \frac{p_{\mathbf{X},A}(\mathbf{x}, A)}{p_{\mathbf{X}}(\mathbf{x})} \tag{B.25}$$

Conditional Probability Density for Two Random Vectors

Let events A and B represent the occurrence of two random vectors in small regions, with probability

$$P(A|B) = P[(\mathbf{x} < \mathbf{X} \leq \mathbf{x} + \Delta\mathbf{x})|(\mathbf{y} < \mathbf{Y} \leq \mathbf{y} + \Delta\mathbf{y})]$$

By using Bayes rule we can write

$$\begin{aligned} P(A|B) &= \frac{P(\mathbf{x} < \mathbf{X} \leq \mathbf{x} + \Delta\mathbf{x}, \mathbf{y} < \mathbf{Y} \leq \mathbf{y} + \Delta\mathbf{y})}{P(\mathbf{y} < \mathbf{Y} \leq \mathbf{y} + \Delta\mathbf{y})} \\ &\approx \frac{p_{\mathbf{X},\mathbf{Y}}(\mathbf{x}, \mathbf{y})\, \Delta x_1 \ldots \Delta x_N\, \Delta y_1 \ldots \Delta y_M}{p_{\mathbf{Y}}(\mathbf{y})\, \Delta y_1 \ldots \Delta y_M} = \frac{p_{\mathbf{X},\mathbf{Y}}(\mathbf{x}, \mathbf{y})}{p_{\mathbf{Y}}(\mathbf{y})}\, \Delta x_1 \ldots \Delta x_N \end{aligned}$$

If we define a conditional probability density for the two random vectors as

$$p_{\mathbf{X}|\mathbf{Y}}(\mathbf{x}|\mathbf{y}) = \frac{p_{\mathbf{X},\mathbf{Y}}(\mathbf{x}, \mathbf{y})}{p_{\mathbf{Y}}(\mathbf{y})} \tag{B.26}$$

then

$$P(\mathbf{x} < \mathbf{X} \leq \mathbf{x} + \Delta\mathbf{x}|\mathbf{y} < \mathbf{Y} \leq \mathbf{y} + \Delta\mathbf{y}) = p_{\mathbf{X}|\mathbf{Y}}(\mathbf{x}|\mathbf{y})\Delta x_1 \ldots \Delta x_N$$

If the events are independent, the conditional density is the same as the density of $\mathbf{X}$ alone, or

$$p_{\mathbf{X}}(\mathbf{x}) = p_{\mathbf{X}|\mathbf{Y}}(\mathbf{x}|\mathbf{y}) = \frac{p_{\mathbf{X},\mathbf{Y}}(\mathbf{x}, \mathbf{y})}{p_{\mathbf{Y}}(\mathbf{y})}$$

and

$$p_{\mathbf{X},\mathbf{Y}} = p_{\mathbf{X}}(\mathbf{x})\, p_{\mathbf{Y}}(\mathbf{y})$$

Other Forms of Bayes Rule

We can derive other useful forms from Bayes rule. From (B.24) and (B.25), it is seen that

$$P[A|(\mathbf{x} < \mathbf{X} \le \mathbf{x} + \Delta\mathbf{x})] = \frac{p_{\mathbf{X},A}(\mathbf{x}, A)}{p_{\mathbf{X}}(\mathbf{x})} = \frac{p_{\mathbf{X}|A}(\mathbf{x}|A)P(A)}{p_{\mathbf{X}}(\mathbf{x})} \tag{B.27}$$

With the use of (B.23) for the denominator of this equation, we get for one event A_m of a set of mutually exclusive and collectively exhaustive events $A_1, A_2, \ldots, A_M$,

$$P[A_m|(\mathbf{x} < \mathbf{X} \le \mathbf{x} + \Delta\mathbf{x})] = \frac{p_{\mathbf{X}|A_m}(\mathbf{x}|A_m)P(A_m)}{\sum_1^M p_{\mathbf{X}|A_j}(\mathbf{x}|A_j)P(A_j)}$$

With the use of (B.26), in the form shown and also interchanging $\mathbf{X}$ and $\mathbf{Y}$, we obtain

$$p_{\mathbf{Y}|\mathbf{X}}(\mathbf{y}|\mathbf{x}) = \frac{p_{\mathbf{X},\mathbf{Y}}(\mathbf{x}|\mathbf{y})p_{\mathbf{Y}}(\mathbf{y})}{p_{\mathbf{X}}(\mathbf{x})} \tag{B.28}$$

This is equivalent, for a random vector, to a special case of Bayes rule given previously for discrete random variables. If (B.18) and (B.26) are used in the denominator of (B.28), it becomes

$$p_{\mathbf{Y}|\mathbf{X}}(\mathbf{y}|\mathbf{x}) = \frac{p_{\mathbf{X}|\mathbf{Y}}(\mathbf{x}|\mathbf{y})p_{\mathbf{Y}}(\mathbf{y})}{\int_{-\infty}^{\infty} p_{\mathbf{X}|\mathbf{Y}}(\mathbf{x}|\mathbf{y})p_{\mathbf{Y}}(\mathbf{y})\,dy}$$

Complex Random Vectors

It is convenient in describing electromagnetic phenomena to use complex vectors to account for the magnitude and phase angle of spatially directed sinusoidally time-varying quantities. This leads naturally to complex random vectors, $\mathbf{X}_c = \mathbf{X}' + j\mathbf{X}''$. Probability distribution and density functions of continuous complex random vectors are found like those for real random vectors, with (B.10) and (B.11). The mean $\mathbf{m}_c$ is complex. The covariance matrix differs slightly from that of a real random vector, (B.12). It is

$$\mathbf{C}_c = E[(\mathbf{X}_c - \mathbf{m}_c)(\mathbf{X}_c - \mathbf{m}_c)^\dagger]$$

The covariance matrix for complex random vectors is positive semidefinite (Urkowitz, 1983, p. 284).

A real random vector, $\mathbf{X}_r$, of $2N$ elements, which corresponds to complex vector $\mathbf{X}_c$, can be defined by

$$\mathbf{X}_r = [\,X_1' \;\; X_1'' \;\; \cdots \;\; X_n' \;\; X_n'' \;\; \cdots \;\; X_N' \;\; X_N''\,]^T$$

If three conditions on the covariances of the vector elements are met, the pdf of the complex vector is equal to the pdf of the real vector. They are (Urkowitz, 1983, p. 337):

1. $\text{cov}\,(X'_n, X''_n) = 0 \qquad \text{all } n$
2. $\text{cov}\,(X'_n, X'_m) = \text{cov}(X''_n, X''_m) \qquad \text{all } n, m$
3. $\text{cov}\,(X''_n, X'_m) = -\text{cov}(X''_m, X'_n) \qquad \text{all } n \neq m$

B.4. PROBABILITY DENSITY FUNCTIONS IN REMOTE SENSING

The Uniform Random Variable. A phase angle measured at random time intervals has a uniform density

$$p_\Phi(\phi) = \frac{1}{2\pi} \qquad 0 \le \phi < 2\pi$$

The Rayleigh pdf. If a dart is thrown repeatedly at a target, the point at which it hits relative to a center for all the impacts is governed by two random variables, R and Θ. We assume without proof that the variables have a joint pdf

$$p_{R,\Theta}(r, \theta) = \frac{re^{-r^2/2b}}{2\pi b} \qquad \begin{array}{l} 0 \le \theta < 2\pi \\ 0 \le r < \infty \end{array}$$

The marginal densities, found by integrating over r and θ respectively, are

$$p_\Theta(\theta) = \frac{1}{2\pi} \qquad 0 \le \theta < 2\pi$$

$$p_R(r) = \frac{re^{-r^2/2b}}{b} \qquad 0 \le r < \infty$$

The second pdf, governing the distance from the target center at which the dart hits, is the Rayleigh pdf. Its maximum value, called the mode of R, occurs at $r = \sqrt{b}$. The expected value of R is $\sqrt{\pi b/2}$.

The Gaussian pdf. In the study of random electromagnetic phenomena, the Gaussian pdf is of great importance. Many electromagnetic variables are the sum of independent variables. A noise voltage may be the sum of voltages from many sources, for example. The central-limit theorem states that the probability density of the sum of a large number of independent quantities approaches the Gaussian density regardless of the densities of the individual quantities, provided that the contribution of any one quantity is not comparable with the sum of all the others. It is reasonable, therefore, to utilize the Gaussian pdf for many problems. Another factor, less cogent but nevertheless important, is that sometimes solutions can be found if Gaussian pdfs are assumed, but not otherwise. Even if the pdf corresponding to the phenomenon is not Gaussian, it may approximate it sufficiently to make the resulting solution useful.

For one variable, the Gaussian pdf is

$$p_X(x) = \frac{1}{\sqrt{2\pi}\,\sigma} e^{-(x-m)^2/2\sigma^2}$$

where m is the mean of X and σ is the *standard deviation* (square root of the variance).

The joint Gaussian pdf of two random variables is (Ziemer and Tranter, 1995, p. 313)

$$p_{X,Y}(x, y) = \frac{\exp\left[-\dfrac{\left(\dfrac{x-m_x}{\sigma_x}\right)^2 - 2\rho\left(\dfrac{x-m_x}{\sigma_x}\right)\left(\dfrac{y-m_y}{\sigma_y}\right) + \left(\dfrac{y-m_y}{\sigma_y}\right)^2}{2(1-\rho^2)}\right]}{2\pi\sigma_x\sigma_y\sqrt{1-\rho^2}}$$

where σ_x and σ_y are the standard deviations of X and Y, and

$$\rho = \frac{E[(X - m_x)(Y - m_y)]}{\sigma_x \sigma_y}$$

ρ is the *correlation coefficient* of X and Y.

For a real random vector $\mathbf{X}$ with N components, the Gaussian pdf is (Therrien, 1989, p. 58)

$$p_{\mathbf{X}}(\mathbf{x}) = \frac{1}{(2\pi)^{N/2}\,(\det(\mathbf{C}))^{1/2}} e^{-\frac{1}{2}(\mathbf{x}-\mathbf{m})^T \mathbf{C}^{-1}(\mathbf{x}-\mathbf{m})}$$

where $\mathbf{m}$ is the mean of $\mathbf{X}$, and $\mathbf{C}$ is the covariance matrix. The pdf of complex Gaussian random vector $\mathbf{X}_c$ is given by a similar equation,

$$p_{\mathbf{X}_c}(\mathbf{x}_c) = \frac{1}{(2\pi)^{N/2}\,(\det(\mathbf{C}_c))^{1/2}} e^{-\frac{1}{2}(\mathbf{x}_c-\mathbf{m}_c)^\dagger \mathbf{C}_c^{-1}(\mathbf{x}_c-\mathbf{m}_c)}$$

B.5. RANDOM PROCESSES

Measurement of a received signal in radar requires a finite time, and the result is not a number but a waveform. If additional measurements are made, noise or relative radar-target motion will cause the waveforms to be different for each measurement. In the general case, the waveform variation from one measurement to another is not predictable. We therefore consider, not a random number associated with each measurement, but a random waveform.

An experiment that gives outcomes that are random waveforms is called a *random process*. The outcomes are *sample waveforms*, and the sample waveforms taken together are an *ensemble*. Figure B.3 shows three sample waveforms from an ensemble.

We previously denoted a random variable as $X(\omega)$, where ω is a sample point taken from a set associated with the outcomes of repeated trials of an experiment. In a similar manner, we denote the random process by $X(\omega, t)$, where ω is a sample point from a set associated with the outcomes (waveforms) of the random process. A sample waveform of the ensemble is $X(\omega_i, t)$.

If the ith waveform is considered at time t_j, it has a numeric value denoted by $X(\omega_i, t_j)$. At time t_j, the ensemble of sample waveforms has a set of values from which we can form a random variable, $X(\omega, t_j)$. Doing so allows us to use the developments of random variable theory to study random processes. We identify the random variable formed at t_j as X_j. Figure B.3 suggests the formation of the random variables by taking values from a "slice" through the sample waveforms. For a specific t_j, the random variables of interest to us are continuous functions of ω. If a set of times $\{t_j,\ j = 1, 2, \ldots, N\}$ is chosen, N random variables can be formed. For convenience, we denote these variables as $X(t)$.

We saw previously that the information about a continuous random variable is contained in its probability density function. The information about a random process is contained in the joint probability density function of the N random variables formed in the manner described above and defined by (B.8) and (B.9).

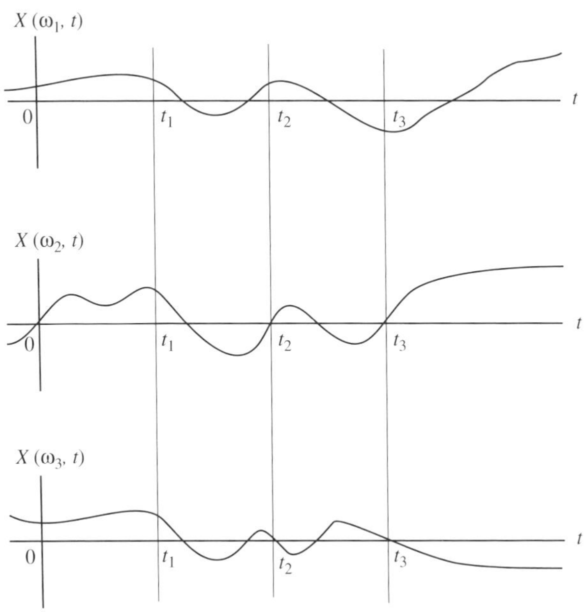

Fig. B.3. Waveforms of a random process.

If a vector $\mathbf{X}$, with N elements $X_1, X_2, \ldots, X_N$, which takes on values $\mathbf{x}$, represents the random process, we can write

$$F_{\mathbf{X}}(\mathbf{x}) = P(\mathbf{X} \leq \mathbf{x})$$

and

$$p_{\mathbf{X}}(\mathbf{x}) = \frac{\partial^N}{\partial x_1 \partial x_2 \ldots \partial x_N} F_{\mathbf{X}}(\mathbf{x})$$

Complete information about a random process can only be obtained from the joint probability density function for any finite set of observation times (Wozencraft and Jacobs, 1965, p. 133).

Time and Ensemble Averages

It may not be possible to obtain the N-fold probability density function for the random process of interest. A partial description can, however, be of value and we consider such.

Our concern is with a random variable $X(t)$, for which t can have any value. At time t, the mean and variance of $X(t)$ can be found, and they will be functions of time,

$$m_x(t) = E[X(t)]$$

and

$$\sigma_x^2(t) = E\left[(X(t) - m_x(t))^2\right]$$

In order to write the correlation and covariance functions, we consider two random variables formed from the random process, $X_1 = X(t_1)$ and $X_2 = X(t_2)$. The correlation of these two random variables is

$$R_{XX}(t_1, t_2) = E\left[X(t_1)X(t_2)\right] = \int_{-\infty}^{\infty} \int_{-\infty}^{\infty} x_1 x_2 p_{X_1,X_2}(x_1, x_2)\, dx_1\, dx_2 \quad \text{(B.29)}$$

where $x_1 = x(t_1)$ and $x_2 = x(t_2)$. This is the *autocorrelation function* of $X(t)$. The substitution that $t_1 = t$, $t_2 = t + \tau$ leads to a commonly seen form,

$$R_{XX}(t, \tau) = E\left[X(t)X(t + \tau)\right] \quad \text{(B.30)}$$

The argument of R_{XX} has a different form in (B.29) and (B.30). Both forms are widely used, and are presented here for that reason, but to avoid confusion we will use the argument form of (B.29). The preferred argument in (B.30) is $t, t + \tau$.

In a similar manner, the covariance can be written as

$$C_{XX}(t_1, t_2) = E\left[(X(t_1) - m_x(t_1))\,(X(t_2) - m_x(t_2))\right]$$

The mean, variance, correlation, and covariance determined in this manner are *statistical* or *ensemble* averages, in contrast to time averages that could be taken for one of the waveforms of the set of waveforms in the random process.

Stationary Random Processes

Consider a moving target seen by a monostatic radar. The received signal is the return from the target with added noise. The relative contributions of target and noise to the received voltage will change as the target moves, with the target having greater importance as it approaches the radar and less as it recedes. It is clear that random variable $X_i = X(t_i)$ will have properties that depend on t_i and that the random vector $\mathbf{X}$, which represents the random process, will have properties that depend on the time origin of Fig. B.3. This is an example of a *non-stationary process*.

Complete information about a random process is contained in the joint probability density function of (B.9). We expect, for this example, that the probability density will change if times $t_1, t_2, \ldots$, at which $x_1, x_2, \ldots$ of (B.9) are determined, are shifted by a fixed amount; that is, if the time origin of Fig. B.3 is changed.

With the example of a non-stationary random process as background, we define a *strict-sense stationary* random process as one for which the joint probability density function of (B.9) is invariant to a time-origin shift. This can be expressed by

$$p_{X_1,\ldots,X_N}[x(t_1), \ldots, x(t_N)] = p_{X_1,\ldots,X_N}[x(t_1+\tau), \ldots, x(t_N+\tau)]$$

A random process may fail to meet this requirement but meet less rigorous tests. If the marginal densities $p_{X_1}[x(t_1)]$ and $p_{X_2}[x(t_2)]$ are equal, the process is *stationary to order one*. The mean of $X(t_1)$ is

$$E[X(t_1)] = \int_{-\infty}^{\infty} x(t_1) p_{X_1}[x(t_1)]\, dx(t_1) = \int_{-\infty}^{\infty} u p_{X_1}[x(t_1)]\, du$$

and that of $X(t_1+\tau)$ is

$$\begin{aligned} E[X(t_1+\tau)] &= \int_{-\infty}^{\infty} x(t_2) p_{X_2}[x(t_1+\tau)]\, dx(t_2) \\ &= \int_{-\infty}^{\infty} x(t_2) p_{X_1}[x(t_1)]\, dx(t_2) = \int_{-\infty}^{\infty} u p_{X_1}[x(t_1)]\, du \end{aligned}$$

It can be seen that

$$E[X(t_1+\tau)] = E[X(t_1)] \tag{B.31}$$

If

$$p_{X_1,X_2}[x(t_1), x(t_2)] = p_{X_1,X_2}[x(t_1+\lambda), x(t_2+\lambda)] \tag{B.32}$$

the process is stationary to order two. For such a process, the autocorrelation of $X(t_1)$ and $X(t_2) = X(t_1 + \tau)$ is

$$R_{XX}(t_1, t_2) = R_{XX}(t_1, t_1 + \tau) = \int_{-\infty}^{\infty} \int_{-\infty}^{\infty} x_1 x_2 p_{X_1,X_2}[x(t_1), x(t_2)]\, dx_1\, dx_2 \tag{B.33}$$

The autocorrelation after time shift λ is

$$\begin{aligned} R_{XX}(t_1 + \lambda, t_2 + \lambda) &= R_{XX}(t_1 + \lambda, t_1 + \tau + \lambda) \\ &= \int_{-\infty}^{\infty} \int_{-\infty}^{\infty} uv p_{X_1,X_2}[x(t_1 + \lambda), x(t_2 + \lambda)] du\, dv \end{aligned} \tag{B.34}$$

where $u = x(t_1 + \lambda)$ and $v = x(t_2 + \lambda)$. If (B.32) is taken into account, comparison of (B.33) and (B.34) shows that

$$R_{XX}(t_1 + \lambda, t_1 + \tau + \lambda) = R_{XX}(t_1, t_1 + \tau) \tag{B.35}$$

The origin shift λ does not affect the autocorrelation function; therefore, R_{XX} is independent of time t_1 and depends only on τ. This property is sometimes expressed as

$$R_{XX}(t, t + \tau) = E[X(t)X(t + \tau)] = R_{XX}(\tau) \tag{B.36}$$

We noted after (B.30) that the arguments of the autocorrelation functions of two equations were expressed in two different forms. Equation (B.36) is still another form which is used. It is valid for random processes satisfying (B.32).

A random process that conforms to (B.31) and (B.35) is a *wide-sense stationary process*. A process that is stationary to order two is wide-sense stationary, but the converse is not necessarily true (Peebles, 1980, p. 129). Strict-sense stationarity implies wide-sense stationarity, but not vice-versa, except for Gaussian random processes (Ziemer and Tranter, 1995, p. 338).

Ergodic Random Processes

Measurements often do not yield an ensemble of time functions for analysis, and a complete description of the random process cannot be obtained. Instead, we may have only one time function, one member of the ensemble $X(\omega_i, t)$. Since we have only one such function, we omit the identification of the ensemble member and use $X(t)$, with measured value $x(t)$, for the sample waveform.

It is of interest to know how well we can describe the random process from one time function. To put this another way, are certain properties for the known time function equal to those of another time function of the ensemble, assuming another member of the ensemble could be obtained?

The time average and the *time autocorrelation function* of $x(t)$ are

$$\overline{x(t)} = \lim_{T\to\infty} \frac{1}{2T} \int_{-T}^{T} x(t)\,dt$$

$$\mathrm{R}_{\mathrm{XX}}(\tau) = \lim_{T\to\infty} \frac{1}{2T} \int_{-T}^{T} x(t)x(t+\tau)\,dt$$

where the time autocorrelation function is designated with Roman subscript rather than the italic used for the ensemble autocorrelation.

In general, the time average and the time autocorrelation differ for different members of the ensemble and are random variables. We may then find the expectation of $\overline{x(t)}$ by statistical averaging,

$$E[\overline{x(t)}] = E\left[\lim_{T\to\infty} \frac{1}{2T} \int_{-T}^{T} x(t)\,dt\right]$$

If the expectation operator can be taken inside the integral, then

$$E[\overline{x(t)}] = \lim_{T\to\infty} \frac{1}{2T} \int_{-T}^{T} E[x(t_1), x(t_2), \ldots, x(t_N)]\,dt = E(X)$$

In the same way,

$$E[\mathrm{R}_{\mathrm{XX}}(\tau)] = R_{XX}(\tau)$$

These equations show that the two time averages considered are equal to the statistical or ensemble averages if the expectation and time integration operations can be interchanged. If all time averages of a member function of the ensemble are equal to the corresponding ensemble averages, the random process is *ergodic*.

It is difficult to prove that a physical random process is ergodic. We may, however, have no choice but to examine one sample function only and to describe the process from our knowledge of it.

The Cross-Correlation Function

Many physical processes involve two random processes that are correlated. The definition of the correlation function of one random process is readily extended to two and we define the *cross-correlation function* of two random processes as

$$\begin{aligned} R_{XY}(t, t+\tau) &= E[X(t)Y(t+\tau)] \\ &= \int_{-\infty}^{\infty} x(t)y(t+\tau)p_{XY}[x(t), y(t+\tau)]\,dx\,dy \end{aligned}$$

If $x(t)$ and $y(t)$ are at least jointly wide-sense stationary, this correlation is independent of time.

The corresponding time average, the time cross-correlation function, is

$$\mathrm{R}_{XY}(\tau) = \lim_{T\to\infty} \frac{1}{2T} \int_{-T}^{T} x(t)y(t+\tau)\,dt$$

If this time cross-correlation function is equal to the statistical cross-correlation function and if the two random processes are individually ergodic, they are said to be *jointly ergodic*.

We can define a *cross-covariance* similarly if the means of the individual random processes are subtracted.

REFERENCES

P. Z. Peebles, *Probability, Random Variables, and Random Signal Principles*, McGraw-Hill, New York, 1980.

H. Urkowitz, *Signal Theory and Random Processes*, Artech House, Dedham, MA, 1983.

J. M. Wozencraft and I. M. Jacobs, *Principles of Communication Engineering*, Wiley, New York, 1965.

R. E. Ziemer and W. H. Tranter, *Principles of Communications*, Fourth ed., Wiley, New York, 1995.

APPENDIX C

THE KENNAUGH MATRIX

Elements of the bistatic Kennaugh matrix are given here in terms of averages of Sinclair matrix elements. The modified Kennaugh matrix elements are also given for backscattering.

The Bistatic Matrix

Equations (7.6) and (7.16) can be combined to give

$$K_{11} = \frac{1}{2}\langle |S_{xx}|^2 + |S_{xy}|^2 + |S_{yx}|^2 + |S_{yy}|^2 \rangle$$

$$K_{12} = \frac{1}{2}\langle |S_{xx}|^2 - |S_{xy}|^2 + |S_{yx}|^2 - |S_{yy}|^2 \rangle$$

$$K_{13} = \mathrm{Re}\langle S_{xx} S_{xy}^* + S_{yx} S_{yy}^* \rangle$$

$$K_{14} = \mathrm{Im}\langle S_{xx} S_{xy}^* + S_{yx} S_{yy}^* \rangle$$

$$K_{21} = \frac{1}{2}\langle |S_{xx}|^2 + |S_{xy}|^2 - |S_{yx}|^2 - |S_{yy}|^2 \rangle$$

$$K_{22} = \frac{1}{2}\langle |S_{xx}|^2 - |S_{xy}|^2 - |S_{yx}|^2 + |S_{yy}|^2 \rangle$$

$$K_{23} = \mathrm{Re}\langle S_{xx} S_{xy}^* - S_{yx} S_{yy}^* \rangle$$

$$K_{24} = \mathrm{Im}\langle S_{xx} S_{xy}^* - S_{yx} S_{yy}^* \rangle$$

Remote Sensing with Polarimetric Radar, by Harold Mott

$$K_{31} = \mathrm{Re}\langle S_{xx}S_{yx}^* + S_{xy}S_{yy}^*\rangle$$

$$K_{32} = \mathrm{Re}\langle S_{xx}S_{yx}^* - S_{xy}S_{yy}^*\rangle$$

$$K_{33} = \mathrm{Re}\langle S_{xy}S_{yx}^* + S_{xx}S_{yy}^*\rangle$$

$$K_{34} = \mathrm{Im}\langle S_{xx}S_{yy}^* + S_{xy}^*S_{yx}\rangle$$

$$K_{41} = \mathrm{Im}\langle S_{xx}S_{yx}^* + S_{xy}S_{yy}^*\rangle$$

$$K_{42} = \mathrm{Im}\langle S_{xx}S_{yx}^* - S_{xy}S_{yy}^*\rangle$$

$$K_{43} = \mathrm{Im}\langle S_{xx}S_{yy}^* - S_{xy}^*S_{yx}\rangle$$

$$K_{44} = \mathrm{Re}\langle S_{xy}S_{yx}^* - S_{xx}S_{yy}^*\rangle$$

Modified Matrix for Backscattering

Note: These matrix elements were formed with the assumption of an x-polarized incident wave.

$$K_{11} = \frac{1}{2}\langle |S_{xx}|^2 + 2|S_{xy}|^2 + |S_{yy}|^2\rangle$$

$$K_{12} = K_{21} = \frac{1}{2}\langle |S_{xx}|^2 - |S_{yy}|^2\rangle$$

$$K_{13} = K_{31} = \mathrm{Re}\langle S_{xx}S_{xy}^* + S_{xy}S_{yy}^*\rangle$$

$$K_{14} = K_{41} = \mathrm{Im}\langle S_{xx}S_{xy}^* + S_{xy}S_{yy}^*\rangle$$

$$K_{22} = \frac{1}{2}\langle |S_{xx}|^2 - 2|S_{xy}|^2 + |S_{yy}|^2\rangle$$

$$K_{23} = K_{32} = \mathrm{Re}\langle S_{xx}S_{xy}^* - S_{xy}S_{yy}^*\rangle$$

$$K_{24} = K_{42} = \mathrm{Im}\langle S_{xx}S_{xy}^* - S_{xy}S_{yy}^*\rangle$$

$$K_{33} = \langle |S_{xy}|^2\rangle - b + \mathrm{Re}\langle S_{xx}S_{yy}^*\rangle$$

$$K_{34} = K_{43} = \mathrm{Im}\langle S_{xx}S_{yy}^*\rangle$$

$$K_{44} = \langle |S_{xy}|^2\rangle - b - \mathrm{Re}\langle S_{xx}S_{yy}^*\rangle$$

where

$$b = \frac{1}{2}\langle |S_{xx}|^2 + |S_{xy}|^2\rangle - \frac{1}{2}\left[\langle |S_{xx}|^2 + |S_{xy}|^2\rangle^2 - 4\left(\langle |S_{xx}|^2\rangle\langle |S_{xy}|^2\rangle - \langle S_{xx}S_{xy}^*\rangle\langle S_{xx}^*S_{xy}\rangle\right)\right]^{1/2}$$

Modified Average Mueller Matrix for Backscattering

The average Mueller matrix is related to the Kennaugh matrix by (7.38), and the modified forms have the same relationship. The modified average Mueller matrix for backscattering and the modified Kennaugh matrix for backscattering are then related by

$$\mathbf{M}_{\mathrm{av}\,b\,\mathrm{mod}} = \mathbf{diag}(1, 1, -1, 1)\mathbf{K}_{b\,\mathrm{mod}}$$

APPENDIX D

BAYES ERROR BOUNDS

The Bayes error probability given here is simple in form but difficult to determine in general.

Error Bounds for Two Classes

The Bayes error probability is the Bayes risk if the cost coefficients are chosen as $c_{11} = c_{22} = 0$ and $c_{12} = c_{21} = 1$. It is, from (9.9),

$$C = \int_{R_1} P(\omega_2) p_{\mathbf{X}|\omega_2}(\mathbf{x}|\omega_2)\, d\mathbf{x} + \int_{R_2} P(\omega_1) p_{\mathbf{X}|\omega_1}(\mathbf{x}|\omega_1)\, d\mathbf{x}$$

The Bayes decision rule is to assign a vector to ω_1 if the integrand in the left integral is less than the integrand in the right integral. The integrals can be combined if we choose the minimum integrand for every region of the integration,

$$C = \int_{R_1+R_2} \min[P(\omega_1) p_{\mathbf{X}|\omega_1}(\mathbf{x}|\omega_1),\ P(\omega_2) p_{\mathbf{X}|\omega_2}(\mathbf{x}|\omega_2)]\, d\mathbf{x}$$

Two real quantities a and b lying between 0 and 1 inclusive satisfy the inequality

$$\min(a, b) \le a^s b^{1-s}$$

where s is some chosen value lying between 0 and 1 inclusive. This gives an upper bound on the Bayes error probability,

$$C \leq [P(\omega_1)]^s [P(\omega_2)]^{1-s} \int_{-\infty}^{\infty} [p_{\mathbf{X}|\omega_1}(\mathbf{x}|\omega_1)]^s [p_{\mathbf{X}|\omega_2}(\mathbf{x}|\omega_2)]^{1-s} \, d\mathbf{x} \qquad \text{(D.1)}$$

This upper bound is the *Chernoff bound*. The limits on the integral show a recognition that the two regions cover all space. If $s = \frac{1}{2}$, the Bayes upper bound simplifies to the *Bhattacharyya* bound.

If the number of components of $\mathbf{x}$ is more than two, numerical integration of (D.1) is difficult (Press et al., 1992, p. 155), and it may not be worthwhile to attempt to find error bounds. For Gaussian density functions, the integration can be carried out analytically (Therrien, 1989, p. 148).

Error Bounds for the Multiclass Problem

Relabel the Chernoff bound for the two-class problem as $\epsilon_{12} = C$. If assignments are to be made to N classes, generalize (D.1) to give the pairwise errors for classes ω_i and ω_j,

$$\epsilon_{ij} \leq [P(\omega_i)]^{s_{ij}} [P(\omega_j)]^{1-s_{ij}} e^{\mu(s_{ij})}$$

with

$$\mu(s_{ij}) = \ln \int_{-\infty}^{\infty} [p_{\mathbf{X}|\omega_i}(\mathbf{x}|\omega_i)]^{s_{ij}} [p_{\mathbf{X}|\omega_j}(\mathbf{x}|\omega_j)]^{(1-s_{ij})} \, d\mathbf{x}$$

The integration is carried out over regions R_i, R_j, and all other regions from which objects would be assigned to classes other than ω_i and ω_j.

It can be shown (Fukunaga, 1972, p. 280) that the probability of total error for the multiclass problem is related to these two-class errors (pairwise errors) by

$$\epsilon \leq \sum_{i>j}^{N} \sum_{j=1}^{N} \epsilon_{ij}$$

Substituting the probabilities and probability densities into this equation gives the multiclass error probability,

$$\epsilon \leq \sum_{i>j}^{N} \sum_{j=1}^{N} [P(\omega_i)]^{s_{ij}} [P(\omega_j)]^{(1-s_{ij})} \int_{-\infty}^{\infty} [p_{\mathbf{X}|\omega_i}(\mathbf{x}|\omega_i)]^{s_{ij}} [p_{\mathbf{X}|\omega_j}(\mathbf{x}|\omega_j)]^{(1-s_{ij})} \, d\mathbf{x}$$

For simplicity, let $s_{ij} = 1/2$, and note that

$$[P(\omega_i)]^{1/2} [P(\omega_j)]^{1/2} \leq \frac{1}{2}$$

With these substitutions, the upper error bound for assignments to N classes becomes

$$\epsilon \leq \frac{1}{2} \sum_{i>j}^{N} \sum_{j=1}^{N} \int_{-\infty}^{\infty} [p_{\mathbf{X}|\omega_i}(\mathbf{x}|\omega_i) p_{\mathbf{X}|\omega_j}(\mathbf{x}|\omega_j)]^{1/2} \, d\mathbf{x}$$

REFERENCES

K. Fukunaga, *Introduction to Statistical Pattern Recognition*, Academic Press, New York, 1972.

W. H. Press, S. A. Teulkosky, W. T. Vetterling, and B. P. Flannery, *Numerical Recipes in Fortran, The Art of Scientific Computing*, 2nd ed., Cambridge University Press, Cambridge, UK, 1992.

C. W. Therrien, *Decision Estimation and Classification*, Wiley, New York, 1989.

INDEX

Remote Sensing with Polarimetric Radar, by Harold Mott